On Permanent Loan

to

Medical Physics

Dielectric and Electronic
Properties of
Biological Materials

Dielectric and Electronic Properties of Biological Materials

Ronald Pethig
*School of Electronic Engineering Science
and the
Laboratory of the National Foundation for Cancer Research,
University College of North Wales,
Bangor, Gwynedd, UK*

JOHN WILEY & SONS
Chichester · New York · Brisbane · Toronto

Library of Congress Cataloging in Publication Data

Pethig, Ronald.
    Dielectric and electronic properties of biological
materials.

    Bibliography: p.
    Includes index.
    1. Electrophysiology.   2. Dielectrics.
3. Macromolecules – Electric properties.   I. Title.
QH517.P47   574.1'9283   78-13694

ISBN 0 471 99728 5

Typeset in IBM Press Roman by
Preface Ltd, Salisbury, Wiltshire.
Printed in Great Britain by
Unwin Brothers, The Gresham Press, Old Woking, Surrey

To
Angela, Richard and Helen

Contents

Foreword

I became convinced at an early date that the wonderful subtlety of biological reactions could not be explained by the function of clumsy and unreactive protein macromolecules without the concurrence of much smaller and more mobile units. These could hardly be anything else than electrons. Since electrons can move only on conductors, I proposed, almost four decades ago, that proteins may be conductors. The idea found but very few defenders. Its rejection was almost unanimous. This was most encouraging because it belongs to the definition of a discovery that it has to be rejected, being at variance with accepted knowledge. Never having had a scientific education myself I was helpless but remained convinced that without extending biology to the electronic dimension the problem of cancer must remain unsolved and the great number of degenerative diseases must be left rampaging freely.

Fate would have it that Dr. Pethig became interested in coloured proteins and I knew ways to produce them. This brought us together. After he had learned the scientific baby-language which I understand, a most fruitful collaboration and warm friendship developed between us, and he agreed to do measurements which might give substance to my theory. I expect this work to lead to an understanding of the 'living state', the difference between 'animate' and 'inanimate'. I see little hope to be able to explain the subtle difference between a normal and a sick cell as long as we do not understand the basic difference between a cat and a stone.

To be able to cope with disease we have to understand both the normal and pathological. We have made great strides towards a better understanding of the nature of life and cancer, and I hope that others will follow our trail. The present book will lend a helping hand to those who want to do so and help to fill a big white territory on the map of knowledge.

Woods Hole,
July, 1978

ALBERT SZENT-GYÖRGYI

xi

Preface

For someone who has been primarily educated as an electrical engineer, the microtechnology that characterizes the sensitivity of living systems provides an unlimited source of wonder. Modern electronic technology cannot compete, for example, with the bat's acoustic holographic radar, the detection of single phero-mone molecules by moths, the sensory power of finger-tips, nor the light adapt-ability of a phototropic sporangiophore and eyes of a horseshoe crab. How has nature enabled some snakes to detect ambient temperature changes as small as 10^{-3} °C, some fishes to respond to electric fields as low as 10^{-6} V m^{-1}, and certain bacteria, bees, and birds to use magnetic fields of 5×10^{-5} tesla for orientation purposes? There is increasing evidence to suggest that the sensilla of some insects act as receivers of weak, high frequency, electromagnetic signals. Coupled with such sensitivities living systems are also characterized by extremely subtle phenomena; as an example one has only to consider the processes of thought and memory. The differences between the normal and pathological, between a healthy and 'rogue' cancerous cell, for example, arise from a variety of very subtle causes. It would seem reasonable to suggest that some of the sensitivities and subtleties that characterize the miracle of the living state have as their basis mechanisms that operate at the submolecular, electronic, level. The intrinsically lifeless macromolecules of which living organisms are basically composed do after all represent a class of condensed matter, and as such will possess solid-state electrical properties.

The study of the dielectric and electronic properties of biomacromolecules should therefore become part of the interdisciplinary armoury of the biological sciences. Advances in such studies should certainly be of benefit to some branches of medicine, and it could well be that progress in cancer research, for example, will be greatly accelerated by an increased understanding of the submolecular processes involved in cellular activity. This book has been written in the hope that those biochemists, biologists, chemists, electrical engineers, physicists, and theoreticians interested in 'bioelectronics' will all find something of use in it. For example, the con-cept of dielectric polarizability and semiconduction may be new to many biochemists and biologists; electrical engineers and physicists may like to learn more about the basic properties of proteins and water; the experimentalist may gain new incentives and also be interested to have some idea of how quantum mathematics is used to derive energy band structures for biopolymers; and in turn theoreticians may wish to use some of the experimental data in their models. It is also hoped that the book will represent a useful source for further reading, and give an idea of the progress that has been made in the study of the dielectric and electronic properties of biological materials.

It gives me pleasure to take this opportunity to express my gratitude to many of those who have greatly helped me. In particular I wish to thank Keith Morgan, my past tutor and research supervisor at Southampton and now family friend, Professor D. D. Eley in whose laboratory at Nottingham I spent four enjoyable and informative years, and Professor T. J. Lewis for his active encouragement and collaboration at Bangor. I have been greatly sustained by the youthful enthusiasm of my research colleagues; Dave Armitage, Mike Arnold, Joy Behi, Stephen Bone, Paul Carnochan, Tom Cross, Joyce Eden, John Fothergill, Peter Gascoyne, Dave Hayward, and John Morgan, and I have received valuable assistance from Susan and Richard Jones. By far the greatest influence regarding the development of my academic interests has been that created by Dr. Albert Szent-Györgyi, beginning by chance when I came across his article, 'The Study of Energy-Levels in Biochemistry', to the time eleven years later when enjoying the kind hospitality of Professor H. A. Pohl's laboratory in Oklahoma the opportunity arose for me first to visit, and then later enjoy working with Dr. Szent-Györgyi and Jane McLaughlin in their laboratory at Woods Hole. This in turn was only made possible by the efforts of Dr. and Mrs. F. C. Salisbury and the support of the National Foundation for Cancer Research with its pioneering concept of an international, interdisciplinary, *Laboratory Without Walls*.

Finally, I wish to thank Dr. Howard A. Jones, Molecular Sciences Editor of John Wiley and Sons Ltd. for his kind invitation to put pen to paper and unceasing enthusiastic encouragement thereafter, and a special thanks must be given to my family who endured the many long weekends and evenings that were consumed in the writing of this book.

Bangor,
May, 1978   R. PETHIG

Chapter 1

Dielectric Theory—An Outline

Although the dielectric properties of matter are primarily of interest to the physicist and electrical engineer, they have also proved to be of importance and use in the understanding and development of other fields of knowledge, including the chemical and biological sciences. As a result the theory of dielectric behaviour has been extensively developed, with the various contributions to the subject often reflecting the requirements and interests of the scientific field in question. Whereas a fully comprehensive account of all the various aspects of dielectric theory will not be attempted here, the following outline will be sufficient for an appreciation of the subject matter of this book. Also, although atoms and molecules can really only be satisfactorily represented by quantum mechanics, only a brief quantum mechanical description of dielectric theory will be included. This is because we shall mainly concern ourselves with dielectric effects associated with molecular move-ment, for which classical theories are often quite adequate. Quantum theories are more appropriate for dielectric mechanisms involving electronic movements and as such will be mentioned in later chapters dealing with the electronic conduction properties of organic and biological polymeric systems. For more detailed and comprehensive accounts of dielectric theory the reader is referred to the excellent books of Debye,[1] van Vleck,[2] Fröhlich,[3] Böttcher,[4] Smyth,[5] von Hippel,[6] Davies,[7] McCrum et al.,[8] Daniel,[9] and Hill et al.[10]

The electrical properties of a material held between two electrodes of area A and separation d can be completely characterized by its electrical conductance G and capacitance C as

$$G = \frac{\sigma A}{d}; \quad C = \frac{\epsilon_0 \epsilon A}{d} \tag{1.1}$$

where σ is the electrical conductivity (expressed in units of mho/m or siemens/m in SI units), ϵ_0 is the dielectric permittivity of free space of value 8.854×10^{-12} farad/m, and ϵ is the material's permittivity relative to free space. The conductivity σ is the proportionality factor between electric current and electric field and is a measure of the ease with which delocalized electric charge can migrate through the material under the field's influence. The permittivity is the proportionality factor between electric charge and field and reflects the extent to

1

which localized charge distributions can be distorted through polarization by an external electric field. Equations (1.1) characterize the linear electrical properties of the material in a uniform electric field. Linearity here means that the conductivity and permittivity are independent of electric field, and is a property exhibited by most dielectrics for field strengths up to about 10^5 volts/m. Uniform fields are readily achieved with parallel electrodes with $A \gg d$.

The practical definition of the static relative permittivity of a dielectric material, also often referred to as the material's dielectric constant, is the ratio of the capacitances of the same capacitor with first the dielectric and then only vacuum between its plates. For non-polar materials, the increase in capacitance results from polarization of the molecular structure due to charges on the capacitor plates. Positive and negative charges in the molecules are pulled in opposing directions and an effective increase in surface charge appears at the plates. In this way the greater the polarizability of the molecular structure of a material, the greater will be its permittivity. For non-polar materials the polarizability arises from two effects, namely electronic and atomic polarization. In electronic polarization the applied electric field causes a displacement of the electron cloud relative to the nuclei in each atom, whereas in atomic polarization the displacement is of the atomic nuclei relative to one another. If the material is an ionic crystal, then apart from the ions being polarized individually, the negative and positive ions will be displaced with respect to each other. This effect is called ionic polarization.

In some molecules, although they are electrically neutral, there exists a distribution of charge so that the centres of positive and negative charge are not coincidental. Such molecules have a permanent electric dipole moment and are termed polar molecules. This concept, that separated positive and negative charges may exist within molecules, can be traced back to the early nineteenth century where it formed the basis for the theory of *electrochemical dualism* formulated by J. J. Berzelius. Such a state was termed *polarization* by association with the concept of the polarity of magnets. It took the discovery of the electron to revive the chemist's interest in the polarity of molecules, and by the year 1900 scientists were qualitatively assessing the importance of some types of molecules, such as H_2O, possessing a permanent electric dipole. The first quantitative theory of molecular dipole moments was given by Peter J. W. Debye in 1912.[11] Debye, born in 1884 in Maastricht, did most to develop the early concepts associated with molecules possessing permanent dipole moments. The magnitude of a dipole moment is equal to the product of the net separated charges and their separation distance, and has historically been measured in debye units, being taken as the order of magnitude of electric charge multiplied by the order of atomic dimensions, so that 1 debye = 3.33×10^{-30} coulomb metre. (The displacement of one electron through 10^{-10} metre gives a dipole moment of 4.8 debye.) One factor that influenced Debye's work was the discovery in 1897 by Drude[12] of an anomalous dispersion in the radio frequency region for substances containing polar atomic groups. Debye reasoned that this effect must have arisen from the orientational response of the polar molecules to the externally applied electric field.

The total polarizability α_T of a molecule is therefore the sum of three terms,

$$\alpha_T = \alpha_e + \alpha_a + \alpha_0$$

where α_e, α_a and α_0 are the electronic, atomic, and orientation polarizations respectively. It will be the molecular orientational processes with which we shall be most concerned when considering the dielectric properties of macromolecular systems.

In outlining a molecular interpretation of dielectric phenomena we will commence with the so-called 'action at a distance' approach where use is made of Coulomb's law describing the force of interaction between electric charges. It is also possible to adopt a procedure, favoured by Faraday and Maxwell for example, where the electric stresses acting across the surfaces of each dielectric medium are considered. These two seemingly distinct approaches can in fact be shown to be mathematically equivalent. Another approach, adopted by some modern textbooks describing electrostatic phenomena, uses a modified form of Coulomb's law to include a factor (the relative permittivity) to account for the specific dielectric medium in which the electric charges are located. This last approach can be criticized in that it does not seem logical to develop a molecular interpretation of the medium's permittivity using an approach where the initial formulae already include its value as an empirical factor.

From the Coulombic force concept, when charges are placed on the plates of an air-filled capacitor the positive and negative charges attract each other. On bringing the plates closer together this attractive force increases so that a given battery is able to maintain an increasingly larger charge density on the capacitor plates. On inserting a dielectric between the plates, the positive and negative charges in its constituent molecules are pulled in opposing directions, with the positive charges moving towards the negatively charged plate and *vice versa*. The negative capacitor plate, for example, now finds itself nearer to the positive than to the negative charges in the dielectric. The plates will therefore experience an increased attractive force just as if the plates had been brought closer together, and the effective charge capacity will increase. This provides a qualitative interpretation of the action of a dielectric material in a capacitor. What follows is an outline of the theory developed to provide a quantitative interpretation of dielectric phenomena at the molecular level.

DIPOLE MOMENTS AND FIELDS OF INTERACTION

According to Coulomb's law the electrostatic force exerted on a charge q_i at vector distance $\bar{r}_i$ by charges q_j at positions $\bar{r}_j$ is given by

$$F_i = \frac{1}{4\pi\epsilon_0} \sum_j \frac{q_i q_j}{r_{ji}^2} \hat{r}_{ji} \tag{1.2}$$

where $\hat{r}_{ji}$ is the unit vector and $\bar{r}_{ji} = r_{ji}\hat{r}_{ji} = \bar{r}_j - \bar{r}_i$. By using the charge q_i as a probe to measure the effect of charges q_j on their surroundings, we are introducing the concept of there being a resultant electric field produced by charges q_j. The field $\bar{K}_i$ experienced by charge q_i is defined as

$$\bar{E}_i = \lim_{q_i \to 0} \frac{\bar{F}_i}{q_i}$$

and is also given by the relationship

$$\bar{E}_i = -\mathrm{grad}_i V$$

with

$$V = \frac{1}{4\pi\epsilon_0} \sum_j \frac{q_i}{r_{ji}}$$

being the electric potential of the charges j at the point i.

If the charges q_j all lie within a sphere of radius 'a' centred at the origin then their potential V at any point P of position r outside the sphere can be expressed as a convergent series in $1/r$ as

$$V = \frac{1}{4\pi\epsilon_0}\left\{ \frac{Q}{r} + \frac{\bar{m} \cdot \hat{r}}{r^2} + \cdots \right\}$$

with the total charge $Q = \Sigma_j q_j$, and the moment $\bar{m}$ being defined as $m = \Sigma_j q_j \bar{r}_j$. If in the set of charges q_j the positive charges exactly cancel out the negative charges then $Q = 0$ and for distances $r \gg a$ the potential V is given approximately as

$$V = \frac{1}{4\pi\epsilon_0} \frac{\bar{m} \cdot \hat{r}}{r^2} \tag{1.3}$$

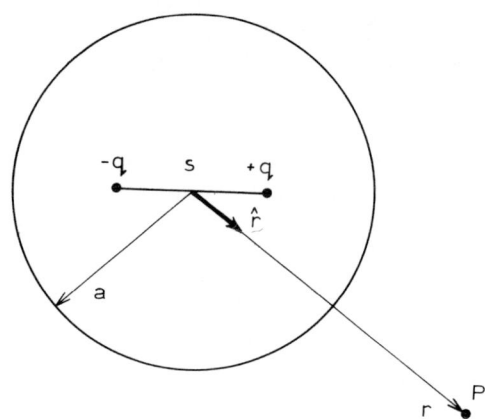

Figure 1.1 The potential of a dipole at point P for $r \gg s$ is given by

$$V = \frac{1}{4\pi\epsilon_0} \frac{\bar{m} \cdot r}{r^2}$$

where $\bar{m} = q\bar{s}$

and the value of $\bar{m}$ does not depend on the choice of origin for the vectors $\bar{r}_j$. The simplest system to consider is that of a dipole composed of a pair of charges $+q$ and $-q$ separated by the vector distance $\bar{s}$ as shown in Figure 1.1. In this case $\bar{m} = q\bar{s}$ and for $r \gg s$ then the potential V is given exactly by equation (1.3). If we let the distance s between the two charges decrease and at the same time let the amount of charge increase in such a way that the moment $\bar{m} = q\bar{s}$ remains constant, then on continuing this process to its limit we obtain an ideal point dipole having a moment $\bar{m}$ in the direction of $\bar{s}$. In this case equation (1.3) is valid everywhere for finite values of r and the dipole's electric field intensity is given by

$$\bar{E} = -\text{grad } V$$

or

$$\bar{E} = \frac{1}{4\pi\epsilon_0} \frac{3(\bar{m} \cdot \hat{r})\hat{r} - \bar{m}}{r^3} \tag{1.4}$$

This result shows that the field created by a point dipole can be resolved into components along the radius vector $\bar{r}$ and parallel to the dipole moment vector $\bar{m}$. The components of $\bar{E}$ are linear functions of $\bar{m}$ with the coefficients being proportional to the components x, y, z of the radius vector $\bar{r}$. Equation (1.4) can therefore be written in tensor notation as

$$\bar{E} = -T \cdot \bar{m}$$

where the tensor

$$T = r^{-3}(1 - 3\hat{r}\hat{r})$$

may be referred to as the dipole field tensor. More commonly it is referred to as the dipole–dipole interaction tensor because the interaction energy of two dipoles is given by the product of one dipole moment and the field of the other dipole. The lines of constant force in the point dipole field, as experienced by a roving positive test charge, are shown in Figure 1.2. In Figure 1.2 the electrostatic field components at any point r are given by

$$\bar{E}_x = \frac{\bar{m}}{4\pi\epsilon_0 r^3}(3\cos^2\theta - 1)$$

$$\bar{E}_y = \frac{3\bar{m}}{4\pi\epsilon_0 r^3}\sin\theta\cos\theta$$

and the electrostatic potential is given by equation (1.3) as

$$V = \frac{\bar{m}}{4\pi\epsilon_0 r^2}\cos\theta$$

If we now consider a spherical region in space containing *none* of the charges q_j, then their potential is given by the convergent series

$$V = V_0 - \bar{r} \cdot \bar{E}_0 + \ldots$$

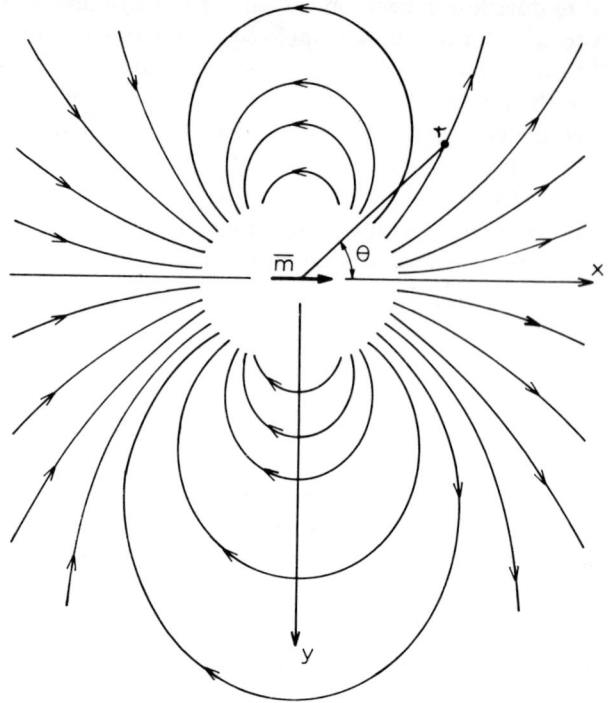

Figure 1.2 The lines of constant force in the field of a
point dipole

where V_0 and $\bar{E}_0$ are the potential and field intensity at the radius vector origin
taken to be the centre of this empty sphere. If now a set of charges q_i are placed
within this sphere, then the potential energy of electrostatic interaction of the
charges q_j with q_i is given by

$$U_q = \sum_i q_i V_i$$

$$= QV_0 - \bar{m} \cdot \bar{E}_0 + \cdots \qquad (1.5)$$

where Q is the total charge and $\bar{m}$ the resultant dipole moment of the charges q_i. If
$Q = 0$ and the sphere containing charges q_i has a radius small compared with the
distance to the nearest charge q_j, then U_q is given approximately by

$$U_q = -\bar{m} \cdot \bar{E}_0 \qquad (1.6)$$

and in the point dipole limit this expression is exact. The force and torque exerted
on the set of charges q_i by the set q_j can be derived from equation (1.2), but they
are more conveniently obtained by differentiation of equation (1.5). For $Q = 0$ and
assuming equation (1.6) is valid, then the force is given by

$$\bar{F} = \bar{m} \cdot \nabla \bar{E}_0$$

and the torque about any point includes, besides the contribution from this force, a couple of magnitude $\bar{m} \times \bar{E}_0$. For a uniform field where $\nabla \bar{E}_0 = 0$, then the only contribution to the torque comes from the couple $\bar{m} \times \bar{E}_0$.

ORIENTATIONAL POLARIZABILITY

When a dielectric material becomes polarized by the application of an external electric field, the dipole moment of the constituent molecules is given by

$$\bar{m} = \alpha_T \bar{E}_1$$

where α_T is the total polarizability of each molecule, and $\bar{E}_1$ is the local field acting on the molecules. In order to distinguish the induced dipole effects, associated with electronic and atomic polarizations, from the permanent dipole effects we will write

$$\bar{m} = \bar{\mu} + \alpha \bar{E}_1$$

where $\bar{\mu}$ is the permanent dipole moment of the molecule and α is the sum of the electronic and atomic polarizabilities. One such material composed of molecules possessing a permanent dipole moment is water, with $\bar{\mu}$ having a value of 1.8 debye units. When subjected to static electric fields or alternating fields of frequency less than 10^9 Hz, water exhibits a relative permittivity ϵ_r value of the order 80. At frequencies higher than 10^{11} Hz, however, ϵ_r falls to a value of the order 4.5. This change in ϵ_r is caused by the orientational polarizability of the water molecules which is effective at low frequencies but is absent at ultra-high frequencies. An expression will now be obtained for the orientational polarizability of permanent dipoles using a method analogous to that first adopted by Langevin for deriving the mean magnetic moment of a gas composed of molecules possessing a permanent magnetic moment.

When an electric field is applied to a material composed of molecules having a permanent dipole moment, the dipoles all experience a torque $\bar{\mu} \times \bar{E}_1$ tending to align them with the field. This orienting tendency is opposed by thermal agitation. From equation (1.6) the potential energy U of a dipole moment $\bar{\mu}$ in a field $\bar{E}_1$ is given by

$$U = -\bar{\mu} \cdot \bar{E}_1 = -\mu E_1 \cos \theta$$

where θ is the angle between the dipole moment and the field direction. According to the Boltzmann distribution law the relative probability of finding a dipole oriented in an element of solid angle $d\theta$ is proportional to $\exp(-U/kT)$, and the thermal average of $\cos \theta$ is given by

$$\langle \cos \theta \rangle = \frac{\int \exp(-U/kT) \cos \theta \, d\theta}{\int \exp(-U/kT) \, d\theta}$$

The integrations are to be carried out over all solid angles, so that

$$\langle \cos \theta \rangle = \frac{\int_0^{\pi} 2\pi \sin \theta \cos \theta \, \exp(\mu E_1 \cos \theta / kT) \, d\theta}{\int_c^{\pi} 2\pi \sin \theta \, \exp(\mu E_1 \cos \theta / kT) \, d\theta}$$

We now let $x = \mu E_1/kT$, and $y = \cos \theta$, then

$$\langle \cos \theta \rangle = \int_{-1}^{1} \exp(xy) y \, dy \Big/ \int_{-1}^{1} \exp(xy) \, dy$$

$$= \frac{d}{dx} \log \int_{-1}^{1} \exp(xy) \, dy$$

$$= \frac{d}{dx} \log(\exp x - \exp(-x)) - \frac{d}{dx} \log x$$

$$= \coth x - \frac{1}{x}$$

$$= L(x)$$

This defines the Langevin function $L(x)$, and is plotted as a function of x in Figure 1.3. The most usual experimental conditions have $\bar{E}_1 < 10^5$ V m^{-1}, and taking the dipole moment value of 1.8 debye (6×10^{-30} C m) for water as being a typical

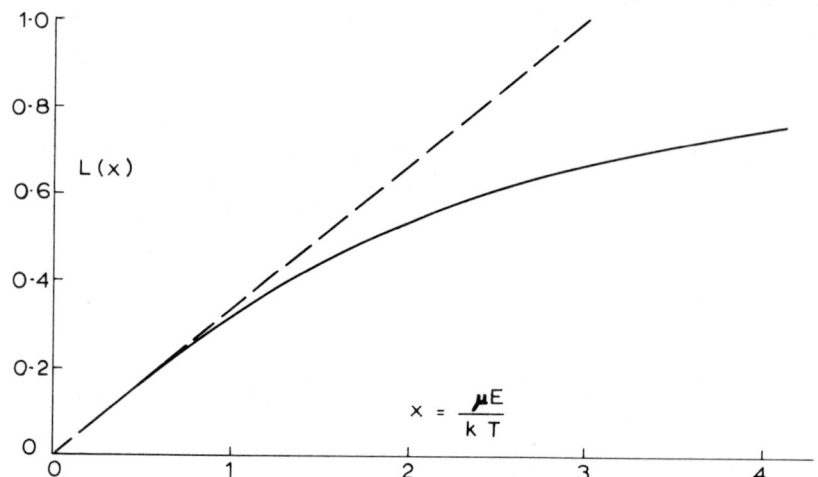

Figure 1.3 Plot of the Langevin function, derived from table given in the *CRC Handbook of Chemistry and Physics*, 53rd ed., p. E102 (1973)

moment, then at room temperature

$$x = \frac{\mu E_1}{kT}$$

$$= \frac{6 \times 10^{-25} \text{ J}}{4.1 \times 10^{-21} \text{ J}}$$

$$= 1.5 \times 10^{-4}$$

In this limit of $x \ll 1$, we have that $L(x) \simeq x/3$, since

$$\coth x = \frac{1}{x} + \frac{x}{3} - \frac{x^3}{45} + \frac{2x^5}{945} -$$

Therefore under normal experimental conditions

$$\langle \cos \theta \rangle = \mu E_1 / 3kT$$

and is of value appreciably smaller than unity. This shows that the polarizing influence of the electric field produces only a slight change in the *average* of $\cos \theta$ for all the dipoles, and does not result in any appreciable change in the direction of the individual dipole moments compared with their original random directions before the field was applied.

The average moment per dipole in the direction of the applied field is given by

$$\bar{\mu}_d = \mu \langle \cos \theta \rangle$$

$$= \mu^2 \bar{E}_1 / 3kT$$

and the magnitude of the orientational polarizability α_0 is

$$\alpha_0 = \mu^2 / 3kT \qquad (1.7)$$

giving

$$\alpha_T = \mu^2 / 3kT + \alpha_e + \alpha_a$$

From the form of the Langevin function shown in Figure 1.3, it is clear that orientational saturation effects are appreciable for $\mu \bar{E}_1 \geqslant kT$. For $\mu E_1 = 5kT$ then 80 per cent of all the molecules in the dielectric will be orientated with their dipole moments in the direction of the applied field. For dipole moments of the order 1 debye such a saturation effect will require a value for E_1 of 6×10^9 V m^{-1} at room temperature, and most materials will have suffered electrical breakdown well before such a field strength is attained. For the situation $\mu E_1 < 0.1kT$ then saturation effects are negligible and equation (1.7) gives an accurate value for the orientational polarizability. For $\mu E_1 > 0.1kT$ then higher order terms must be considered for the Langevin function to give, for example,

$$\alpha_0 = \frac{\mu^2}{3kT} \left(1 - \frac{1}{15} \left(\frac{\mu E_1}{kT} \right)^2 \right)$$

showing that the dielectric response is no longer linear with respect to the applied field.

THE LOCAL ELECTRIC FIELD AND THE MACROSCOPIC PERMITTIVITY

In the last few paragraphs care was made to distinguish the externally applied electric field acting on the dielectric medium from the local field E_1 acting on molecules within the material. Why should we have to make such a distinction?

Consider a parallel-plate capacitor, of electrode separation d, filled with a dielectric material of relative permittivity ϵ. If a voltage V is applied then the average field strength is $E = -V/d$. A charge q moving from the negative electrode to the positive one will experience an average force of $-qE$. We can say that the average field describes the force acting on *extra* charges brought into the dielectric material averaged over all *possible positions* of the charge. The local field E_1, however, refers to that acting on the *constituent* charges in the atoms or molecules located at specific sites in the material. The local *microscopic* field acting on a molecule will be a function of the *macroscopic* average field together with fields of the form of equation (1.4) arising from its own induced dipole moment and those of neighbouring polarized molecules.

An electric field can be represented by the electric field strength E, expressed in volts per meter, and by the electric displacement D, expressed in coulombs per square meter. In a vacuum, $\bar{D} = \epsilon_0 \bar{E}$ and in the dielectric material $\bar{D} = \epsilon \epsilon_0 \bar{E}$. According to well-known theorems of electrostatics,

$$\bar{D} = \epsilon \epsilon_0 \bar{E}$$
$$= \epsilon_0 \bar{E} + \bar{P} \tag{1.8}$$

where the polarization P is defined as the induced dipole moment per unit volume of the dielectric material. The macroscopic field E and the polarization P are related to the relative dielectric susceptibility χ by the definition

$$\bar{P} = \epsilon_0 \chi \bar{E} \tag{1.9}$$

If the material contains N dipoles per unit volume, then

$$\bar{P} = N\alpha_T \bar{E}_1$$

and from equations (1.8) and (1.9) the relative dielectric susceptibility is given by

$$\chi = \epsilon - 1$$
$$= \frac{N\alpha_T}{\epsilon_0} \frac{\bar{E}_1}{\bar{E}} \tag{1.10}$$

Therefore, in order to be able to quantitatively relate the macroscopic permittivity to molecular quantities such as the polarizability α_T, the relationship between the microscopic local field intensity E_1 and the macroscopic field intensity E must be obtained. This represents one of the most difficult problems of dielectric theory since in only a few exceptional cases can such a simple relationship be derived.

The way in which this problem has been approached for the various classes of dielectric materials will now be briefly outlined.

Low Pressure Gases

In gases whose pressures are sufficiently low, the constituent molecules can be considered to be far enough apart for their interactions with each other to be negligible. In other words, at the position of any molecule the electrostatic field created by all the surrounding molecules can be neglected in comparison with the macroscopic field intensity E. As described earlier, the dipolar field intensity experienced by a neighbouring molecule will be of the order $m/4\pi\epsilon_0 r^3$ where m is the molecular dipole moment and r is a representative intermolecular spacing. In this way, since $m = \alpha_T E$, then in order for $E_1 = E$ we must have that

$$\alpha_T/4\pi\epsilon_0 r^3 \ll 1 \tag{1.11}$$

If this inequality can be shown to be valid, then from equation (1.10) the relative dielectric susceptibility of the gas is given by

$$\epsilon - 1 = \frac{N\alpha_T}{\epsilon_0} \tag{1.12}$$

Liquids and Dense Gases

When the inequality of equation (1.11) is not satisfied, then $E_1 \neq E$. The local field acting on a specific molecule will be greatly influenced by the electric fields of other molecules. Following earlier deductions by Mossotti (1850) and Clausius (1879), H. A. Lorentz[13] in 1902 attempted the first quantitative analysis of the effect of such molecular electrostatic interactions. His method for calculating the local field strength E_1 is as follows. The dielectric material is considered to be composed of two regions; a small spherical region centred around the molecule for which the local field is to be calculated, and the remaining part of the dielectric. The radius of the sphere is considered to be large in comparison with the mean intermolecular distance but small in comparison with the dimensions of the dielectric material, as shown in Figure 1.4. The molecules within the spherical cavity are considered to be located at their average lattice sites and interact electrostatically like parallel point dipoles. The molecules outside the sphere are considered to form a continuous medium.

The local field is given by

$$\bar{E}_1 = \bar{E}_1 + \bar{E}_2$$

where $\bar{E}_1$ is the field intensity due to the material outside the sphere and $\bar{E}_2$ is the field intensity due to dipoles inside the sphere. The field $\bar{E}_1$ is comprised of both the field due to distant molecules and the externally applied field $\bar{E}$. The field due to the distant molecules can be represented by apparent charges on the boundary of the

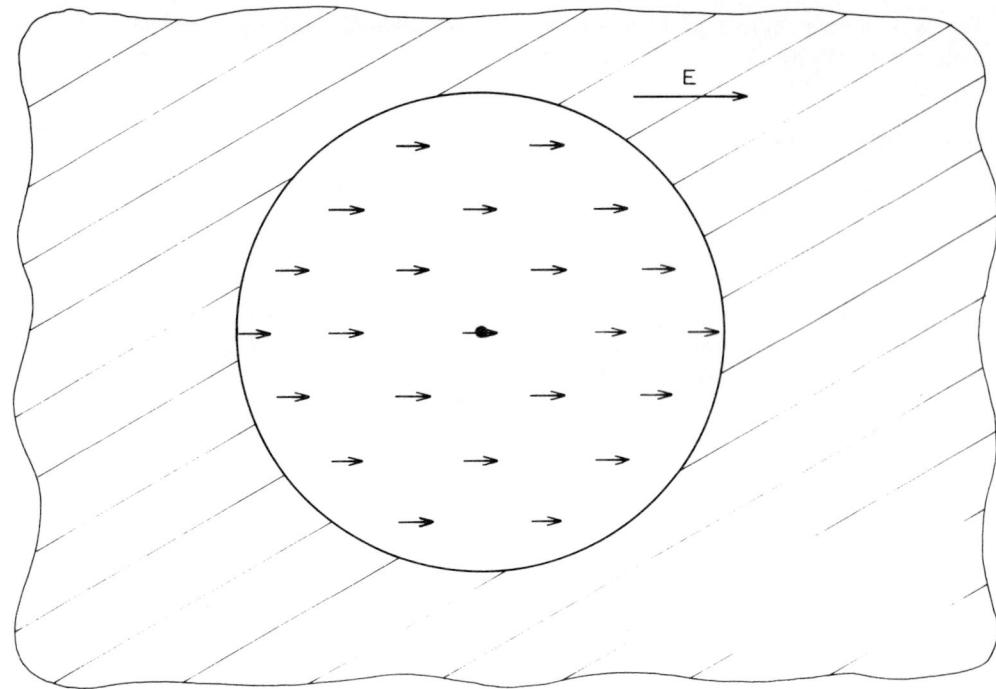

Figure 1.4 Model used for the calculation of the Lorentz local field intensity

spherical cavity, and by standard electrostatic theory it is easily shown that

$$\bar{E}_1 = \bar{E} + \frac{\bar{P}}{3\epsilon_0} = \left(\frac{\epsilon + 2}{3}\right)\bar{E}$$

The calculation of $\bar{E}_2$ is not so straightforward. It can be shown that if the molecules within the sphere form a simple cubic, face-centred cubic, or body-centred cubic lattice about the central molecule of interest, then $\bar{E}_2 = 0$. This result also holds for a completely random arrangement of molecules within the sphere. For isotropic materials such as liquids, then $\bar{E}_2$ can be considered to be small by comparison with $\bar{E}_1$, and to a good *approximation*

$$\bar{E}_1 = \left(\frac{\epsilon + 2}{3}\right)\bar{E} \tag{1.13}$$

so that equation (1.10) can be written as

$$\frac{3(\epsilon - 1)}{\epsilon + 2} = \frac{N}{\epsilon_0}\alpha_T \tag{1.14}$$

On rearranging equation (1.14) we obtain

$$\epsilon - 1 = \frac{N\alpha_T/\epsilon_0}{1 - N\alpha_T/3\epsilon_0} \tag{1.15}$$

which predicts that as the density of the material approaches the value corresponding to $N = 3\epsilon_0/\alpha_T$, then the permittivity becomes infinite so that a very small applied field strength would result in an infinite polarization of the material. Saturation effects would keep the polarization finite in value, but instead the material would become spontaneously polarized and correspond to the ferroelectric state. As no such ferroelectricity has been observed to arise from dipolar effects, then equation (1.14) may not be considered to be completely valid.

Onsager[14] was the first to provide a satisfactory alternative to the Mossotti–Clausius–Lorentz formulation in which the molecule of interest is imagined to be at the centre of a large spherical cavity within the dielectric medium. Onsager replaced this cavity with one being the same size as the molecule whose dipole moment $\bar{m}$ is assumed to occupy only a small region located at the centre of the molecule. According to this scheme the local field $\bar{E}_l$ takes the form

$$\bar{E}_l = \alpha\bar{E} + \beta\bar{m}$$

where $\alpha\bar{E}$ represents the *cavity* field and $\beta\bar{m}$ the *reaction* field. The *cavity* field is that part of the local field that remains unaltered if the molecule in question has its dipole moment removed. The *reaction* field is the component of the local field that now results if the dipole moment is restored to the molecule, for although this moment cannot contribute to its own local field directly it can do so indirectly by inducing polarizations in the neighbouring molecules which in turn modify the local field. This reaction field will have no orienting effect on the molecule since it will always be parallel to the moment $\bar{m}$. It is this fact that is overlooked in the formulation of the Lorentz local field and as a result there is the implication that the dipole moment can partly orient itself which in turn leads to such possibilities as the dielectric 'catastrophe' implicit in equation (1.14). Following Onsager's treatment, the local field is given by

$$\bar{E}_l = \frac{3\epsilon\bar{E}}{(2\epsilon + 1)} + \frac{2(\epsilon - 1)}{(2\epsilon + 1)}\frac{N\mu}{3\epsilon_0} \qquad (1.16)$$

Dilute Solutions

In dilute solutions of a dipolar solute in a non-polar solvent, the polar molecules are far apart so that the interactions between them can be neglected. There will, however, be an interaction between the polar molecules and the molecules of the solvent, and this interaction can be taken into account approximately by using the form of the Lorentz local field given by equation (1.13). In this way equation (1.14) can easily be generalized to describe the permittivity of a mixture of molecules of various kinds by the expression

$$\frac{3\epsilon_0(\epsilon - 1)}{\epsilon + 2} = \sum_k N_k \alpha_{kT} \qquad (1.17)$$

where the subscript k refers to the kth type of molecule. The mole fraction of the kth component is defined as

$$f_k = \frac{N_k}{\sum\limits_{l} N_l}$$

and its density

$$\rho_k = \sum_{k} N_k M_k / N_A$$

where M_k is the molecular weight of component k and N_A is Avogadro's number. Equation (1.17) then can be written as

$$\frac{3\epsilon_0(\epsilon - 1)}{\epsilon + 2} \frac{1}{\rho_k} \sum_{k} f_k M_k = N_A \sum_{k} f_k \alpha_{kT}$$

For a polar solute 1 in a non-polar solvent 2,

$$\frac{3\epsilon_0(\epsilon - 1)}{\epsilon + 2} \frac{1}{\rho_1} [f_1 M_1 + (1 - f_1)M_2] = N_A [f_1 \alpha_{1T} + (1 - f_1)\alpha_{2T}] \qquad (1.18)$$

By plotting the left-hand side of this equation against f_1, and by obtaining α_{2T} from measurements on the pure solvent ($f_1 = 0$), the value for α_{1T}, and hence the dipole moment of the polar molecules, can be obtained. The dipole moment values obtained in this way are found to differ from the value calculated for the molecules in the gas phase and they are also dependent on the solvent used. This is known as the 'solvent effect'.

Solids

Solids composed of non-polar molecules behave qualitatively like non-polar liquids. The permittivity is practically independent of temperature and changes very little when the solid melts. Such temperature changes as there are normally relate to changes of density. The permittivity value differs little from the square of the optical refractive index n^2 since the contribution of atomic displacements to the polarizability is small compared with the electronic displacements. From equation (1.14) we then have that

$$\frac{n^2 - 1}{n^2 + 2} = \frac{N\alpha_e}{3\epsilon_0}$$

In solids composed of polar molecules, it is to be expected that the molecules will have lost their rotational freedom and with it their orientational polarizability. On freezing a polar liquid the permittivity should therefore fall abruptly to a value typical of non-polar materials. This does happen for most polar liquids, but a few others (e.g. camphor, t-nitrobutane and 2,2-dichloropropane) show virtually no change in permittivity when the liquid freezes. In these 'rotator phase' solids, the molecules are usually of a spherical shape and retain a high degree of rotational motion.

In solid polymeric materials containing polar chemical groups, these polar groups often have rotational freedom even at low temperatures. Such polymers can be considered to behave more like viscous liquids than crystalline solids, especially if the polymer contains a non-volatile molecular 'lubricant' used as the plasticizer. The dielectric behaviour of such polymeric solids is discussed in more detail later in this chapter.

DIELECTRIC RELAXATION AND LOSS

The rates with which polarizations can occur are limited, so that as the frequency of the applied electric field is increased some polarizations will no longer be able to attain their d.c. or low frequency values. As might be expected the slowest polarization mechanism is often that of dipolar reorientation, and so this is usually the first polarization term to disappear as the frequency is increased. The dipole moments are just not able to orient fast enough to keep in alignment with the applied electric field and the total polarizability falls from α_T to $(\alpha_T - \alpha_0)$. This fall, with its related reduction of permittivity, and the occurrence of energy absorption, is referred to as dielectric relaxation, or dispersion. The frequency at which this fall in orientation polarization occurs can vary from very low frequencies of the order 10^{-1} Hz and below for large hindered macromolecules to frequencies up to 10^{12} Hz for small molecules. In this frequency range α_a and α_e remain unchanged, since the dispersion due to the fall-off of the atomic polarization occurs at frequencies comparable with the natural frequencies of vibrations of the atoms in a molecule (i.e. in the infrared spectrum around 10^{14} Hz), and that for electronic polarization occurs at still higher frequencies corresponding to electronic transitions between different energy levels in the atom (visible, u.v., and X-ray frequencies). The frequencies at which electronic and atomic dispersions occur are determined by the inertial properties of the molecules or atoms and have the form of a resonance dispersion. By contrast, the dipolar orientation process gives rise to relaxation dispersions where both the frequency and shape of the dielectric loss characteristic depend mainly on the immediate environment of the molecular dipole, and the corresponding changes in permittivity are very different from that of the resonance dispersion (see Figure 1.8).

To derive a mathematical model for such an orientational relaxation process we shall assume that the polarization P is composed of two parts P_1 and P_2, where P_1 arises from atomic and electronic displacements and P_2 arises from the much slower process of dipolar reorientation. For the frequency range of interest it is assumed that P_1 responds instantly to the applied field E, and has the constant value $P_1 = \chi_1 E$. The part P_2 lags behind E in such a way that at any instant P_2 approaches its final value of $\chi_2 E$ at a rate proportional to $(\chi_2 E - P_2)$, so that

$$\frac{dP_2}{dt} = \frac{1}{\tau}(\chi_2 E - P_2)$$

If the field $\bar{E}$ is applied as a step-function at $t = 0$ when $P_2 = 0$, then on integration of the above equation

$$P = P_1 + P_2$$

$$= \left[\chi_1 + \chi_2 \left(1 - \exp -\frac{t}{\tau} \right) \right] E$$

which shows that the polarization approaches its final value exponentially with a time constant τ. In other words the dielectric material gradually responds or relaxes to the polarizing influence of the applied field with a characteristic relaxation time τ. In an alternating field $\bar{E} = \bar{E}_0 \exp(j\omega t)$, the corresponding solution for P_2 gives

$$P = P_1 + P_2 = \left(\chi_1 + \frac{\chi_2}{1 + j\omega\tau} \right) E$$

This result corresponds to their being a complex permittivity of the form

$$\epsilon^* = \epsilon_\infty + (\epsilon_s - \epsilon_\infty)/(1 + j\omega\tau) \tag{1.19}$$

where ϵ_∞ is the permittivity measured at a sufficiently high frequency for the orientational polarization to have disappeared, ϵ_s is the limiting low frequency permittivity, ω is the angular frequency $2\pi f$, j is $\sqrt{-1}$ and τ is the characteristic macroscopic dielectric relaxation time for the relaxation mechanism.

For small and relatively simple molecular structures there is often only a single orientational process, but for polymers the dielectric dispersion can consist of several components associated with small side-chain movements ranging up to whole macromolecular motions. Other complications can arise for macromolecular systems associated with its heterogeneity, where as a result of the so-called Maxwell–Wagner effect apparent dielectric relaxation can occur at low frequencies. This phenomenon will be treated in more detail in Chapter 5.

The real and imaginary components of the complex permittivity are given by

$$\epsilon^* = \epsilon' - j\epsilon''$$

This notation reminds us that the real and imaginary parts of the permittivity are $90°$ out of phase with each other. The real part ϵ' represents the permittivity already discussed (equation 1.1) and is given by

$$\epsilon' = \epsilon_\infty + \frac{\epsilon_s - \epsilon_\infty}{1 + \omega^2 \tau^2} \tag{1.20}$$

The imaginary component is given by

$$\epsilon'' = \frac{(\epsilon_s - \epsilon_\infty)\omega\tau}{1 + \omega^2\tau^2} \tag{1.21}$$

The variations of ϵ' and ϵ'' with frequency are shown in Figure 1.5 for a τ value of $10^{-6}/2\pi$ seconds. Equations (1.20) and (1.21) are commonly known as the Debye dispersion formulas, and it is to be remembered that they refer specifically to the situation where equilibrium is attained exponentially with time when a constant external field is imposed on a dielectric. The transition from the low frequency to

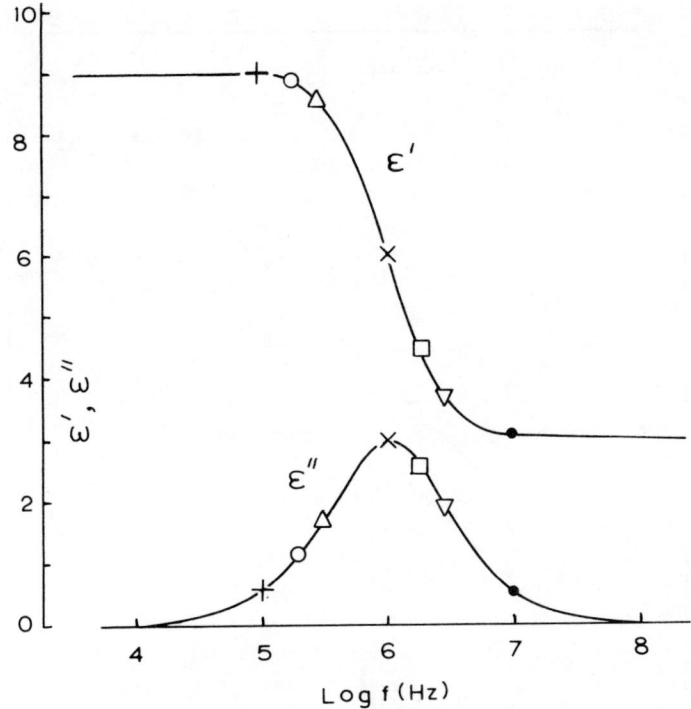

Figure 1.5 Variation of the dielectric parameters ϵ' and ϵ'' with frequency for a Debye-type relaxation process, with $\tau = 10^{-6}/2\pi$ seconds

the high frequency dielectric behaviour extends roughly over about four decades in frequency, and it is also useful to note that for a single Debye-type relaxation process, the width of the ϵ'' peak at the half-height value is 1.14 decades in frequency. By simple mathematical manipulation, equations (1.20) and (1.21) can be modified to give the dielectric parameters in the form of two equations representing straight lines, namely

$$\epsilon''\omega = (\epsilon_s - \epsilon')/\tau \tag{1.22}$$

and

$$\epsilon''/\omega = (\epsilon' - \epsilon_\infty)\tau \tag{1.23}$$

These plots are given in Figure 1.6 for the same data plotted in Figure 1.5, where it can be seen that equation (1.22) is more useful for the low frequency results and equation (1.23) for the higher frequencies. Graphical representation of the dielectric results such as those given in Figure 1.6 can be useful if the full dispersion region lies outside the available frequency range of the measurement apparatus. For example, if measurements for the particular data of Figure 1.5 were only possible up to 300 kHz and if a reasonable value for ϵ_s could be determined, then from the

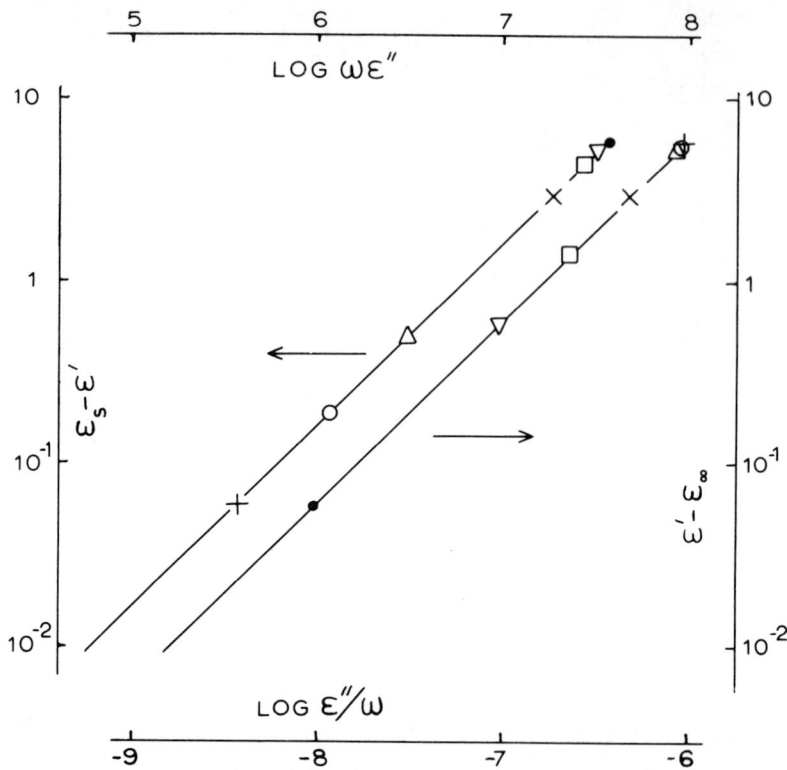

Figure 1.6 The dielectric parameters of equations 1.22 and 1.23 plotted for the same data as Figure 1.5

plot of $\log(\epsilon_s - \epsilon')$ versus $\log \omega \epsilon''$ the magnitude of the relaxation time τ could be obtained (e.g. for $\epsilon_s - \epsilon' = 0.1$ then from equation (1.22), $\omega \epsilon'' = (10\tau)^{-1}$).

The relaxation time τ represents the reciprocal of the mean rate coefficient of the dielectric relaxation process, and since such molecular processes usually follow an Arrhenius temperature law, then we can write

$$\tau = A \, \exp(\Delta H/RT) \tag{1.24}$$

where ΔH is the Arrhenius activation enthalpy per mole and A is a constant. From equation (1.23)

$$\frac{d(\ln \tau)}{d(1/T)} = \frac{\Delta H}{R} \tag{1.25}$$

so that a plot of $\ln \tau$ against $1/T$ gives a straight line of slope $\Delta H/R$. The relaxation process can be further developed in terms of a chemical rate process[15,16] and equation (1.24) becomes

$$\tau = \frac{h}{kT} \exp\left(-\frac{\Delta S}{R}\right) \exp\left(\frac{\Delta H}{RT}\right) \tag{1.26}$$

where h is Planck's constant and ΔS is the molar entropy of activation for the relaxation process.

The phase lag between the polarization and the applied field leads to an absorption of energy and joule heating. The rate of conversion of electrical energy to heat in the material is represented by the imaginary component ϵ'', and ϵ'' is in this way referred to as the dielectric loss factor. The ratio $\epsilon''/\epsilon' = \tan \delta$ is often used to characterize dielectrics for electrical engineering purposes, and can have values ranging from less than 10^{-4} for low-loss polymers up to 1 for very lossy materials. Other synonyms used for $\tan \delta$ are the dissipation factor defined as the ratio of energy lost to energy stored, and the reciprocal of the Q-factor, $1/Q$.

The relationship between the relaxation time τ, the frequency of maximum loss, and the maximum loss value can be found by differentiating equation (1.21) with respect to ω and equating to zero.

$$\frac{d\epsilon''}{d\omega} = 0 \text{ at } \omega_{max} = 1/\tau \tag{1.27}$$

and

$$\epsilon''_{max} = (\epsilon_s - \epsilon_\infty)/2 \tag{1.28}$$

So from equation (1.26) a plot of $\ln(\omega_{max})$ against $1/T$ gives a straight line of slope $-\Delta H/R$.

Deviations from an ideal Debye-type single relaxation are likely to occur for macromolecular systems as a result, for example, of such effects as combinations of cooperative and isolated movements of side-chain molecular groups. This gives rise to a spread of relaxation times each contributing to a Debye-type dispersion. For a set of closely spaced relaxation times the resultant ϵ'' loss curve as a function of frequency will be much broader than that given in Figure 1.5. A method of checking for a single relaxation time can be obtained by rearranging equations 1.20 and 1.21 to eliminate $\omega\tau$ giving

$$[\epsilon' - (\epsilon_s - \epsilon_\infty)/2]^2 + (\epsilon'')^2 = [(\epsilon_s - \epsilon_\infty)/4]^2 \tag{1.29}$$

Equation (1.29) is of the form $x^2 + y^2 = r^2$, i.e. the equation for a circle of radius r, but as only positive values for ϵ'' are possible, then a plot of ϵ' against ϵ'' produces a semicircle of radius $(\epsilon_s - \epsilon_\infty)/2$ with the centre at $\{(\epsilon_s - \epsilon_\infty)/2,0\}$. Such a plot of ϵ' against ϵ'' is referred to as a Cole–Cole plot,[17] and an example is shown in Figure 1.7 for the same data as Figure 1.5. If a symmetrical distribution of relaxation times occurs about a mean relaxation time, then a depressed semicircle is found in the Cole–Cole plot of ϵ' against ϵ''. To account for such a symmetrical distribution of relaxation times, Cole and Cole[17] modified equation (1.19) to the form

$$\epsilon^* = \epsilon_\infty + \frac{\epsilon_s - \epsilon_\infty}{1 + (j\omega\tau)^\beta} \tag{1.30}$$

where β is the parameter $0 < \beta \leqslant 1$. For a single relaxation time $\beta = 1$ and β tends to zero as the distribution tends to an infinite one. Davidson and Cole[18] later

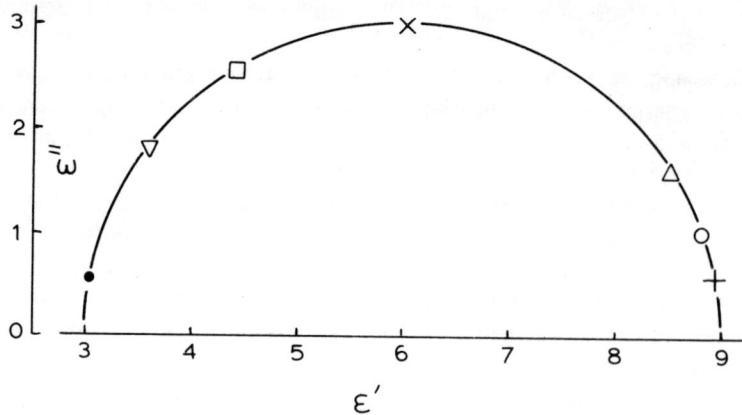

Figure 1.7 The Cole–Cole plot for the data of Figure 1.5

proposed another modification of equation (1.19) to

$$\epsilon^* = \epsilon_\infty + \frac{\epsilon_s - \epsilon_\infty}{(1 + j\omega\tau)^\gamma}, \quad 0 < \gamma \leqslant 1 \tag{1.31}$$

which gives a skewed arc-shaped ϵ', ϵ'' curve and is characteristic of a non-uniform distribution of relaxation mechanisms where the distributions on the high frequency side of the principal relaxation time decreases more rapidly than those on the low frequency side. More generally

$$\epsilon^* = \epsilon_\infty + (\epsilon_s - \epsilon_\infty) \int_0^\infty \frac{G(\tau)}{1 + j\omega\tau} \cdot d\tau \tag{1.32}$$

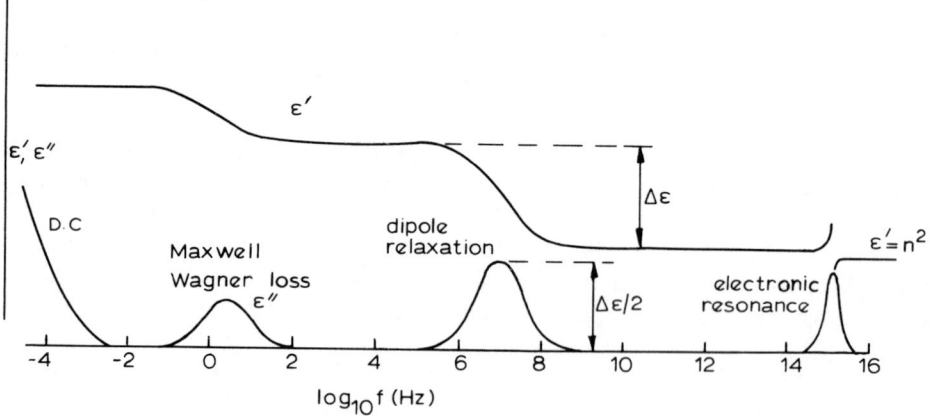

Figure 1.8 The variations of the dielectric parameters ϵ' and ϵ'' with frequency for various dielectric dispersion processes

where $G(\tau)\,d\tau$ is the fraction of molecules at a particular instant associated with a relaxation time between τ and $(\tau + d\tau)$.

Another factor to be considered is that since dielectric absorption is a measure of the energy dissipated in the material, then processes which are usually related to the d.c. conductivity σ can also contribute to the total dielectric absorption. At a frequency f the total dielectric loss ϵ''_T will be given by

$$\epsilon''_T = \epsilon'' + \sigma/2\pi f \epsilon_0 \tag{1.33}$$

from which it can be seen that the d.c. conductivity could contribute significantly to the absorption at low frequencies, but decreases in influence as the frequency is increased.

A schematic outline of how ϵ' and ϵ'' vary with frequency for the various dielectric dispersions mentioned is shown in Figure 1.8. The permittivity ϵ' tends to the square of the refractive index at optical frequencies and higher.

If a dielectric material exhibits a dispersion due to dipolar relaxation, then assuming the local field is given by the Lorentz formulation the low frequency permittivity ϵ_s is given by equation (1.14),

$$\frac{3(\epsilon_s - 1)}{\epsilon_s + 2} = \frac{N}{\epsilon_0}\left(\frac{\mu^2}{3kT} + \alpha_e + \alpha_a\right) \tag{1.34}$$

At the high frequency side of the dipolar dispersion the corresponding permittivity ϵ_∞ is given by

$$\frac{3(\epsilon_\infty - 1)}{(\epsilon_\infty + 2)} = \frac{N}{\epsilon_0}(\alpha_e + \alpha_a) \tag{1.35}$$

On subtracting equation (1.35) from (1.34)

$$\frac{\epsilon_s - 1}{\epsilon_s + 2} - \frac{\epsilon_\infty - 1}{\epsilon_\infty + 2} = \frac{N\mu^2}{9\epsilon_0 kT}$$

or

$$\frac{\epsilon_s - \epsilon_\infty}{(\epsilon_s + 2)(\epsilon_\infty + 2)} = \frac{N\mu^2}{27\epsilon_0 kT} \tag{1.36}$$

This equation can be used for gases, dilute solutions in non-polar solvents, and pure liquids of low polarity to obtain quite accurate values for the dipole moment μ from measurements of the dielectric dispersion. For materials where the Onsager formulation for the local electric field is more appropriate, then it can easily be shown (e.g. reference 7, p. 73) that

$$\frac{(\epsilon_s - \epsilon_\infty)(2\epsilon_s + \epsilon_\infty)}{\epsilon_s(\epsilon_\infty + 2)^2} = \frac{N\mu^2}{9\epsilon_0 kT} \tag{1.37}$$

Equations (1.36) and (1.37) are based on the assumption that no correlations or chemical associations occur between the polar molecules. This is not true for the hydroxylic compounds, for example, which form molecular complexes involving intermolecular and intramolecular hydrogen bonds. This often leads to an increased effective polarity and hence an increased dipole moment value. Following the theories of Kirkwood[19] and Fröhlich,[3] the most satisfactory modification for equation (1.37) to take into account such effects is to introduce the 'correlation parameter' g so that the factor $g\mu^2$ replaces μ^2. A value for g of unity implies no molecular associations or correlations and for hydrogen-bonded liquids values for g have generally been found to be in the region 1.5—4.0. The generally accepted value for g for water at 20 °C is 2.82, which falls with increasing temperature to a value of 1.64 at 300 °C.[20]

In polymeric systems, steric hindrances and dipole—dipole interactions often result in a decrease of the effective dipole moment value in interaction with the electric field. For such systems the correlation parameter g has a value of less than unity, and for many polymers in dilute solution values for g have been found to range from 0.5 to 0.8.[21,22] Values for g do not appear yet to have been determined for any polymers in the bulk solid state or viscoelastic fluid state.

The relation between ϵ' and ϵ'' (i.e. between dispersion and absorption) is quantitatively given by the so-called Kramers—Kronig relations:

$$\epsilon'(f) - \epsilon_\infty = \frac{2}{\pi} \int_0^\infty \frac{\epsilon''(f)F \, dF}{F^2 - f^2} \tag{1.38}$$

$$\epsilon''(f) = -\frac{2}{\pi} \int_0^\infty \frac{\epsilon'(f) - \epsilon_\infty \, dF}{F^2 - f^2} \tag{1.39}$$

where F is a dummy variable supplementing the frequency f. For example if the difference in permittivities on either side of a complete dispersion is $\Delta\epsilon$, then

$$\Delta\epsilon = \frac{2}{\pi} \int_{-\infty}^\infty \epsilon'' \, d(\ln f) \tag{1.40}$$

indicating that the total area under the dielectric loss curve of ϵ'' versus $\log_{10} f$ should equal $(\pi\Delta\epsilon)/(2 \times 2.3)$. Such a relationship represents a useful check for the consistency of experimental data (the factor 2.3 arises when comparing theory, such as equation 1.40 involving natural logarithms, with the experimental data more conveniently plotted as $\log_{10} f$) and shows that the total area under the ϵ'' loss curve is proportional to the total concentration of dipoles in the material and their dipole moment, irrespective of their distribution of relaxation times. This has been emphasized by Sillars[23] and Gevers[24] who show that using the theories of Debye and Onsager

$$\int_{-\infty}^\infty \epsilon'' \, d(\ln f) = \frac{\pi}{54\epsilon_0 kT} (\epsilon_s + 2)^2 \mu^2 N$$

$$= \frac{\pi}{2} (\epsilon_s - \epsilon_\infty) \tag{1.41}$$

As the distribution of relaxation times increases from that of the single relaxation time case, the ϵ'' versus $\log f$ curve becomes broader in extent, but the peak height (ϵ''_{max}) reduces so as to maintain the total area under the curve at an unchanging value. Reddish[25] has made use of this fact to define a parameter α' to represent the spread of relaxation times as

$$\alpha' = \frac{\pi \epsilon''_{max}}{\text{Area of } \epsilon'' \text{ loss curve}} \qquad (1.42)$$

where $\pi\epsilon''_{max}$ would equal the area under a single-relaxation Debye-type loss curve, so that α' falls from unity and tends to zero as the distribution tends to an infinite one. It can be said that the distribution factor α' of equation (1.42) has a more readily understandable physical meaning than the purely mathematical and empirical parameters β and γ in equations (1.30) and (1.31).

CALCULATION OF DIPOLE MOMENTS AND PERMITTIVITY

When a bond is formed between two dissimilar atoms, it will usually be the case that one of the atoms will be more electronegative than the other. The important consequence of this is that most covalent bonds constitute electrical dipoles; the bonding electrons having been drawn more to one atom than to the other. This results in the electrical centre of the electrons not being the same as that of the nuclei, the latter being virtually at the mass centre of the molecule. The magnitude of the bond dipole moment will depend primarily on the difference in electronegativity of the two elements forming it. The order of electronegativity is

$$H < C < N < O < F \quad \text{and} \quad I < Br < Cl < F$$

and similarly for other series and groups in the periodic table. In the formation of a bond the magnitude of the dipole moment increases with increasing separation of the atoms in the electronegativity series, the atom to the left becoming positive with respect to the other.

The dipole moments of bonds can be represented, approximately at least, as additive vector quantities acting in the direction of the chemical bond, and so the moment of the whole molecule is the vector sum of the constituent moments. A perfectly symmetrical molecule will therefore be non-polar, although it may contain polar linkages. A simple case of this is carbon tetrachloride in which the moments of four tetrahedrally directed C—Cl bonds cancel each other. Unsymmetrical compounds, however, are almost invariably polar; the presence of oxygen, nitrogen, or a halogen atom in a carbon compound makes this polarity very marked. It may be noted that in addition to polar and non-polar linkages, and polar and non-polar molecules, it is often the practice to speak of polar and non-polar groups. An alkyl radical (C_nH_{2n+1}), for example, is generally regarded as non-polar, although it has a small polarity, but —OH, —CN, —COOH, —NO$_2$, etc., are said to be polar groups. These ideas will be of great assistance to us in the next chapter when we begin to discuss the properties of polypeptide side chains.

Values for some bond and group dipole moments are given in Table 1.1. These values must only be treated as approximate ones representing the average taken over a whole range of chemical structures in which these bonds or chemical groups occur. The value of 0.4 debye given for the C–H bond is often quoted, but in fact is the one designation most open to question in that not only can this value approach zero, it can also be of reversed sign according to the degree of hybridization and electron delocalization of the bonding orbitals. When in doubt, the designation of a zero dipole moment for the C–H bond will not cause too serious an error.

Table 1.1   Some chemical bond and group dipole moment values, given in debye units

+ ← −	+ ← −	+ ← −	+ ← −
H—C	H—N	H—O	C—Cl
0.4[a,b]	1.31[a]	1.51[a]	1.46[a], 1.7[b]
+ ← −	+ ← −	+ ← −	+ ← −
C—N	C≡N	C—O	C=O
0.22[a]	0.9[a]	0.74[a]	2.3[a], 2.7[b]
− → +	− → +	− → +	+ ← −
C—OH	C—OCH$_3$	C—NH$_2$	C—NO$_2$
1.65[b]	1.3[b]	1.2[b]	3.1[b]

[a]C. P. Smyth[5]
[b]M. Davies[7]

As already indicated, the resultant dipole moment of a molecule or molecular group may be evaluated by adding vectorially the various group or bond moments. This can be done graphically, or by means of the equation

$$\mu_r^2 = \mu_1^2 + \mu_2^2 + 2\mu_1\mu_2 \cos \theta \tag{1.43}$$

where μ_r is the resultant of two groups or bond moments μ_1 and μ_2 (the values including the appropriate sign) which make an angle θ with each other. An example of this is shown in Figure 1.9(a) for the bond moments in the free water molecule. Figure 1.9(b) shows how the resultant dipole moment for the planar formamide molecule can be calculated from the constituent bond moments. Formamide is of considerable importance from the biological viewpoint, since it is the simplest molecule containing an N–C–O linkage characteristic of amides and polypeptides. The bond angles given in Figure 1.9(b) were obtained from the work of Hirota *et al.*[26] The calculations used to obtain the resultant dipole moment of formamide, as given in Figure 1.9(b), were:

$$\mu_x = 1.31 \cos(180 - 118.4) + 1.31 \cos(180 - 119.6) - 0.22$$
$$- 0.4 \cos(180 - 114.6) + 2.7 \cos(180 - 124.5) = 2.41 \text{ debye}$$

$$\mu_y = 1.31 \sin(180 - 119.6) - 1.31 \sin(180 - 118.4) + 2.7 \sin(180 - 124.5)$$
$$+ 0.4 \sin(180 - 114.6) = 2.58 \text{ debye}$$

$$\mu_r = (\mu_x^2 + \mu_y^2)^{\frac{1}{2}} = \underline{3.53 \text{ debye}}, \text{ with } \Theta = \tan^{-1} \frac{2.58}{2.41} = 46.95°$$

From Stark effect measurements, Kurland and Wilson[27] obtained the dipole moment for formamide as 3.71 debye at an angle of 39.6° from the C—N bond, so that the moment derived in Figure 1.9(b) differs by less than 5 per cent from this experimental value. If a molecule contains two or more dipoles, which as a result of rotation about bonds between atoms can change their relative orientations, then the

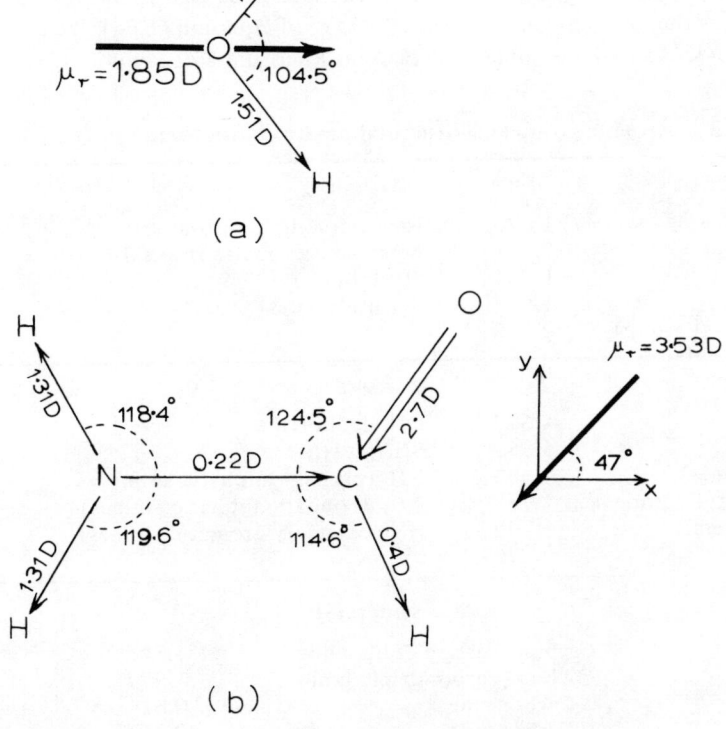

Figure 1.9 The bond moments and total dipole moment for (a) the free water molecule, and (b) the planar formamide molecule

calculation of the total dipole moment requires the averaging over all possible rotations (see, for example, Smyth,[5] p. 234). If rotation about any bond is hindered, by a potential barrier for example, then the problem is even more difficult, and the total effective dipole moment tends to increase with increasing temperature. Tables of experimental molecular dipole moment values are given in several books.[28-31]

It is also useful to be able to calculate the permittivity ϵ of non-polar solids, or the limiting high frequency permittivity ϵ_∞ for polar solids. The relationship between the molecular permittivity and total polarizability is given by equation (1.34). If ρ is the density of the material and M its molecular weight, then the number of molecules per unit volume is given by

$$N = \frac{N_A \rho}{M}$$

where N_A is the Avogadro number. Equation (1.34) can therefore be written as

$$P = \frac{\epsilon_\infty - 1}{\epsilon_\infty + 2} \cdot \frac{M}{\rho} = \frac{N_A}{3\epsilon_0}(\alpha_a + \alpha_e) \tag{1.44}$$

where P is the molar polarization or refraction, and the high frequency permittivity ϵ_∞ replaces the low frequency or static value ϵ_s of equation (1.34). We can see from equation (1.44) that the molar polarization P has units of volume; it is, in fact, a

Table 1.2   Atomic, group and structural polarizabilities (cm^{-3})

Atomic			
H(in CH_2)	1.028	N(tertiary aliphatic amines)	2.744
C(in CH_2)	2.591	N(tertiary aromatic amines)	4.243
O(ethers)	1.764	S(sulphides)	7.921
O(acetals)	1.607	S_2(disulphides)	16.054
O(carbonyl group)	2.122	Cl	5.844

Group			
CH_2	4.647	OH(alcohols)	2.546
CH_3	5.653	SH(thiols)	8.757
CO(ketones)	4.601	NH_2(primary aliphatic amines)	4.438
CO(methyl ketones)	4.758	NH(secondary aliphatic amines)	3.610
COO(esters)	6.200	NH(secondary aromatic amines)	4.678
COOH	7.226		

Structural	
Carbon–carbon double bond	1.575
Carbon–carbon triple bond	1.977
3-Carbon ring	0.614
4-Carbon ring	0.317
5-Carbon ring	−0.19
6-Carbon ring	−0.15

Table 1.3 Examples of the application of equation 1.44 and Table 1.2

$+CH_2—CH_2+_n$
Polyethylene

$+CH_2—C+_n$ (with phenyl group)
Polystyrene

$+CH_2—CH+_n$ (with Cl)
Polyvinylchloride

$+CH_2—CH_2—O—C(=O)—C_6H_4—C(=O)—O+_n$
Polyethylene terephthalate

Polymer	Density (g/cm^3)	P (cm^{-3})	M	ϵ_∞ (eqn 1.44)	ϵ (10 GHz)	ϵ (a)
Polyethylene	0.92	$9.29n$	$28.06n$	2.32	2.29	2.30
Polystyrene	1.04	$33.53n$	$104.14n$	2.51	2.53	2.49–2.55
Polyvinylchloride	1.54	$14.11n$	$62.50n$	2.60	2.76	—
Polyethylene terephthalate	1.39	$49.63n$	$192.18n$	2.68	2.78	2.80

(a) From Polymer Handbook, Eds. J. Brandrup & E. H. Immergut, J. Wiley & Sons, Inc., New York (1975)

molar volume and provides an approximate measure of the actual total volume (without free space) of the electronic clouds of the molecules in one gram mole, as distinct from the apparent volume M/ρ. As such P should be independent of temperature. In fact, the first reliable values for the radii r of ions in crystals were obtained from ionic refraction measurements using the identity

$$(\alpha_a + \alpha_e) = r^3$$

$$= \frac{n^2 - 1}{n^2 + 2} \cdot \frac{3M\epsilon_0}{N_A \rho}$$

Like other forms of molar volume, the molar polarization is an additive and constitutive property for a molecular system, although the value for any particular atom will depend on how it is bonded to another atom. For a very large number of compounds the molar polarization is approximately additive for the atoms, chemical groups and bondings present in the molecular structure. Table 1.2 summarizes various atomic, chemical group and structural contributions to the molar polarization, and is based on Vogel's[32] very extensive measurements of the refractive dispersion of the sodium D-line emission. The sodium D-line occurs at a frequency of 5.1×10^{14} Hz, which is low enough to ensure that P includes both atomic and electronic polarizabilities.

Examples of the application of equation 1.44 and Table 1.2 are given in Table 1.3 for one non-polar and three polar polymers, where the theoretically derived ϵ_∞ values can be compared with permittivity values obtained experimentally at 10 GHz using a resonating microwave cavity technique,[33] and with the literature values obtained at lower frequencies.[34] Equation (1.44) predicts quite accurately the high frequency permittivity of the non-polar polyethylene and the weakly polar polystyrene, but as might have been expected, the prediction is not quite so accurate for the more polar polymers, although the error for polyvinylchloride is still tolerable at 5.8 per cent. This error could result from either the frequency value of 10 GHz not being sufficiently high to avoid the dipolar dispersion, or in using an underestimated molar polarization for chlorine. Fletcher,[35] for example, quotes a polarization value of 5.97 for chlorine, as compared with the value 5.84 given in Table 1.2.

MODELS FOR DIELECTRIC ABSORPTION

In liquids we can generally assume that the molecular dipoles can point in any direction and are continually changing direction as a result of thermal agitation. Debye[1] has provided a very simple model where we may consider dipolar molecules as spheres whose rotation is opposed by the viscosity of the surrounding medium. The relevant relaxation time for such a system is

$$\tau = \zeta/2kT$$

where ξ is a molecular friction constant relating the torque produced by the external electric field to the molecule's angular velocity. Debye interpreted ξ by assuming the molecular dipole to respond to torque as would a rigid sphere of radius a turning in a hydrodynamic fluid of macroscopic viscosity η. For such a model Stokes' classical calculation gives $\xi = 8\pi\eta a^3$ hence

$$\tau = 4\pi\eta a^3 /kT \tag{1.45}$$

We shall see later in Chapter 3 that this simple model is particularly useful for interpreting the dielectric properties of dilute aqueous solutions of amino acids, peptides, and proteins. We shall also be reminded of the complicated nature of the structure of water involving hydrogen-bond associations. In this way it is of considerable interest to note that equation (1.45) predicts a surprisingly accurate τ value for water. Taking the molecular radius of water to be one-half the inter-oxygen distance at 1.4×10^{-10} m and the 293 K viscosity value as 10^{-3} kg/m s, then the 293 K value for τ is calculated as 8.5×10^{-12} s, in good agreement with the experimental value of 9.3×10^{-12} s (Hasted,[20] p. 47). Furthermore, the ratio of the τ's, of value 1.27, for D_2O and H_2O is very close to the ratio of their viscosities at 1.25.

In solids the dipoles will be more hindered than in the liquid model, and again, as in so much of dielectric theory, Debye was the first to introduce a useful physical model. His model[1] is that of a single type of dipole for which two

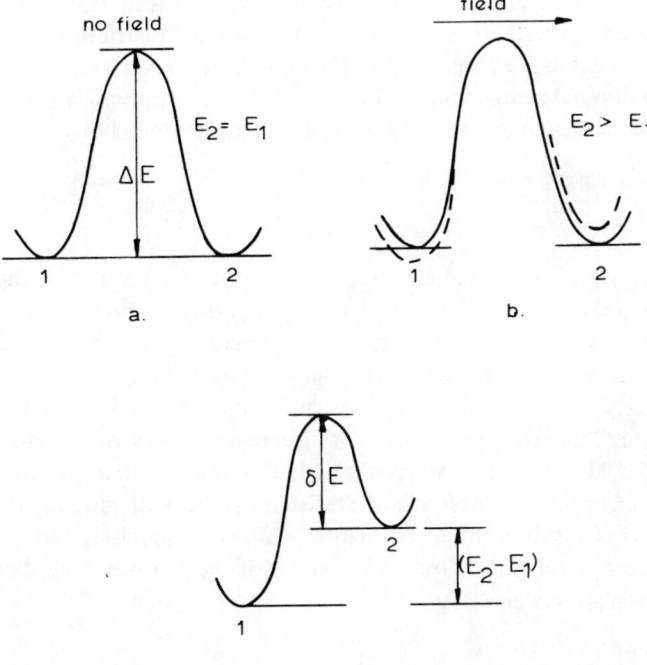

Figure 1.10 (a) The two-site potential energy barrier model for dipolar orientation, showing the effect (b) of application of an electric field, and the situation (c) where the two equilibrium sites are not equal

orientations are possible, these being alignments parallel and antiparallel to the external electric field. As a result of the crystalline structure of the dielectric material, these two orientations will be separated by a potential energy barrier ΔE, as shown in Figure 1.10(a).

For no electric field, the two sites are taken to be of identical energy, and hence equally populated, with identical probabilities P_{12} and P_{21} for transitions between them. Application of a constant field E at the molecule lowers the energy of parallel alignment relative to the other, as indicated by the dashed line in Figure 1.10(b). The new equilibrium state, with more molecules in site 1 is reached by a net excess of transitions from site 2 to 1. Fröhlich[3] discusses this two-site model in more detail, assuming transition probabilities for no field of the form

$$(P_{21})_0 = (P_{12})_0$$

$$= f_0 \exp(-\Delta E/kT)$$

where ΔE is the barrier height (per mole) and f_0 is the oscillation frequency in the potential minimum. On computing site populations, and net polarization as a function of time for a sinusoidal electrical field, this leads directly to a Debye-type relaxation equation with the relaxation time given by $1/\tau = 2(P_{21})_0$, and hence is determined by an Arrhenius type function of the form of equations (1.24) and (1.25). If there is a distribution of such barrier heights arising from fluctuations in molecular arrangements, then the result will be a broader distribution of relaxation times.[16] In many solids we might expect the equilibrium positions of the dipoles to be unequal, as illustrated in Figure 1.10(c). In this case, applying absolute reaction rate theory, the molecular relaxation time is given approximately by

$$\tau = \frac{h}{kT} \exp\left(\frac{\delta E}{kT}\right) \tag{1.46}$$

where δE is equal to the smaller energy barrier. As we have seen the value of τ can be found from dielectric loss measurements by finding the frequency f_m ($f_m = 1/2\pi\tau$) at which the loss factor ϵ'' reaches a maximum. Figure 1.11 can be used to obtain estimates of the potential energy barrier associated with the Debye-type relaxation processes such as that described by equation (1.24).

An interesting consequence arises for a barrier model when the zero-field two-site energies, E_1 and E_2, are not equal. For the situation shown in Figure 1.10(c) then according to Boltzmann statistics, there will always be a greater number of dipoles residing in site 1 than site 2. More precisely, for the general barrier model, the net average dipole moment residing in a site after relaxation has occurred between sites is given by

$$\langle \mu \rangle = \sum_i^z {}^0P_i \mu_i \tag{1.47}$$

where ${}^0P_i$ is the equilibrium probability of site i and μ_i is the dipole moment in site i.[36] This moment can only be relaxed if the reference coordinate frame is allowed to reorientate. Whereas such reorientation would occur for a liquid model, it cannot

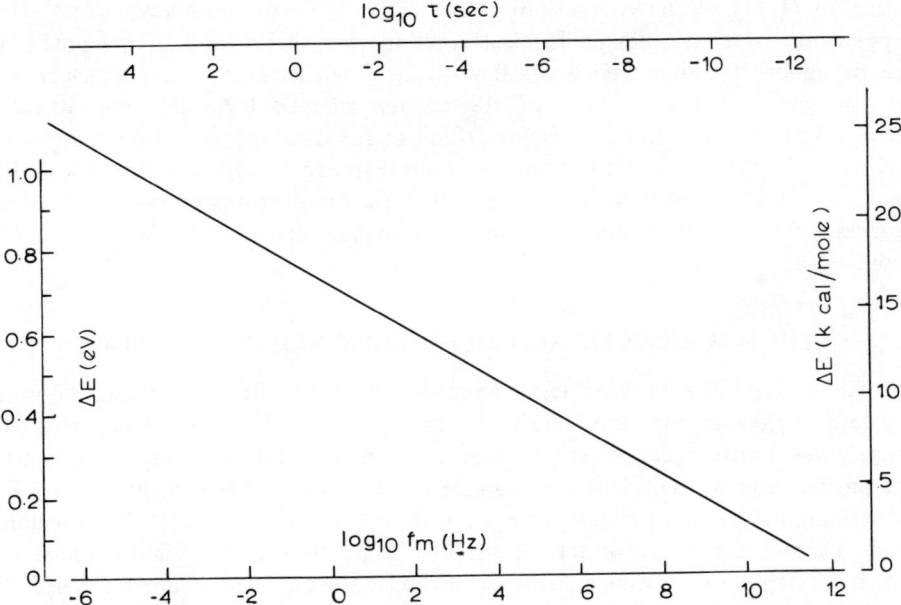

Figure 1.11 Variation of relaxation time τ and frequency of maximum dielectric loss f_m with barrier energy height for a temperature of 300 K

occur for solids. The factor μ^2 of equations (1.36) and (1.37), for example, represents the mean square of the total moment, more correctly written as $\langle \mu^2 \rangle$. For a barrier model, then the only observable dipolar relaxation or dispersion processes will be associated with those dipoles actually undergoing relaxation. The experimentally observed moment will have a total magnitude given by $\langle \mu^2 \rangle - [\langle \mu \rangle]^2$, where $[\langle \mu \rangle]^2$ is the unrelaxed component given by equation (1.47). As the temperature increases, then the magnitude of this unrelaxed component will decrease, and for the two-site model, the effective moment will increase by an amount related to the Boltzmann factor $\exp(E_1 - E_2)/kT$. Such an effect can tend to negate the prediction given by equations (1.36) and (1.41), that the dipolar dispersion will decrease with increasing temperature.

Extensions of the two-site barrier model have been developed by several authors[37-39] to include both two- and three-dimensional multisite models giving more than just one single relaxation time.

Many polar solids, e.g. aliphatic long-chain ketones and most ethers, exhibit no sizeable dielectric absorption even at temperatures approaching their melting point. This indicates that the energy barriers between the dipole equilibrium positions are very large (>40 kcal/mole). Other materials, such as long-chain esters and some ethers, do exhibit dielectric absorption at room temperature and have activation energies of the order 10 kcal/mole. On the other hand, many amorphous polymers, supercooled liquids, and solutions, exhibit low frequency (~1 kHz) room temperature dielectric relaxation processes resembling the Davidson–Cole function of equation (1.31), but have large activation energies of the order 65 kcal/mole.[40] The application of conventional chemical reaction rate theory and of the potential barrier model does not seem applicable for such high activation energies, which are of the same order as many bond dissociation energies (typically 60–150 kcal/ mole). For such materials we cannot think of the observed activation energies in terms of the energy required to reorient a solute molecule, for example. Instead the molecular process must be viewed in terms of a wholly cooperative reorientation process between solute and solvent molecules, which may even be defect diffusion controlled.

THE MACROSCOPIC AND MICROSCOPIC RELAXATION TIME

So far the symbol τ has been termed as the macroscopic dielectric relaxation time as defined through equation (1.19). In deriving equation 1.19 it can be appreciated that τ represents a decay time or time constant determined as an experimental parameter, and is equivalent for example to the circuit time-constant factor 'CR' describing the decay of charge from a capacitor C through a resistor R. How does this macroscopic decay or relaxation time relate to the intrinsic molecular or microscopic relaxation time τ_i, which is obviously the quantity required if dielectric relaxation phenomena are to be interpreted at the molecular level? The microscopic relaxation time τ_i is related to the mean square angular deviation of a molecule over a given period of time. It is of the order of magnitude of the time required for a given molecule, if fixed and then released, to revert to random orientation in the absence of any macroscopic polarizations which might result from external fields, or surrounding polar molecules, for example. If the molecule is envisaged to change positions by jumps over an energy barrier, then τ_i can be defined as one-half the average time a molecule waits before jumping (cf. Fröhlich's model for deriving equation (1.46)). A further, more comprehensive discussion of the distinction between the macroscopic and microscopic relaxation times can be found in Böttcher's book.[4]

For materials exhibiting a single decay or macroscopic relaxation time τ, Powles[41] has shown that the intrinsic microscopic relaxation time τ_i is given by

$$\tau_i = \frac{(2\epsilon_s + \epsilon_\infty)}{3\epsilon_s} \tau \tag{1.48}$$

This implies that using the reaction rate equation (1.26) involving τ, and neglecting any variations of ϵ_s with temperature, then the error in deriving the value of the entropy of activation is quite small and gives, at the most, an underestimate of only 0.8 cal/°C per mole. Furthermore, this not very large difference between τ and τ_i decreases as the distribution of total relaxation processes increases, and for an infinite distribution τ and τ_i have equal values.

Since the difference between the macroscopic and microscopic relaxation time results from local field effects and orientation correlations between dipoles, it is not immediately clear that we should expect that a single microscopic relaxation time will always lead to a process governed by a single macroscopic relaxation time. In fact, such a prediction is valid, as Williams[42] shows in his excellent review of the use of dipole correlation functions in the study of dielectric relaxations.

STATISTICAL MECHANICAL AND QUANTUM THEORIES

The macroscopic behaviour of a dielectric material under the influence of an external field reflects the response of *each* individual molecule, so that the observed dielectric relaxation times represent the average value for typically 10^{25} molecules per m^3. Statistical mechanics can provide the means by which to calculate the macroscopic behaviour of such an ensemble of dipoles. Cole[43] approached this problem using the so-called Kubo formalism and derived an expression for the average polarization of an ensemble of dipole moments in terms of a dipole moment autocorrelation function

$$\phi(t) = \frac{\langle \mu(0)\mu(t) \rangle}{\langle \mu(0)\mu(0) \rangle}$$

This function describes the way in which the dielectric responds to an applied electric field by denoting the time-varying change of the projection of the average dipole moment to an original direction at time $t = 0$. Detailed accounts of this method are given in the literature[44,45] and a general review has been given by Williams.[42]

The derivation of equation (1.7) describing the orientational polarizability used classical Maxwell—Boltzmann statistics where it is assumed that the angle θ between a dipole axis and the field can vary continuously. The use of quantum mechanics overcomes this assumption. Earlier theories, notably those of Pauli and Kramers using what can be called the older quantum theory, gave a different result to that of equation (1.7). However, the new quantum theory leads to essentially the same result as that originally derived by Debye. Details of such quantum theories can be found in several books.[2,46,47] The relevance of quantum theories is also emphasized by the existence of low temperature (<4 K) mechanical dispersions in polymers containing methyl groups which have been explained in terms of the quantum mechanical tunnelling of the methyl groups through rotational potential barriers.[48] Also, accurate calculations of the dipole moment value for molecules can really only be approached in terms of the Hamiltonian operator describing the nuclear and electronic charge interactions for the whole molecule. Unfortunately such calculations are extremely complex and at present can only be attempted for relatively small molecules if semi-empirical assumptions are to be excluded. In this way, the semiclassical approach based on Pauling's Valence Band Theory[49] still often provides the only practicable method.

Finally, we should add that the theory of dielectric processes continues to be developed. Some of the current topics involve extending the Debye equations to include inertia effects,[50] and in clarifying the situation arising from conflicting theories of dielectric relaxation.[51-53] Modifications of equations (1.30) and (1.31) have also been suggested to describe the dielectric behaviour of polymeric systems, and interpretations given of such general Debye functions in terms of distributions of relaxation times.[54,55] From the biological viewpoint, the use of dielectric measurements has had, and will continue to have, a very important rôle to play in the investigation and elucidation of the electrochemical, physical, and dynamic aspects of biomacromolecular systems. An increased awareness of the processes of interaction of electromagnetic power with biological tissue is also of considerable importance to the medical sciences. These various aspects of the dielectric properties of biological materials will form the subject matter of the next six chapters.

References

1. P. Debye, *Polar Molecules*, The Chemical Catalog Co. (1929) and Dover, New York (1947).
2. J. H. van Vleck, *Electric and Magnetic Susceptibilities*, Oxford, London (1932).
3. H. Fröhlich, *Theory of Dielectrics*, Oxford, London (1949).
4. C. J. F. Böttcher, *Theory of Electric Polarization*, Elsevier, Amsterdam (1952).
5. C. P. Smyth, *Dielectric Behaviour and Structure*, McGraw-Hill, New York (1955).
6. A. R. von Hippel, *Dielectrics and Waves*, Wiley, New York (1954).
7. M. Davies, *Some Electrical and Optical Aspects of Molecular Behaviour*, Pergamon, Oxford (1965).
8. N. G. McCrum, B. E. Read, and G. Williams, *Anelastic and Dielectric Effects in Polymeric Solids*, Wiley, London (1967).
9. V. V. Daniel, *Dielectric Relaxation*, Academic Press, New York (1967).
10. N. E. Hill, W. E. Vaughan, A. H. Price, and M. Davies, *Dielectric Properties and Molecular Behaviour*, Van Nostrand, London (1969).
11. P. Debye, *Z. Physik.*, **13**, 97 (1912).
12. P. Drude, *Z. Physik. Chem.*, **23**, 267 (1897).
13. H. A. Lorentz, *The Theory of Electrons*, Dover, New York (1952).
14. L. Onsager, *J. Am. Chem. Soc.*, **58**, 1486 (1936).
15. S. Glasstone, K. J. Laidler, and H. Eyring, *The Theory of Rate Processes*, McGraw-Hill, New York (1941).
16. W. Kauzmann, *Rev. Mod. Phys.*, **14**, 12 (1942).
17. R. H. Cole and K. S. Cole, *J. Chem. Phys.*, **9**, 341 (1941).
18. D. W. Davidson and R. H. Cole, *J. Chem. Phys.*, **18**, 1417 (1950).
19. J. G. Kirkwood, *J. Chem. Phys.*, **7**, 911 (1939).
20. J. B. Hasted, *Aqueous Dielectrics*, p. 90, Chapman and Hall, London (1973).
21. M. V. Volkenstein, *J. Polym. Sci.*, **29**, 441 (1958).
22. M. V. Volkenstein, *Configurational Statistics of Polymer Chains*, Interscience, New York (1963).
23. R. W. Sillars, *Proc. Roy. Soc. (London)*, **169**A, 66 (1939).
24. M. Gevers, *Philips Research Repts.*, **1**, 298 (1945/6).
25. W. Reddish, *Trans. Faraday Soc.*, **46**, 459 (1950).
26. E. Hirota, R. Sugisaki, C. J. Nielsen, and G. O. Sorensen, *J. Molec. Spectrosc.*, **49**, 251 (1974).
27. R. J. Kurland and E. B. Wilson, *J. Chem. Phys.*, **27**, 585 (1957).
28. J. W. Smith, *Electric Dipole Moments*, Butterworth, London (1955).
29. O. A. Osipov and V. I. Minkin, *Dipole Moments* (in Russian), Izv. Vys. Skola, Moscow (1965).
30. M. McClellan, *Tables of Experimental Dipole Moments*, Freeman, San Francisco (1965).
31. V. I. Minkin, O. A. Osipov, and Yu. A. Zhdanov, *Dipole Moments in Organic Chemistry* (in Russian) Izv. Khimia, Leningrad (1968).

32. A. I. Vogel, *J. Chem. Soc.*, **607–674**; 1804–1825 (1948).
33. T. E. Cross and R. Pethig, unpublished data.
34. *Polymer Handbook*, Eds. J. Brandrup and E. H. Immergut, J. Wiley & Sons, Inc., New York (1975).
35. K. A. Fletcher, *British Plastics*, Dec. 1959.
36. G. Williams and M. Cook, *Trans. Faraday Soc.*, **67**, 990 (1971).
37. J. D. Hoffman, *J. Chem. Phys.*, **23**, 1331 (1955).
38. B. M. Axilrod, *J. Res. NBS.*, **56**, 81 (1956).
39. R. J. Meakins, *Trans. Faraday Soc.*, **51**, 953 (1955).
40. M. F. Shears, G. Williams, A. J. Barlow, and J. Lamb, *J.C.S. Faraday II*, **70**, 1783 (1974).
41. J. G. Powles, *J. Chem. Phys.*, **21**, 633 (1953).
42. G. Williams, *Chem. Rev.*, **72**, 55 (1972).
43. R. H. Cole, *J. Chem. Phys.*, **42**, 637 (1965).
44. J. P. Boon and S. A. Rice, *J. Chem. Phys.*, **47**, 2480 (1967).
45. P. Resibois, *J. Chem. Phys.*, **41**, 2979 (1964).
46. D. W. Davies, *The Theory of the Electric and Magnetic Properties of Molecules*, J. Wiley & Sons, London (1967).
47. F. K. Fong, *Theory of Molecular Relaxation*, J. Wiley & Sons, New York (1975).
48. S. Reich and A. Eisenberg, *J. Chem. Phys.*, **53**, 2847 (1970).
49. L. Pauling, *The Nature of the Chemical Bond*, 3rd Ed., Cornell, Univ. Press, New York (1960).
50. B. Quentrec and P. Bezot, *Molec. Phys.*, **27**, 879 (1974).
51. U. M. Titulaer and J. M. Deutch, *J. Chem. Phys.*, **60**, 1502, 2703 (1974).
52. R. L. Fulton, *Molec. Phys.*, **29**, 405 (1975).
53. D. E. Sullivan and J. M. Deutch, *J. Chem. Phys.*, **62**, 2130 (1975).
54. B. K. Scaife, *Proc. Phys. Soc. (London)*, **81**, 124 (1963).
55. J. V. Bertelsen and A. Lindgard, *J. Polym. Sci: Polym. Phys. Ed.*, **12**, 1707 (1974).

Chapter 2

Intrinsic Dielectric Properties of Biopolymers

In later chapters we shall consider the properties of hydrated proteins, and will deal particularly with the contributions that the various forms of absorbed water can make to their dielectric and electronic properties. In this and the next chapter we shall consider those dielectric and electrochemical properties that are specifically related to the intrinsic compositional and structural properties of biopolymers in both the solvated and, what we might term, the 'dry' state. Also, we shall consider almost exclusively the most abundant and perhaps most remarkable class of biopolymers, namely the proteins.

THE α-AMINO ACIDS

Proteins, which play essential structural and catalytic roles in living systems, are built up from one or more linear chains of amino acids. In principle, the term *amino acid* could be used to refer to any compound that contains amino and acidic groups. From the standpoint of the biological sciences, this term is in fact restricted to the α-amino carboxylic acids isolated from natural sources. Well over 100 have in fact been isolated, but only 20 are commonly obtained upon hydrolysis of proteins. These acids invariably contain an amino group in a position alpha to a carboxyl group and can be represented by the following chemical structure

$$R-\underset{\underset{NH_2}{|}}{\overset{\overset{H}{|}}{C}}-COOH$$

where R is the variable side chain characterizing the particular amino acid. We will see that it is these side chains that determine to a very large part the dielectric, electronic, and electrochemical properties of protein systems. The chemical structures of these side chains for the 20 α-amino acids that are the building blocks of proteins are listed in Table 2.1. Residues of most of these amino acids occur in all proteins, and the composition of some proteins is shown in Table 2.2. More than a

36

Table 2.1 The naturally occurring α-amino acids

Name and structural formula	Abbreviation	Mol. wt.	Side-chain (R) property
Alanine —CH₃	Ala	89.1	Non-polar
Arginine —(CH₂)₃—NH—C	Arg	174.2	Polar; basic
Aspartic acid —CH₂—C	Asp	133.11	Polar; acidic
Asparagine —CH₂—C	Asn	132.12	Polar; neutral
Cysteine —CH₂SH	Cys	121.16	Polar; neutral
Glutamic acid —(CH₂)₂—C	Glu	147.13	Polar; acidic
Glutamine —(CH₂)₂—C	Gln	146.15	Polar; neutral
Glycine —H	Gly	75.07	Non-polar
Histidine —CH₂—	His	155.16	Polar; basic

Table 2.1   (continued).

Name and structural formula	Abbreviation	Mol. wt.	Side-chain (R) property
Isoleucine $-CH-CH_2-CH_3$ (with CH_3 on CH)	Ilu	131.18	Non-polar
Leucine $-CH_2-CH-CH_3$ (with CH_3)	Leu	131.18	Non-polar
Lysine $-(CH_2)_4-NH_2$	Lys	146.19	Polar; basic
Methionine $-(CH_2)_2-S-CH_3$	Met	149.21	Non-polar
Phenylalanine $-CH_2-$ (phenyl ring)	Phe	165.19	Non-polar
Proline (ring structure)	Pro	151.13	Polar; neutral
Serine $-CH_2OH$	Ser	105.10	Polar; neutral
Threonine $-CH-CH_3$ (with OH)	Thr	119.12	Polar; neutral
Tryptophan $-CH_2-$ (indole ring)	Trp	204.23	Polar; neutral
Tyrosine $-CH_2-$ (phenol ring, OH)	Tyr	181.19	Polar; acidic
Valine $-CH-CH_3$ (with CH_3)	Val	117.15	Non-polar

Table 2.2 Amino-acid residue content of some protein molecules

Residue	Equine cytochrome-c	Egg white lysozyme	Bovine ribonuclease	Bovine chymotrypsinogen	Bovine serum albumin
Ala	6	12	12	22	46
Arg	2	11	4	4	23
Asp	3	8	5	9	54
Asn	5	13	10	14	–
Cys	2	8	8	10	35
Glu	9	2	5	5	76
Gln	3	3	7	10	–
Gly	12	12	3	23	16
His	3	1	4	2	17
Ilu	6	6	3	10	13
Leu	6	8	2	19	61
Lys	19	6	10	14	58
Met	2	2	4	2	4
Phe	4	3	3	6	27
Pro	4	2	4	9	28
Ser	–	10	15	28	27
Thr	10	7	10	23	34
Trp	1	6	–	8	2
Tyr	4	3	6	4	19
Val	3	6	9	23	35
Total	104	129	124	245	575
Mol. wt.	13 400	14 600	13 700	25 000	66 700

century spans the time between the original discovery of glycine by Braconnot in 1820, and the isolation of threonine from casein by McCoy, Meyer, and Rose in 1935. To this list of amino acids should be added hydroxylysine and hydroxyproline, which appear only in collagen, the major protein of bone, tendon, skin, and other connective tissue, and thyroxine which appears to occur only in thyroglobulin.

THE POLYPEPTIDE CHAIN

The way in which linear chains of amino acids, called polyamino-acid or polypeptide chains, are built up by the formation of peptide bonds is shown in Figure 2.1(a). These peptide bonds take the form of an amide linkage between α-amino and α-carboxyl groups of adjacent amino acids. For each linkage formed

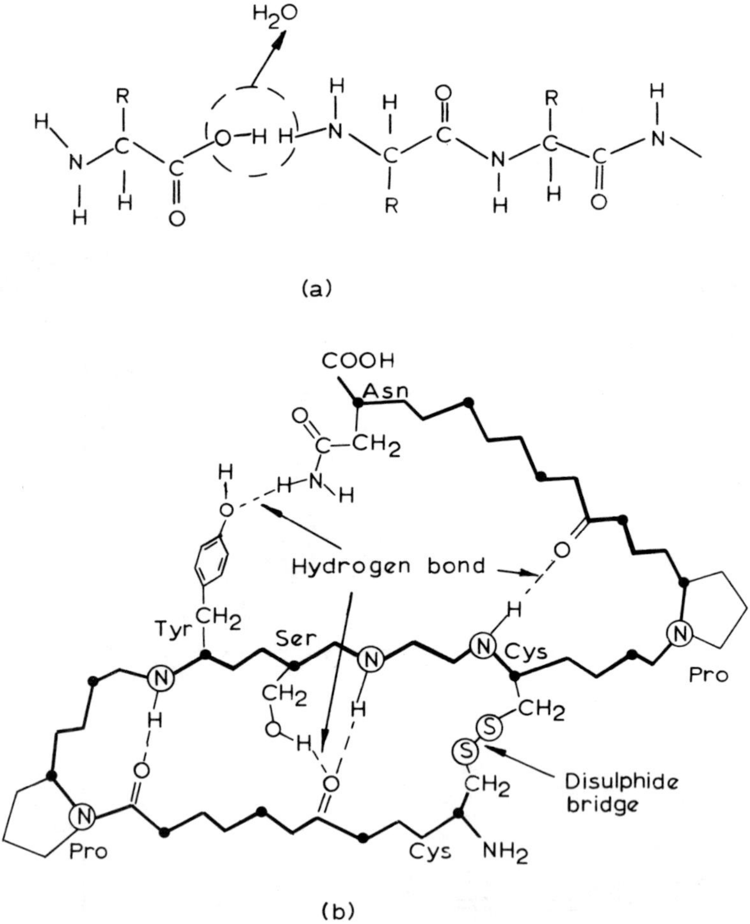

(a)

(b)

Figure 2.1 (a) The formation of a peptide bond in a polypeptide chain; (b) a typical polypeptide chain with various cross-linkages

there is the loss of one water molecule, so polypeptide chains are in fact formed of amino-acid residues. The three-dimensional structure of proteins is influenced by the way some residues cross-link with other residue members of the same, or of another, polypeptide chain, to form regions of α-helical or pleated β-sheet configurations for example. By far the most important linkage is formed by hydrogen bonds involving either the carbonyl C=O and amino N−H groups of the main chain, or the side chains of asparagine, glutamine, serine, threonine, and tyrosine. Another important linkage is the covalent disulphide bridge between the side chains of two cysteine residues. Disulphide bridges have bond strengths of the order 50 kcal/mole whereas the hydrogen bonds are much weaker at about 6 kcal/mole. The number of disulphide bridges in a particular protein can vary quite markedly. For example, the bovine serum albumin molecule has 17 disulphide bridges, whereas horse cytochrome-c has none. The two possible chain cross-linkages are demonstrated in Figure 2.1(b). A proline residue is included in Figure 2.1(b) to illustrate its property (as a result of its being the only amino acid where the side chain loops back to rejoin the main chain) of forcing bends in an otherwise straight chain. The relationship between the structure and function of biomacromolecules is most interesting and obviously of fundamental importance. More of this subject can be learnt from the enjoyable and informative books by Dickerson and Geis,[1] and Wold,[2] for example.

The polypeptide backbone of proteins consists, then, of a repeated sequence of three atoms and their associated substituents. In principle we might expect that rotation could occur about any of the three bonds, as is shown in Figure 2.2. However, X-ray crystallographic studies of simple peptides show that whereas the N−C_α bond length is 1.46 Å, as expected for a single bond, the C−N peptide bond is 1.33 Å. This length is only a little longer than the value of 1.27 Å for the average C=N bond length in model compounds, and is taken as strong evidence for double-bond character in the peptide bond. Also, X-ray studies show that the six atoms $C_{\alpha'}$ NHCOC$_\alpha$ are, or very close to being, coplanar. The basic dimensions, geometry, and coplanar nature of the peptide bond are shown in Figure 2.3.

The picture of the peptide group given by Figure 2.2 should therefore be modified to that of Figure 2.4, which shows the delocalization of an electron from the carbonyl double bond. The result is that the C−O bond is less than double, and the C−N bond has partial double-bond character, with the bonding electrons occupying a π-like molecular orbital that extends over all three atoms. This delocalized electronic structure of Figure 2.3 gives the peptide unit an increased energy of stabilization of around 20 kcal[3] compared with the structure of Figure 2.2. Energy is therefore required to rotate the amido link about the C−N axis, a

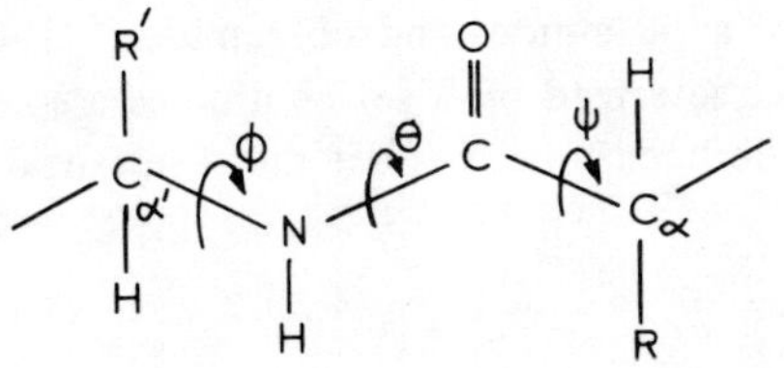

Figure 2.2 Bonds in the peptide unit and rotations which could possibly occur

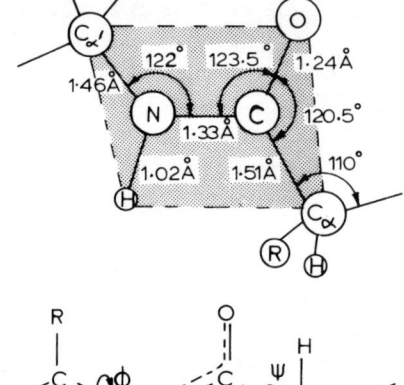

Figure 2.3 The geometry and coplanar nature of the peptide unit. The dimensions and angles are as those given by Ramachandran and Sasikekharen[3]

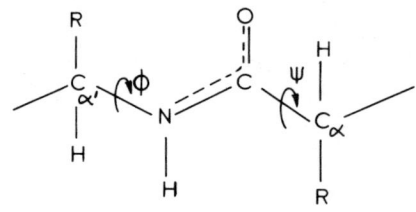

Figure 2.4 Rotation about the N–C bond is greatly hindered as a result of the delocalization of bonding electrons in the N–C–O atomic system

rotation through Θ degrees requiring an energy equal to $20 \sin^2 \Theta$ kcal/mole. As a result of this restriction, only two unique rotatable bonds are allowed per peptide in the polypeptide backbone, defined by the angles ϕ and ψ in Figure 2.4. These rotation angles have become the basis for modern conformational analyses of polypeptides.

THE PEPTIDE DIPOLE MOMENT

The study of interatomic forces has also been extensively developed in calculating the forces stabilizing biopolymers. Of considerable relevance to us is the estimation of electrostatic contributions to the interatomic interaction potential energies. From the concept of electronegativity it will be evident that different atoms in a peptide share differently in the ground state valence electron distribution. The interaction between these partially charged atoms is characterized by an electrostatic potential energy given by

$$V_{es} = \frac{-e_i e_j}{4\pi\epsilon_0 \epsilon r_{ij}} \tag{2.1}$$

where e_i and e_j are the residual charges separated by distance r_{ij}. No really satisfactory method is available for estimating an appropriate molecular permittivity ϵ in the region of the atoms, but a value of 3.5 is commonly used, based on high frequency dielectric measurements of solid amides and polyamides.[4] The net result of the charge separation can be characterized by a set of dipoles associated with the polar amide groups of the peptide backbone, so that the resultant dipole moment μ_p is defined by

$$\mu_p = \Sigma e_i e_j r_{ij}$$

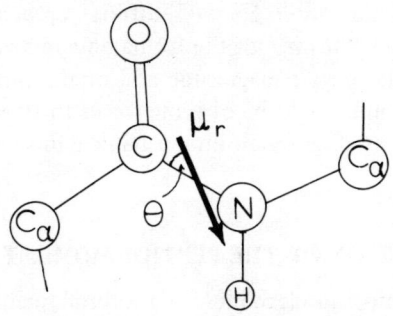

Figure 2.5 The permanent dipole moment vector of the peptide unit. Various values for μ_r and θ are given in Table 2.3

It is reasonable to assume that the permanent dipole moment for the peptide unit will have a magnitude of the same order as the value 3.71 D obtained by Kurland and Wilson for the formamide molecule described in Chapter 1 (Figure 1.9(b)). The value 3.7 D has in fact been adopted by several workers. For example, Brant et al.[5] have described the electronic charge distribution within the peptide groups in terms of a point dipole moment vector of magnitude 3.7 D, located in the plane of the peptide unit at the midpoint of the peptide bond, and making an angle of 56° with this bond. This assigned magnitude is also consistent with the observed[6] dipole moments of the order 3.8 debye for a number of alkyl amides such as the methyl and dimethyl formamides and acetamides.

The bond moment values that have been used by Brant et al.[5] and Scheraga et al.[7] as a simplified description of the resultant dipole moment vector μ_r of the peptide group, as shown in Figure 2.5, are given in Table 2.3. The convention used in Figure 2.5 to show the direction of μ_r is the same as that employed in Figure 1.9 and Table 1.1, where the moment vector is directed away from the negative charge centre towards the positive centre. Using this convention, which we note is not always the one adopted in the literature, the dipole vectors will tend to be aligned with the applied electric field direction. Chemists tend to adopt the opposite convention to describe dipole moment directions. Physicists use the convention adhered to here, and needless to say they consider it to be the more correct method! Also included in Table 2.3 are bond moment values derived from Table 1.1, which can be seen to give a resultant dipole moment of similar value and orientation as that obtained from the empirical values of Brant and Scheraga. This

Table 2.3 Constituent bond moment values (debye units) for the peptide group of Figure 2.5, and the resultant moment μ_r and orientation θ

Bond	References		Table 1.1
	5	7	
μ(C–O)	2.35	2.48	2.7
μ(C–N)	Zero	0.21	0.22
μ(C$_\alpha$–N)	Zero	Zero	0.22
μ(H–N)	1.35	1.31	1.31
μ_r	3.67	3.67	3.8
θ	55.5°	58.5°	55.4°

resultant dipole moment orientation θ, of the order $55-58°$, differs appreciably from that of $39.6°$ obtained by Kurland and Wilson[9] for the formamide molecule. Some doubt should then be attached to the precise magnitude and orientation of the dipole moment for the peptide group, and it will be of value for us to attempt an estimation independently for ourselves, based on quantum mechanical molecular orbital theory.

QUANTUM MECHANICAL CALCULATION OF THE PEPTIDE MOMENT

To begin with, we note that not even the most modern molecular orbital quantum theories are capable of reproducing either theoretical or experimental dipole moments consistently.[10] It will, therefore, be sufficient for us to use a relatively simple method, devised by Del Re,[11] that is capable of deriving dipole moments to within 10 per cent accuracy, compared with observed values, for many saturated organic molecules. Del Re's method is based essentially on the Hückel method[12,13] used for π-electron systems, but instead of treating π-electron-type delocalized orbitals, electrons forming σ-bonds are considered as strictly localized units. Also, inductive effects are introduced to describe the influence of neighbouring atoms for each bond of the molecule.

Without considering the theoretical details in depth, we will just note that in conventional molecular orbital theory (LCAO—MO theory) the bonding between two atoms i and j is described in terms of the appropriate atomic orbitals X_i and X_j as

$$\Psi = C_1 X_i + C_2 X_j$$

For such a localized bond, and considering the normalization requirement, the associated net excess of bond charge appearing on atom i (assumed of equal magnitude to the net deficit on atom j) is given by

$$qi(j) = 1 - 2C_1^2 \tag{2.2}$$

Using the so-called variation method ($\partial E/\partial C_1 = 0$, etc.), the minimum bond energy E is expressed in terms of the linear system of equations

$$\left. \begin{aligned} C_1(H_{ii} - E) + C_2(H_{ij} - ES_{ij}) = 0 \\ C_1(H_{ij} - ES_{ij}) + C_2(H_{ij} - E) = 0 \end{aligned} \right\} \tag{2.3}$$

with the permitted values of E corresponding to the roots of the secular determinant

$$\begin{vmatrix} H_{ii} - E & H_{ij} - ES_{ij} \\ H_{ij} - ES_{ij} & H_{jj} - E \end{vmatrix} \tag{2.4}$$

The terms H_{ii}, H_{jj}, and H_{ij} are the coulomb and resonance integrals respectively. S_{ij} is the overlap integral, and is of no importance for our method. The H terms are conventionally written as

$$\left.\begin{array}{l} H_{ii} = \alpha + \delta_i\beta \\ H_{jj} = \alpha + \delta_j\beta \\ H_{ij} = \mathscr{E}_{ij}\beta \end{array}\right\} \tag{2.5}$$

with α and β being two basic parameters whose values are not required, and $\mathscr{E}_{ij}$ is the so-called exchange integral. Del Re introduces two other parameters, δ_i^0 and $\gamma i(j)$, defined according to

$$\delta_i = \delta_i^0 + \sum_{j \text{ adj. to } i} \gamma i(j)\delta_j \tag{2.6}$$

with $\gamma i(j)$ being the inductive parameter indicating the fraction of the coulomb integral of j to be added to that of i. From equations (2.3), (2.4), and (2.5), we obtain for the bond charge

$$qi(j) = \frac{\delta_j - \delta_i}{2\mathscr{E}_{ij}}\left(1 + \left(\frac{\delta_j - \delta_i}{2\mathscr{E}_{ij}}\right)^2\right)^{-\frac{1}{2}} \tag{2.7}$$

In his paper, Del Re quotes the approximate result

$$qi(j) = \frac{\delta_j - \delta_i}{2\mathscr{E}_{ij}}$$

We will use equation (2.7). To obtain the bond charges the procedure is simply one of obtaining a series of equations of the form of equation (2.6) for each atom in the molecule, solve for the various unknowns and apply equation (2.7). The total net charge of an atom is then given by

$$Qi = \sum_j qi(j) \tag{2.8}$$

and the bond moment μ_{ij} is obtained from

$$\mu_{ij} = qi(j) \times r_{ij} \tag{2.9}$$

where r_{ij} is the bond length between atoms i and j. The values of the empirical parameters of equations (2.6) and (2.7) are obtained from Table 2.4 as derived by Del Re.

Table 2.4   Some of the Del Re[11] parameters for bond charge calculations

	C–H	C–C	C–N	C–O	N–H
$\mathscr{E}_{ij}$	1.00	1.00	1.00	0.95	0.45
$\gamma i(j)$	0.3	0.1	0.1	0.1	0.3
$\gamma j(i)$	0.4	0.1	0.1	0.1	0.4
δ_i^0	0.07	0.07	0.07	0.07	0.24
δ_j^0	0.00	0.07	0.24	0.40	0.00

We will now apply this method to calculate the various σ-bond charges and dipole moments for the peptide unit of polyglycine, as shown in Figure 2.6. The system of equations corresponding to equation (2.6) take the form, using the parameters of Table 2.4, of:

$$\delta H_1 - 0.4\delta N = 0$$

$$\delta H_2 - 0.4\delta C_\alpha = 0$$

$$\delta H_3 - 0.4\delta C_\alpha = 0$$

$$\delta C - 0.1\delta O - 0.1\delta N - 0.1\delta C_\alpha = 0.07$$

$$\delta C_\alpha - 0.1\delta N - 0.3\delta H_2 - 0.3\delta H_3 - 0.1\delta C = 0.07$$

$$\delta N - 0.3\delta H - 0.1\delta C_\alpha - 0.1\delta C = 0.24$$

$$\delta O - 0.1\delta C = 0.4$$

Solving for the 7 unknown values of these equations, we obtain:

$$\delta H_1 = 0.123; \quad \delta H_2 = \delta H_3 = 0.061; \quad \delta C = 0.157;$$
$$\delta C_\alpha = 0.152; \quad \delta N = 0.307 \quad \text{and} \quad \delta O = 0.416.$$

The σ-bond fractional electronic charges, derived using equations (2.7) and (2.8) and the data of Table 2.4, are shown for the peptide group in Figure 2.6(a). Figure 2.6(b) shows the associated σ-bond dipole moments, derived using equation (2.9) and the bond dimensions given in Figure 2.3. Using the same method as that outlined for the peptide group, the σ-bond charges and bond moments have also been derived for the formamide molecule, and these are shown in Figures 2.6(c) and (d).

Having considered the σ-bond orbitals, consideration must now be given to the most important feature of the peptide group and formamide, namely the double-bond character of the C—N bond. This structural feature can be described in terms of a resonance between the two valence-bond structures:

Our problem is two-fold; to assign dipole moments for the C—O and C—N bonds in accordance with the extent of their double-bond character, and to derive an estimate of the degree of ionic character for the peptide group.

Pauling[14] has derived the empirical formula

$$D_n = D_1 - (D_1 - D_2)\frac{1.84(n-1)}{0.84n + 0.16} \tag{2.10}$$

in which D_n is the interatomic distance for a bond of intermediate type, D_1 and D_2 that for a single and double bond respectively, and n the bond number. Using the idealized bond data of Cox and Jeffrey,[15] namely C—N(1.47 Å), C=N(1.28 Å), C—O(1.43 Å), and C=O(1.18 Å), and the formamide and peptide bond data of

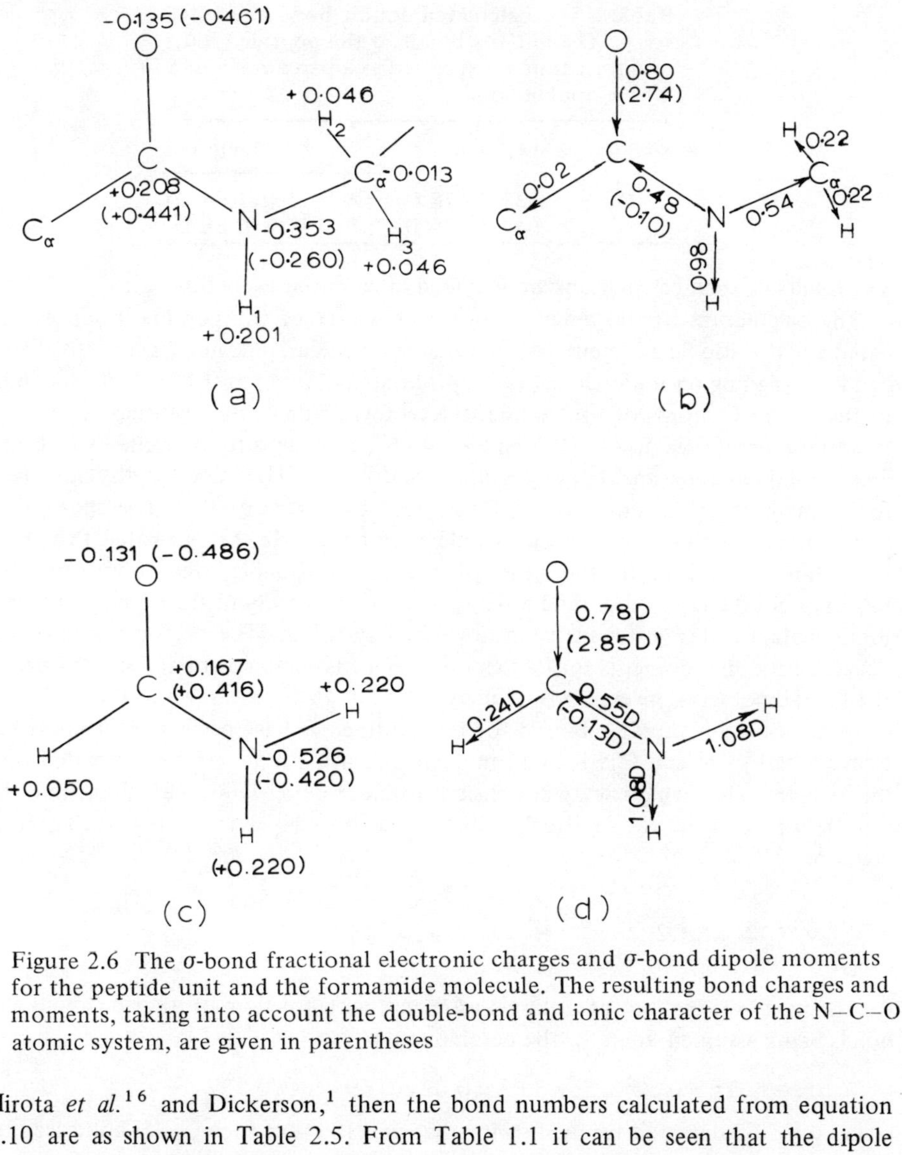

Figure 2.6 The σ-bond fractional electronic charges and σ-bond dipole moments for the peptide unit and the formamide molecule. The resulting bond charges and moments, taking into account the double-bond and ionic character of the N—C—O atomic system, are given in parentheses

Hirota *et al.*[16] and Dickerson,[1] then the bond numbers calculated from equation 2.10 are as shown in Table 2.5. From Table 1.1 it can be seen that the dipole moments for double-bonded atoms are considerably greater than those for the corresponding single-bonded atoms, the increase being 1.56 D and 0.68 D for the C=O and C=N bonds respectively.

To a first approximation we will assume that the moments for the intermediate type bonds are given in terms of the σ-bond moments and bond numbers according to

$$\mu(C\!\cdots\!O) = \mu(C\!-\!O) + 1.56(n - 1), \quad \text{and}$$

$$\mu(C\!\cdots\!N) = \mu(C\!-\!N) + 0.68(n - 1).$$

Table 2.5   Calculated double-bond character
for C—O and C—N bonds of the peptide group
and formamide, expressed as a percentage of a
whole double bond

Bond	Peptide	Formamide
C—O	63% (1.78 D)	75% (1.94 D)
C—N	64% (0.92 D)	47% (0.88 D)

The results of such calculations are included in parentheses in Table 2.5.

The calculation of the extent of ionic character of the peptide group will be based on the dipole moment vector obtained by Kurland and Wilson[9] for formamide. Assigning an ionic charge (of appropriate polarity) of 0.155 electronic units to the N and O atoms of formamide, the vectorial sum of the resulting constituent bond moments gives μ_r = 3.71 D and θ = 43.6°, compared to the values 3.71 D and 39.6° obtained experimentally by Kurland and Wilson. The bond lengths and angles for formamide are assumed to be those given by Hirota et al.[16] The appropriate ionic charge for the peptide group would then be 0.161e, being related to that of formamide according to the ratio of their total double-bond characters. The resulting bond charges and bond moments taking into account the double-bond and ionic character, are given in parentheses in Figure 2.6. The vectorial sum of the constituent bond moments for the peptide group, assuming planarity and neglecting the C_α—H moments, gives μ_r = 3.63 debye and θ = 46.7°, as shown in Figure 2.7.

In deriving this dipole moment for the peptide, we have neglected the contribution of the C_α—H and C_α—R bond moments. Leaving aside effects associated with the side-arm (R), which we will consider in more detail later, we will assume the C_α—H dipole can rotate freely about the C_α—N axis. Using the general equation[17,18]

$$\mu^2 = \sum_{j=1}^{n} \mu_j^2 + 2 \sum_{j=1}^{n} \sum_{s<j} \prod_{k=j}^{s+1} \cos \theta_k \mu_j \mu_s$$

for the mean-square sum of n dipole moments with freedom of rotation about the bonds being assumed, then for the peptide group

$$\mu^2 = 3.63^2 + 0.22^2 - 2 \times 3.63 \times 0.22 \cos 69° \cos 70.5°$$

giving μ = 3.62 debye, a result showing that the effects of such C_α—H bonds can be ignored.

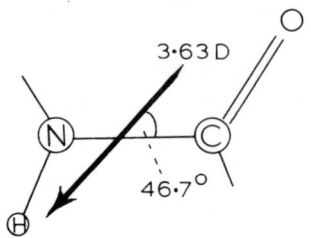

Figure 2.7 The permanent
dipole moment vector of the
peptide unit, as derived using
the quantum mechanical
molecular orbital theory of
Del Re[11]

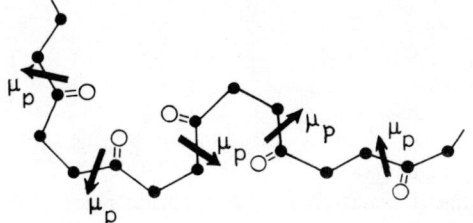

Figure 2.8 A schematic representation of a polypeptide chain as a string of peptide dipole moments

Finally, we note that we have only considered the peptide dipole moment arising from the electron distributions in their ground states. As electrons are excited from their ground states to higher energy levels, a redistribution of charge occurs and the magnitude and orientation of the dipole moment will alter. Such dipole variations are obviously of fundamental relevance to the optical spectroscopic properties of biopolymers.

CONFORMATIONAL EFFECTS

With the peptide unit possessing a permanent dipole moment μ_p, polypeptide chains can be considered as strings of connected dipole moments, as shown schematically in Figure 2.8.

From Chapter 1 (e.g. equation 1.7) we know that a molecule's dipole contribution to the permittivity of a medium is proportional to μ^2. For a completely rigid linear polymer chain of n regularly spaced dipolar units of moment μ, fixed and directed normal to the chain, the total contribution to the permittivity will either be zero or of the order of $(n\mu)^2$ depending upon whether the dipoles vectorially cancel or are additive. If these dipolar units are perfectly free to rotate, the contribution will be $n\mu^2$. For most polypeptide systems n will be a large number, typically of the order 10^2-10^3, so that the peptide unit dipole contribution to the permittivity will be extremely sensitive to the polypeptide configuration and to any hindering of internal rotations.

Debye and Bueche[19] have considered the dipole moments of polar polymers in relation to their structure, and show that the total mean-square dipole moment is given by

$$\langle \mu_t^2 \rangle = 3 \left\langle P \left\{ \sum_1^n \sum_1^n (\mu_n f)(\mu_n f) \right\} \right\rangle_{AV}$$

In this equation μ_n is the vector magnitude of the nth dipole of the chain and f is a unit vector in the direction of the applied electric field E. The sums extend over all of the n dipoles on the chain and the average is to be taken over all of the possible chain configurations and orientations, with P being the probability of occurrence of any particular chain configuration. In the case of a carbon–carbon chain with a dipole of moment μ coming off every other atom, at an angle β with the preceding C–C bond and angle γ with the following C–C bond, then for the case of free rotation about each carbon atom

$$\langle \mu_t^2 \rangle = n\mu^2 \left[1 + 2 \cos \theta \cos \beta \cos \gamma / (1 - \cos^2 \theta) \right] \qquad (2.11)$$

with θ being the C—C valence angle.

With there being restriction of rotation about the C—N bond in the peptide unit, and by modifying the peptide geometry slightly to give $\angle C_\alpha NC = \angle NCC_\alpha = 120°$, then equation (2.11) can be used as an approximation for polypeptide chains, so that for a completely random polypeptide chain the resultant dipole moment is given by

$$\langle \mu^2 \rangle \simeq 1.1 \, n\mu_p^2$$

We have ignored again, for the present, the possibility that the side chains (R) may possess dipole moments. If rotations about the peptide group linkages are restricted, by steric hindrances for example, then the experimentally determined dipole moment will not be that given by equation (2.11). The measured value will depend on the detailed nature of the potential energy barriers restricting molecular rotations, and the result could be an enhancement, partial cancellation, or even complete cancellation of the total dipole moment. A good example where the peptide group moments are totally additive is the extended, rigid, α-helix configuration, as demonstrated in Figure 2.9(a). Homopolypeptides such as poly-γ-benzyl-L-glutamate (PBLG) are known to have extended α-helix configurations. From measurements of the molecular weight dependence of the dipole moment of PBLG polymers in the helix-stabilizing solvent 1,2-dichloroethane, Wada[20] concluded that the individual peptide residues contribute 3.4 debye units each, in a direction parallel to the helix axis. This contribution of 3.4 D will represent not only the peptide moment μ_p contribution, but will also include solvent, hydrogen bonding, and polar side-chain effects. We shall see in the next chapter that solvent effects will tend to enhance the measured magnitudes. The α-helix configuration is stabilized

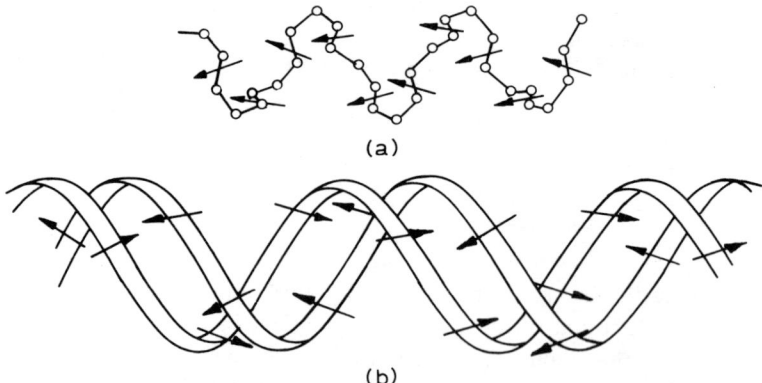

(a)

(b)

Figure 2.9 (a) Diagram illustrating the way in which the peptide moments are additive in an α-helical configuration; (b) the double helix of the DNA molecule, to show schematically how the peptide moments cancel to give no net permanent dipole moment

by a regular network of intramolecular hydrogen bonds involving the carbonyl C=O and amino N—H groups of the main chain. Each peptide residue is successively connected by such hydrogen bonding to the residue four units further along the chain, and we can envisage the possible valence-bond structures[21,22] of such hydrogen bonds to include the following:

C=O $\cdots$ H$^+$N$^-$ (ionic, no charge separation)
C=O$^+$—HN$^-$ (covalent OH bond, with charge separation)
C=O $\cdots$ H$^-$N$^+$ (ionic, no charge separation)
C=O$^+$ H$^-$N (covalent ON bond, with charge separation)
C=O H—N (covalent HN bond)

Four of these structures involve charge transfer, with three of them of such polarity as to enhance the effect of the residue moment of PBLG, as has already been indicated by Wada.[23]

The axis of a rigid polypeptide α-helix is directed at an angle of about $56°$ to the C—N bond of the constituent peptide residues, so that the peptide contribution to the total moment parallel to the helix axis will be

$$\mu_{/\!/} = n\mu_{\mathrm{p}}.\cos(56° - \theta) \text{ Debye units}$$

From Figure 2.7, then

$$\mu_{/\!/} = 3.63n \cos 9.3°$$
$$= 3.58n \text{ D}$$

Wada's result of 3.4n D, which we have seen will tend to overestimate the contribution of the residue moments, would suggest that the effect of the dipolar side chains of PBLG is such as to oppose the main-chain residue moments.[23]

The calculation of the 'vacuum' dipole moment value of macromolecules such as PBLG can be made using a modification[24] of a theory first derived by Buckingham.[25] According to this theory the 'vacuum' dipole moment can be calculated for an ellipsoidal molecule in an infinitely dilute solution to take into account its molecular shape and the action of the Onsager-type reaction field on the surrounding medium. The corresponding 'vacuum' dipole moment value μ_{res} per peptide residue of rod-shaped molecules such as PBLG is then given by

$$\mu_{\mathrm{res}} = \langle \mu^2 \rangle^{1/2}(\alpha/\beta)^{1/2} \qquad (2.12)$$

where $\langle \mu^2 \rangle$ is the mean-square dipole moment per peptide residue derived from dielectric measurements such as those of Wada.[20] In equation (2.12)

$$\alpha = \frac{4\epsilon_1^2 + 4n_1^2\epsilon_1 - 2n_1^4 + 3n_1^2}{(2\epsilon_1 + n_1^2)^2}$$

and

$$\beta = \frac{\epsilon_1(2\epsilon_1 + 1)}{3} \left(\frac{1 + (n_2^2 - 1)A}{\epsilon_1 + (n_2^2 - \epsilon_1)A} \right)^2$$

where $A = [\ln(2p) - 1]/p^2$ for macromolecules having an axial ratio p of $p > 10$. In these equations ϵ_1 is the permittivity of the solvent, and n_1, n_2 are the refractive indexes of the solvent and solute, respectively. On the basis of measurements by Doty et al.[26] the axial ratio for PBLG can be taken as roughly one-tenth of the value of the weight-average degree of polymerization. Using equation (2.12), then for PBLG in dichloroethane solution the following values for μ_{res} have been determined; 4.0 debye,[23] 4.2 debye,[27] and 4.11 debye.[28] For low molecular weight PBLG in m-cresol solution, the value of 4.7 debye was derived for μ_{res}.[29] Recent measurements for PBLG in m-cresol by Nakamura et al.,[30] using a newly developed pseudo-random noise dielectric spectrometer, indicate that the residue dipole moment value decreases with an increasing degree of polymerization as shown by the following list of molecular weights and corresponding dipole moment values; 74 000 (4.5 D), 95 000 (4.4 D), 150 000 (4.0 D), and 210 000 (3.7 D). This result was assumed to be associated with the flexibility of the helices, and can possibly be taken as evidence to show that the contribution of the side-chain dipole moments is to oppose that of the main-chain residue moments.

Following the work of Wada and Kihara[31] the dipole moment per residue of a helical chain can be written as

$$\mu_{res} = \mu_0 \{ [\exp(-2DL) - 1 + 2DL]/2D^2 L^2 \}^{1/2} \qquad (2.13)$$

where L is the helix chain length, and D is related to the flexial rigidity of the helix being equal to the reciprocal of twice the persistance length characterizing the conformation of the helical chain. From the dielectric data obtained at 70 °C for PBLG in m-cresol,[30] the value for μ_0 in equation 2.13 is determined to be 5.3 ± 0.5 debye and the persistance length value is 540 ± 5 Å. The dielectric results obtained by Nakamura et al.[30] also show that the rotational relaxation time τ for PBLG in m-cresol is related to the size and shape of the macromolecules very accurately according to Perrin's equation[32] for prolate ellipsoids;

$$\frac{\tau T}{\eta} = \frac{\pi L^3}{6k[\ln(K/b) - 1/2]}$$

where η is the viscosity of the solvent at temperature T, b is the semi-minor axis of the equivalent ellipsoid of revolution for the molecule of length L, and k is Boltzmann's constant.

Before considering the effects of polar side chains in more detail, we will note that the α-helix configuration contains 3.6 peptide residues per turn, so even a modest helix of just ten turns will have a dipole moment of the order 120 D. One of the most important biopolymers of life, DNA, which is commonly found to have a molecular weight in the range $10^6 - 10^9$, is composed of two α-helices. But these two helices point in opposite directions,[33] so we have the interesting result that the dipole moments of one α-helix chain exactly counterbalance those in the other, as shown in Figure 2.9(b). This feature of the DNA macromolecule in having no net helix-contributed dipole moment has not always been fully appreciated, although the nature of this essentially complete dipolar antiparallelism must provide a major contribution to the energy stabilization of the double helix. For molecules like PBLG, on the other hand, the resultant dipole moment can easily exceed a value of 2000 Debye units. In this way the condition $\mu E_1 > kT$ can readily be achieved and with it the onset of the orientational saturation effect described in terms of the Langevin function in Figure 1.3. Practically complete field alignment of PBLG molecules in dioxan solution has been demonstrated by Jones et al.[34] By observing saturation effects for PBLG solutions in pulsed fields of amplitudes up to 6×10^6 V m^{-1}, Gregson et al.[35] were able to determine both the dipole moment and the anisotropy of polarizability. The electric anisotropy of polarizability is given by $(\alpha_{/\!/} - \alpha_\perp)$ where $\alpha_{/\!/}$ is the component of polarizability along the symmetry axis of the molecule (i.e. the same direction as the permanent dipole moment) and $\alpha_\perp$ is the value perpendicular to the axis of the molecule. A positive anisotropy value (i.e. $\alpha_{/\!/} > \alpha_\perp$) corresponds to a prolate macromolecule whilst for oblate macromolecules the anisotropy value is negative. Measurements on similar preparations of PBLG in 1,2-dichloroethylene and dioxan gave dipole moment and effective molecular anisotropy values, as determined at a frequency of 6 MHz, of 1930 debye, 8.3×10^{-21} cm^3 and 1410 debye, 10.5×10^{-21} cm^3, respectively. The authors were not able to decide whether the different values for the two solvents were due to solvent—helix or helix—helix (i.e. aggregation) interactions. As quoted by Gregson et al.,[35] values for $(\alpha_{/\!/} - \alpha_\perp)$ for PBLG in dichloroethane fall in the range 1.4 to 1.5×10^{-21} cm^3 when determined by optical methods. At optical frequencies, only the electronic polarizabilities will contribute to $(\alpha_{/\!/} - \alpha_\perp)$ whereas at 6 MHz both the atomic and electronic polarizabilities will be present. The atomic polarizability determines the total intensity of the normal mode infrared absorption bands and usually has a value of only 15 per cent of that of the electronic polarizability. The values of $(\alpha_{/\!/} - \alpha_\perp)$ determined at 6 MHz are therefore of much greater magnitude than might have been expected. This result could possibly reflect the presence of low frequency vibrational modes involving the helical chain. As the helix is only stabilized by relatively weak hydrogen bonds such vibrations could involve several turns of the helix and hence quite large dipole moment orientations which could contribute significantly to the value for $\alpha_{/\!/}$.

POLAR SIDE-CHAIN EFFECTS

In Table 2.1 we see that 13 of the 20 common α-amino acids have polar side chains. The presence of such polar groups attached to the polypeptide chains will obviously influence their overall dielectric properties.

It will be advantageous at this stage to consider some of the knowledge gained from the study of ordinary organic polymers, a useful review of which has been given by Block and North.[36] We will also use the now commonly adopted nomenclature, originally proposed by Reddish and his colleagues,[37] for identifying the locations of observed dielectric loss peaks. The dielectric dispersion observed at the lowest frequency (for constant temperature), or alternatively at the highest temperature if the frequency is maintained constant, is called the α-relaxation. The terms β, γ, etc. are then used to identify any remaining relaxation regions in order of increasing frequency (or decreasing temperature).

In amorphous polymers a loss region (α-relaxation), associated with the glass—rubber transition, is observed at or above the glass-transition temperature T_g. This glass—rubber transition in polymers is associated with hindered rotations of the main backbone of the polymer chain. At temperatures below T_g, these main backbone chain motions are not possible, and subsequent β, γ, etc. relaxations can only be attributed to side-chain rotations or to localized movements in the main chain. Matters are a bit more complicated for crystalline polymers, but in general the α-relaxation is usually associated with backbone movements in the crystalline regions, and the β-relaxation with movements of relatively large main-chain portions in the disordered or amorphous regions. For crystalline polymers this β-relaxation is often associated with the glass—rubber transition, while the γ and succeeding relaxations are thought to involve either side-chain or limited main-chain movements. Dielectric dispersions associated with polar side chains can therefore be classified broadly into two main types. The first is where the dipoles are attached to the main chain in such a way that the chain must move before orientation can occur. An example of this is polyvinylchloride $-(CH_2-CHCl)_{\overline{n}}$ which exhibits a relatively small dispersion until it is heated to near its glass-transition point and the main chains are 'unfrozen'. The second is where the dipoles can rotate about a single bond even if the main chain remains immobile. Polymeric resins containing OH groups such as phenolics or polyvinyl alcohol $-(CH_2-CH)_n$ fall into this

$$\overset{|}{OH}$$

category, and exhibit room temperature dielectric dispersions typically at frequencies around 100 kHz. Large, bulky, side chains tend to decrease the flexibility of the backbone chain, resulting in an increase in activation energy and lowering of frequency of the α-relaxation. If, now, a large bulky group is attached to the original side chains, the backbone chain mobility will tend to be decreased further, but at the same time adjacent polymer chains will be pushed away tending to increase the backbone mobility. The overall effect of adding additional groups to the side chains thus depends on two opposing factors and in some cases there is not likely to be a very major change in the dielectric relaxation characteristics. These side-chain effects are summarized in Table 2.6, where values for τ and ΔH, as defined by equation (1.26), are given for some representative polymers. The values for the relaxation time τ and enthalpy of activation ΔH were determined from dilute solution studies,[36] so the results in fact are representative of cooperative α–β transition processes. The influence of the steric hindrance resulting from an increase in bulk of groups directly attached to the backbone chain is clearly illustrated by the increase in relaxation time and activation energy in going from PMA to PMMA (H group replaced with CH_3). The relatively negligible effect in increasing the length of the flexible side chain is also clearly demonstrated by the results for PMA, PnBMA, and PnNMA. The effect of making measurements on dilute solutions will give the same result as adding plasticizer to the polymer. Plasticizers tend to decrease the packing density of the main polymer chains, so that the mobility of these main chains increases and the relaxation time for the α-process decreases,

Table 2.6   Dielectric relaxation times and enthalpies of activation for some organic polymers in dilute solutions at 298 K

Polymer	Structure	τ (s)	ΔH (kcal/mole)
Polymethylacrylate (PMA)	$\begin{array}{c} H \\ | \\ \text{+CH}_2\text{—C+}_n \\ | \\ \text{O=C—O—CH}_3 \end{array}$	8.84×10^{-11}	5.5
Polymethylmethacrylate (PMMA)	$\begin{array}{c} CH_3 \\ | \\ \text{+CH}_2\text{—C+}_n \\ | \\ \text{O=C—O—CH}_3 \end{array}$	4.1×10^{-9}	6.5
Poly-n-butylmethacrylate (PnBMA)	$\begin{array}{c} CH_3 \\ | \\ \text{+CH}_2\text{—C+}_n \\ | \\ \text{O=C—O—C}_4\text{H}_9 \end{array}$	4.7×10^{-9}	6.5
Poly-n-nonylmethacrylate (PnNMA)	$\begin{array}{c} CH_3 \\ | \\ \text{+CH}_2\text{—C+}_n \\ | \\ \text{O=C—O—C}_9\text{H}_{19} \end{array}$	5.3×10^{-9}	6.5

often by as much as ten decades. Solvation, or the addition of plasticizer, has a smaller effect on the relaxation of side chains, since side chains tend to act as their own plasticizer.

We will expect the same general tendencies of dielectric behaviour for simple polypeptide chains. The situation for protein molecules, though, will tend to be more complicated, since the polypeptide chain conformation is determined by interchain hydrogen bonds and disulphide bridges (see Figure 2.1(b)). Such interchain interactions will modify both main-chain and side-chain relaxations.

Of the biopolymers, poly-γ-methyl-L-glutamate (PMLG) and poly-γ-benzyl-L-glutamate (PBLG) are the two most investigated regarding dielectric effects associated with polar side chains. PMLG, in which the side chain R is $CH_2 CH_2 CO \cdot OCH_3$ was the first synthetic polypeptide to provide evidence for the existence of the α-helix,[38,39] and PBLG, in which R is $CH_2 CH_2 CO$ $OCH_2 C_6 H_5$, was the first synthetic polypeptide to be obtained in a highly orientated form.[39] PBLG is still one of the most easily orientated of the synthetic polypeptides, and as such is the one to have been most extensively studied. The side chains of PMLG and PBLG are shown in Figure 2.10.

Using the same quantum mechanical molecular orbital method outlined earlier in this chapter, σ-bond moments have been calculated for these two side chains, and the values derived are given in Figure 2.10, with the carbonyl double bond being given the value 2.3 D (Table 1.1). Assuming planarity of the carbon and oxygen

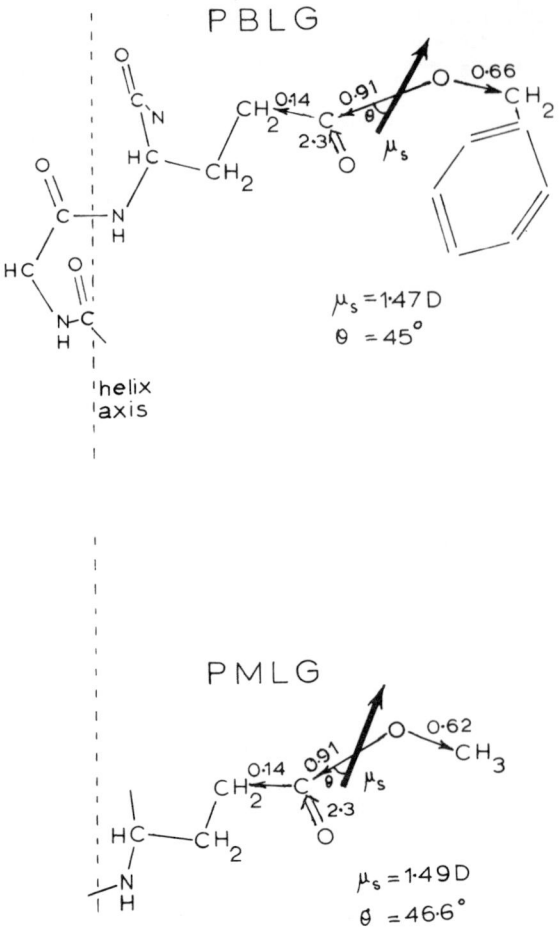

Figure 2.10 The side chains of PBLG and PMLG, showing their bond moments and resultant permanent dipole moment vectors. For complete freedom of rotation of the carbonyl group, μ_s = 2.8 D

atoms in these side chains, then the vectorial sum of the bond moments gives a resultant dipole moment μ_s of the order 1.5 D aligned in roughly the same direction as the C=O bond. The orientation of these side-chain moments μ_s is such as to be at an angle of about 14° with the main α-helical axis. The side-chain bond dimensions and angles required for these calculations were based on those given by Tanaka and Ishida.[40] Using the bond moments given by Smyth, as given in Table 1.1, then μ_s = 1.65 D. If complete freedom of rotation is assumed for the C=O bond about the carbon atom (which in fact is unlikely) then the resultant dipole moment is increased to the value 2.8 D for both the side chains of PMLG and PBLG. Other bond rotations do not produce such large changes in μ_s.

The resultant dipole moment for such helical polypeptide chains as PMLG and PBLG can be written as

$$\mu_r = n(\mu_{m\parallel} \pm \mu_{s\parallel})$$

where $\mu_{m\parallel}$ and $\mu_{s\parallel}$ are the axial components of the main-chain and the side-chain moment, respectively. The positive and negative signs correspond to whether the side-chain moment is parallel or antiparallel to the main-chain moment. We have already decided that for PBLG the side-chain component opposes that of the main chain, in the same manner as shown in Figure 2.10. From Wada's measurements[20] we know that $\mu_r = 3.4$ D, and we can now estimate $\mu_{s\parallel}$ for the rigid side chain to be of the order 1.5–2.0 D. This gives $\mu_{m\parallel}$ as 5.1 ± 0.2 D. We have also determined the peptide residue component μ_p to be 3.6 D, so we can now estimate that the solvent (dichloroethane) and hydrogen-bond effects contributed about 1.5 D per peptide residue to Wada's results for PBLG.

Hikichi et al.[41] have reported detailed dielectric measurements for PBLG. They made measurements on solid films of PBLG in the frequency range $0.3-10^6$ Hz and

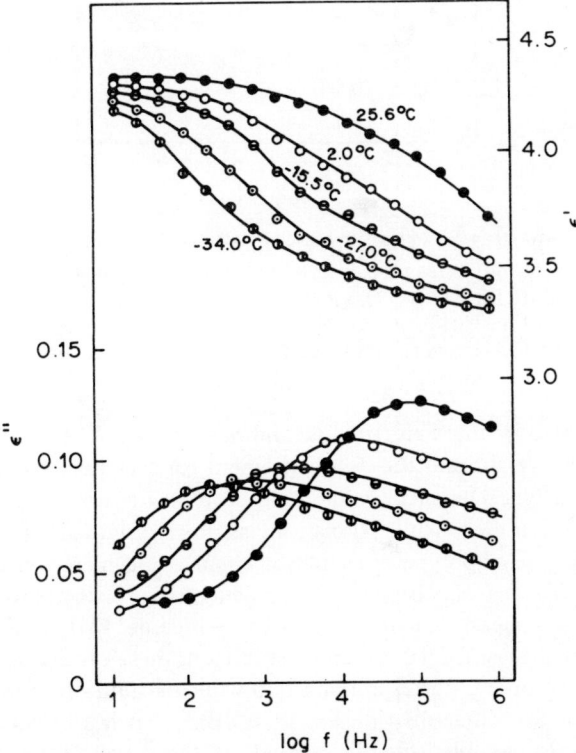

Figure 2.11 The frequency and temperature dependencies of ϵ' and ϵ'' for poly-γ-methyl-L-glutamate (PMLG). (Reproduced from A. Tanaka and Y. Ishida, *J. Polym. Sci. Polym. Phys.*, **11**, 1117 (1973) by permission of John Wiley & Sons, Inc.)

58

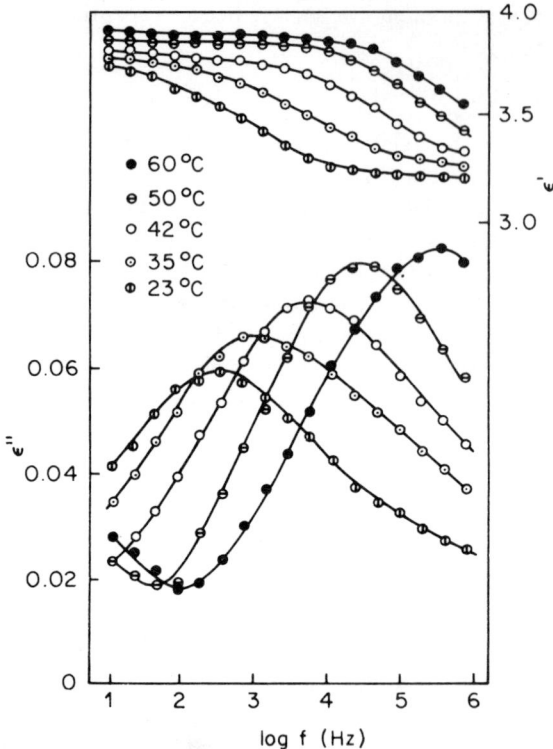

Figure 2.12 The frequency and temperature
dependencies of ϵ' and ϵ'' for poly-γ-benzyl-L-
glutamate (PBLG). (Reproduced from A. Tanaka
and Y. Ishida, *J. Polym. Sci. Polym. Phys.*, **11**, 1117
(1973) by permission of John Wiley & Sons, Inc.)

observed low values for the dielectric loss and permittivity. The PBLG samples were
of high molecular weight and the dipole moment for each helical chain would have
exceeded 2000 debye. The small value, of the order 2.5, obtained for the permittivity was taken to indicate that relaxations associated with the polar side chains,
and not with the main polypeptide helical chain, were being observed. In more
recent work by Tanaka and Ishida[40] these dielectric investigations were repeated
for PBLG and the measurements extended to include PMLG. Their results are
shown in Figures 2.11 and 2.12. It can be seen from these results that the dielectric
loss peaks are very broad, corresponding to a wide distribution of relaxation times,
with PMLG having a wider distribution than PBLG. Such a broad distribution of
relaxation times is possibly directly related to the α-helical form of the main
polypeptide chain. As a result of there being 3.6 amino-acid residues per turn of the
α-helix, the polar side chains will occupy 18 different locations that are repeated
after every five turns of the helix. The less bulky PMLG side chain will not be as
restricted regarding possible steric hindrances to differing conformations, and so

apart from a distribution of relaxation times corresponding to the 18 different locations, we might also expect a wider distribution for PMLG than PBLG.

Using infrared, nuclear magnetic resonance (n.m.r.) and crystallographic data for these two polypeptides, Tanaka and Ishida[40] computed potential energy maps for the various internal rotations of the side chains. The total potential energies were taken to include torsional energy, non-bonded interactions of the van der Waals type, and electrostatic potential energy as described by equation (2.1), with the value 4 being adopted for the molecular permittivity. These calculations were made for all the side groups in seven α-helices, to include the effect of six neighbouring helices about a central one. Convincing evidence was obtained to show that the side-chain dipole moment orientations were distributed about the potential energy minima according to Boltzmann statistics. The dielectric dispersion was calculated assuming a Barrier model with dipolar transitions occurring between potential energy minima separated by potential barriers, as described in Chapter 1. The theory assumed a completely random conformation for the main-chain axis, a Boltzmann distribution for the dipole energies, and the Onsager model (as used to formulate equation 1.16) to describe the local field. Using the Barrier model, with barrier heights calculated from the potential energy maps, good agreement between theory and experiment was achieved. For such a model, where atomic rotations are restricted by various potential barriers, the rotations of the side-chain atoms will be expected to be temperature sensitive. N.m.r. measurements by Tanaka and Ishida[40] confirm this and indicate that whereas the methyl group of the PMLG side chain rotates freely even at −173 °C, the protons attached to the carbon atom nearest the helical main chain are only able to rotate appreciably at temperatures above 100 °C.

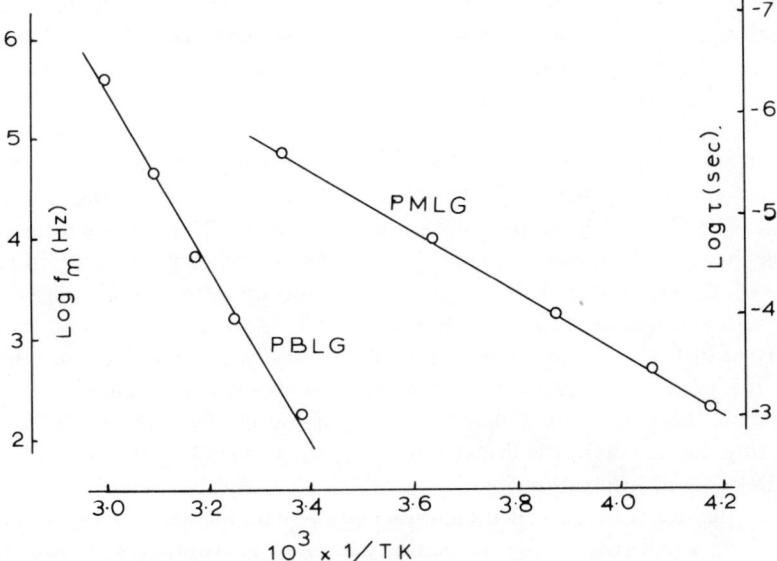

Figure 2.13 Arrhenius plots of the frequency of maximum dielectric loss f_m and relaxation time τ for PBLG and PMLG

Using the dielectric data of Figures 2.11 and 2.12, the variations of the frequency of maximum loss with temperature have been plotted in Figure 2.13. Using equation (1.26), values have been calculated for the enthalpies and entropies of activation for the dielectric dispersions of PMLG and PBLG, and these are given below:

	ΔH kcal/mole	ΔS cal/deg mole	$N \langle \mu^2 \rangle g$ SI units	$\langle \mu^2 \rangle g$ debye units
PMLG	14	-2×10^{-2}	2.87×10^{-32}	0.47
PBLG	40	$+0.14$	1.95×10^{-32}	0.50

Also included in this table are values derived for the product $N\langle \mu^2 \rangle g$, with N the dipole density and g a dipole–dipole correlation parameter to take account of possible side-chain interactions. The calculations were based on equation (1.37) where the full Onsager internal field expression given by equation (1.16) was used, and a correlation parameter g has been included. The resulting equation (in SI units) is:

$$N\langle \mu^2 \rangle g = \frac{9kT(\epsilon_s - \epsilon_\infty)(2\epsilon_s + \epsilon_\infty)\epsilon_0}{\epsilon_s(2 + \epsilon_\infty)^2} \tag{2.14}$$

The side-chain dipole densities were calculated using the specific density values of 1.31 g/ml[39] and 1.27 g/ml[42] for PMLG and PBLG, respectively. This gives $N = 5.5 \times 10^{27}$ m^{-3} for PMLG and $N = 3.5 \times 10^{27}$ m^{-3} for PBLG.

Careful examination and extrapolation of the dielectric results of Figures 2.11 and 2.12 indicates that with increasing temperature, the effective distribution of relaxation times, given by the width of the loss peaks and the total dispersion $(\epsilon_s - \epsilon_\infty)$, decreases. The rate of decrease of the total dispersion is greater than that predicted by the theoretical $1/T$ relationship, as given by equation (1.37) for example, indicating that the effective product $N\langle \mu^2 \rangle g$ decreases. These temperature effects are also evident in the more recent results of Hikichi et al.[43] for PBLG, where the total dispersion falls in value by 44 per cent in the temperature range 40–120 °C. The $\langle \mu^2 \rangle g$ values, derived using equation (2.1) and the dielectric data of Hikichi et al. are shown in Figure 2.14. In deriving these values account has been taken of the 10 per cent fall in the low frequency permittivity value ϵ_s in the temperature range 40–120 °C,[43] and in the 3.7 per cent reduction of the specific density of PBLG[42] due to thermal expansion over this temperature range. In Figure 2.14, the product $\langle \mu^2 \rangle g$ is seen to decrease with increasing temperature. At the same time, the Cole–Cole parameter β of equation (1.30) tends to the value $\beta = 1$, indicating that the dielectric dispersion is tending towards that of a single relaxation time Debye-type mechanism.

The effective reduction of the product $\langle \mu^2 \rangle g$ with increasing temperature could result from a reduction of the correlation parameter g. Assuming a side-chain dipole moment μ_s of 1.5 D and that all side chains are equally effective, then the corresponding values for g in Figure 2.14 are $g = 0.38$ at 40 °C reducing to $g = 0.28$

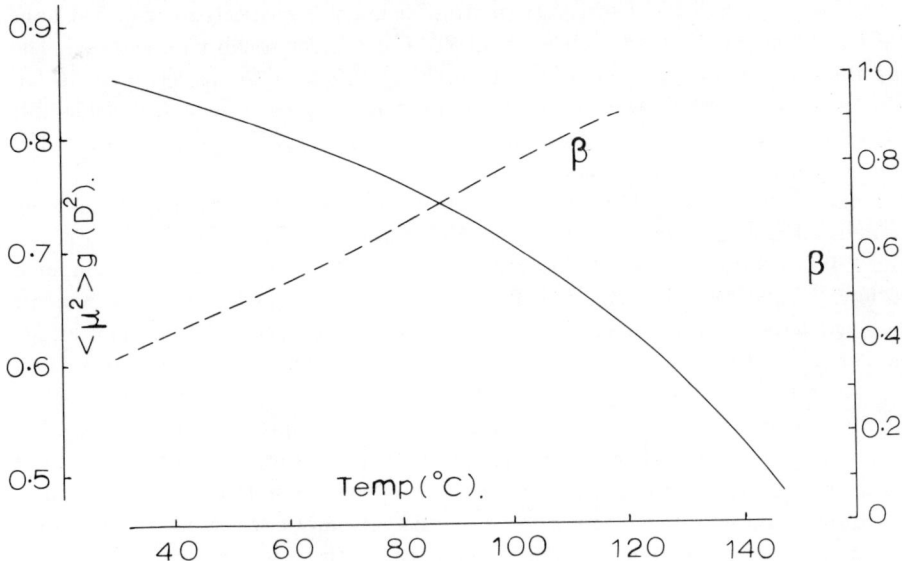

Figure 2.14 Temperature variation of the product $\langle \mu^2 \rangle g$ and the Cole–Cole parameter β for PBLG

at 120 °C. In Chapter 1, a similar type of reduction in g was described for water, with g varying from the value 2.82 at 20 °C to 1.64 at 300 °C. The value of g greater than unity for water arises from intermolecular hydrogen bonds which tend to additively correlate the dipole moments of neighbouring water molecules. With increasing temperature these hydrogen bonds are disrupted and the value for g falls. A value for g of less than unity for PBLG implies that somehow the effective side-chain dipole moments are reduced.

The spacing between neighbouring α-helical main chains is of the order 15 Å for crystalline PBLG, and allowing for the radius of the α-helix at 2.5 Å and that the side chains will have an effective length of the order 9 Å (taking account of the van der Waals radius of the most extreme H atom of the benzyl group), then considerable interaction between side chains of neighbouring helices is to be expected. Adopting the value for g of 1.0, then the dielectric dispersion at 40 °C measured by Hikichi et al.[43] would correspond to relaxations involving only 17 per cent of the total side-chain density, taking $\mu_s = 1.5$ D. Tanaka's and Ishida's results[40] for PBLG correspond to only a 10 per cent contribution from the total side-chain density. With an increase of temperature to 120 °C we would expect the number of relaxing side chains to increase considerably, especially since broad-line n.m.r. studies[44,45] for solid PBLG have shown that the side chains exhibit considerable motion even at 20 °C. Assuming a value for g of unity, then at 120 °C the observed dispersion still only corresponds to a 28 per cent relaxation of the side chains. A mechanism involving motions of only parts of the side chain, as a result of steric hindrances, would not be satisfactory either, as this would tend to lead to an increase in dispersion with increasing temperature as the side-chain mobility increased.

On the other hand, if contributions from main-chain motions are assumed to be effective, then the dielectric behaviour of PBLG can be reasonably understood. The crystal structure, and side-chain length, for PBLG is such that apart from considerable 'entanglements' between side chains of neighbouring α-helices, considerable interaction must also exist between side chains and helical main chains. The motions of the side chains could in this way be directly governed by α-helical motions and interactions between neighbouring helices. The thermal expansion and modulus of elasticity measurements by McKinnon and Tobolsky[42] indicate that solid PBLG exhibits a glass—rubber transition at 17 °C. This indicates that hindered motions of the helical chains will have occurred in the dielectric measurements described here for PBLG. Such main-chain motions involving the peptide residue moment μ_p will tend to reduce the effective dipole moment μ_s of the side chains, since they are directed antiparallel to each other. This overcomes problems regarding the low effective density of relaxing side chains. This negative contribution from the main-chain moment will increase with increasing temperature, and the activation energy of 40 kcal/mole, deduced from Tanaka and Ishida's work, and of 44 kcal/mole quoted by Hikichi et al.,[43] could certainly correspond to the potential barrier restricting main-chain motions. Also, with increasing temperature, the α-helix motions will tend to control completely the side-chain motions and the dominant relaxation times can be expected to tend to a single value. Using the terminology of Deutsch et al.,[37] the dielectric dispersion of solid PBLG appears to be adequately described in terms of a cooperative α and β process.

Another way of investigating side-chain effects in polypeptides is by measurements on an α-helix constituted of optical antipodes. The C_α atom in the peptide unit has four different groups attached tetrahedrally to it, and these can be bonded in two different ways related by a mirror reflection as shown in Figure 2.15. The chemical and physical properties of these two isomers are identical, except they will rotate incident polarized light in opposite senses, and as such they have come to be called D and L optical isomers from the Latin *dextro* (right) and *laevulo* (left). It has been shown that all of the α-amino acids that are the constituents of proteins take the form of the L-configuration. The D-amino acids are usually considered as 'unnatural' but nevertheless many of them are components of biologically impor-

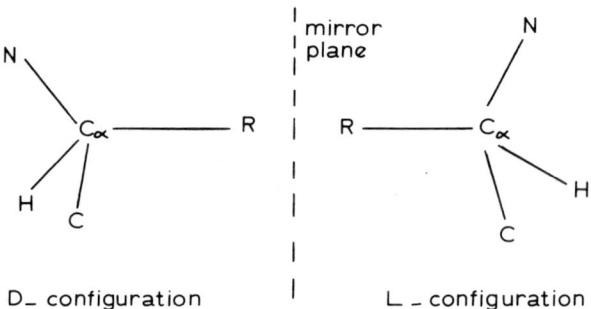

Figure 2.15 The D- and L-configurations for the C_α atom in the peptide unit

tant substances other than proteins. The photosynthesis of glucose by plants from CO_2 and H_2O gives the D-isomer only. The L-form is 'unnatural' and, furthermore, is not even metabolized by animals. Such stereospecificity is probably imperative for the high efficiency of enzymes. For example, a protein molecule composed of 100 peptide units not having such stereospecificity, would have 2^{100} or 10^{30} possible optical isomers. A vessel with a capacity of the order 10^7 litres would be required to contain all such possible protein molecules. An organism so constituted as to be able to deal specifically with each one of these isomers would have to be truly gigantic!

Wada[23] has predicted that incorporating D-residues into an α-helix of PBLG (composed of L-isomers as the letter 'L' indicates) will tend to increase the effective dipole moment, since the D-residue side chains will have their dipole moments directed against those of the L-residues. Although associated helical conformational changes will tend to complicate such a procedure, Wada's results[23] for poly-γ-benzylglutamate L—D copolymers give support to the viewpoint that the side-chain moments of PBLG are directed against the main-chain moment. Hikichi *et al.*[43] carried out dielectric measurements for mixtures of PBLG and poly-γ-benzyl-D-glutamate (PBDG), and a reduction in the dispersion strength was found, compared to that of pure PBLG. This effect can be interpreted as resulting from the formation of regular stacks of benzyl groups between adjacent chains of opposite helical pitch sense, so reducing the number of relaxing side chains. At about 100 °C the dielectric properties of PBLG—PBDG mixtures exhibited a transition, to follow the dielectric properties of pure PBLG at higher temperatures. Such behaviour is consistent with the breaking down of the benzyl group stacks, so freeing the side groups to allow them to undertake normal motions.

Clearly, the investigation of polar side-chain effects and interactions in polypeptides and other biopolymers by dielectric measurements remains an area of great potential and considerable importance. Side-chain effects in protein molecules will be expected to be more complex than those in helical polypeptides, but the knowledge gained for such polypeptide systems will still be of considerable value.

MOLECULAR PERMITTIVITY OF POLYPEPTIDE SYSTEMS

In calculating the electrostatic potential energy contribution to the total conformational energy of polypeptide systems, a value for the molecular permittivity ϵ is required (e.g. see equation 2.1). A value for ϵ of the order of 3.5—4.0 has often been used for such calculations, based on the microwave permittivity value for some solid amides and polyamides. For example, Brant and Flory[46] have used a value for ϵ of 3.5, and Scott and Scheraga[47] adopted a value of 4.0. Ramachandran[3] considers that a value of ϵ in the range 3—5 is quite reasonable for electrostatic potential energy calculations, whereas a value of $\epsilon \simeq 1$ is more appropriate for van der Waals and similar interaction energies. Tanaka and Ishida[40] adopted a value for ϵ of 4 for their calculations of the potential energy maps for PBLG and PMLG, even though their dielectric measurements for PBLG gave $\epsilon \simeq 3.2$ at 1 MHz.

Table 2.7   The polarizability α and molar
polarization P values for the peptide group
atoms

Atom	$\alpha \times 10^{24}\,(cm^{-3})$	$P(cm^{-3})$
H	0.42	1.06
O	0.84	2.12
N	1.15	2.90
C	1.30	3.28

As there appears to be no adequate basis upon which to calculate an appropriate value for the molecular permittivity of proteins and polypeptides, it will be of value for us to derive an estimate value based on the method outlined in Chapter 1 for simple organic polymers. The polarizability values α for the various atoms of the peptide group shown in Figure 2.3 are given in Table 2.7. The values are those given by Ramachandran and Sasisekharan[3] and take into account the partial double-bond character of the carbonyl and the C—N bond. The molar polarization values P given in Table 2.7 were calculated from the polarizability values using the formula given by equation (1.44),

$$P = \frac{4\pi}{3} N_A \alpha$$

where N_A is the Avogadro number. Using the side-chain structural formulae of Table 2.1 and the atomic molar polarization values of Tables 1.2 and 2.7, values

Table 2.8   The molar polarizability values for the 20
α-amino-acid residues

Residue	Molecular weight	$P(cm^{-3})$
Ala	71.1	18.6
Arg	156.2	42.2
Asp	115.1	24.9
Asn	114.1	28.4
Cys	103.2	26.4
Glu	129.1	29.5
Gln	128.1	36.1
Gly	57.1	14.0
His	137.2	39.5
Ilu	113.2	32.6
Leu	113.2	32.6
Lys	128.2	36.0
Met	131.2	35.9
Phe	147.2	37.8
Pro	133.1	26.9
Ser	87.1	20.2
Thr	101.1	24.8
Trp	186.2	54.1
Tyr	163.2	40.3
Val	99.1	27.9

have been calculated for molar polarizations of the 20 α-amino-acid residues, and these are given in Table 2.8. Recalling equation (1.44) the limiting high frequency molecular permittivity ϵ will be given by

$$\frac{\epsilon - 1}{\epsilon + 2} = \frac{P\rho}{M}$$

where ρ is the density of the material and M its molecular weight. In Chapter 3 we shall see that the so-called 'oil-drop model' for protein molecules will lead us to expect most protein molecules to consist of a core of hydrophobic amino-acid residues surrounded by polar hydrophilic ones. Kuntz[48] has recently estimated the density of the hydrophobic and non-hydrophobic regions of carboxypeptidase, and obtains a density of 0.93 g/cm³ for the hydrophobic regions, 1.55 g/cm³ for the remainder of the protein molecule, and an overall density value of 1.39 g/cm³. The overall density value for proteins in the past has commonly been taken[49] as 1.17 g/cm³.

From Table 2.8, the seven hydrophobic residues (Ala, Gly, Ilu, Leu, Met, Phe, and Val) have an average molecular weight of 104 g units, and an average value of 28.5 for the molar polarization. The remaining 13 α-amino-acid residues have an average molecular weight and molar polarization of 129 and 33, respectively. Assuming carboxypeptidase to be typical of the globular proteins in general, then, ignoring amino and carboxylic peptide chain ends, the high frequency molecular permittivity in the low density hydrophobic interior of such proteins will have a value given approximately by

$$\frac{\epsilon - 1}{\epsilon + 2} = \frac{n \times 28.5 \times 0.93}{n \times 104}$$

$$= 0.255$$

The resulting permittivity value is $\epsilon = 2.03$. For the non-hydrophobic regions surrounding this less dense non-polar core, the corresponding permittivity value is $\epsilon = 2.97$, and the mean permittivity for the whole protein molecule has the value $\epsilon = 2.63$. More exact calculations, based on the amino-acid compositions given in Table 2.2 and on the molecular weight and molar polarization values given in Table 2.8, give the permittivity of the hydrophobic regions for BSA and lysozyme as 2.03 and 2.02, respectively, assuming a density value of 0.93 g/cm³. The non-hydrophobic regions of assumed density 1.55 g/cm³, have permittivities of 2.85 and 2.97, and the mean permittivity values are 2.64 and 2.7 for BSA and lysozyme, respectively. From these estimations based on the density values for carboxypeptidase, it would appear that using a permittivity value $\epsilon \simeq 1$ for van der Waals interactions would not be as satisfactory as choosing $\epsilon \simeq 2$ for the interaction of groups not in immediate contact with each other. Also, the value $\epsilon \simeq 5$ would appear too large, especially for electrostatic interactions involving atoms in the non-polar hydrophobic protein interior. A good compromise would be to adopt the value $\epsilon \simeq 2$ for the protein interior, and $\epsilon \simeq 3$ for the outer regions of protein molecules. Where peptide residues or side chains protrude into an aqueous surrounding medium, then the appropriate molecular permittivity for electrostatic interactions could approach the free-water value $\epsilon \simeq 80$, although the value $\epsilon \simeq 2$ would still be appropriate for van der Waals interactions since these involve the high frequency permittivity. It should be remembered that, when considering the interaction of charge centres separated by small molecular distance in a condensed phase, the concept of an effective permittivity is not clear. When dealing at the atomic level the negative charge centres cannot be thought of as being locally fixed in space, but instead should be represented as electron 'probability' clouds. It is the polarizability of these electron clouds that will determine the microscopic permittivity when the value of r_{ij} in equation (2.1) is of the order of nearest neighbour interatomic distances. The permittivity values determined above can be considered to represent the macroscopic permittivity appropriate for calculations involving charged groups not in immediate contact with each other.

The molecular weights of the peptide residue and side chain for PMLG and PBLG are 143.2 and 219.3, respectively, and the corresponding molar polarizations are 36 and 60. Using the specific density values of 1.31 and 1.27 g/cm^3, then the corresponding molecular permittivity values for solid PMLG and PBLG are $\epsilon = 2.5$ and $\epsilon = 2.6$, respectively. The 1 MHz permittivity values obtained by Tanaka and Ishida[40] are a little higher than these estimates, at a value of the order 3.2. This would suggest the possibility that for the frequency region above 10^6 Hz at room temperature, other dipolar dispersions exist, possibly related to side-chain motions not coupled to motions of the helical main chain. This conclusion is also consistent with the results obtained by Gregson et al.[35] for the polarizability anisotropy of PBLG, where the unexpectedly large value for $(\alpha_{\parallel} - \alpha_{\perp})$ was considered to possibly arise from helical motions occurring in the frequency range between 6 MHz and the near infrared. The search for, and possible identification of, such additional orientational dipolar dispersions would be of obvious interest and value.

References

1. R. E. Dickerson and I. Geis, *The Structure and Action of Proteins*, Harper and Row, New York (1969).
2. F. Wold, *Macromolecules: Structure and Function*, Prentice-Hall, New Jersey (1971).
3. G. N. Ramachandran and V. Sasisekharan, *Advan. Protein Chem.*, **23**, 283 (1968).
4. W. O. Baker and W. A. Yager, *J. Am. Chem. Soc.*, **64**, 2171 (1942).
5. D. A. Brant, W. G. Miller, and P. J. Flory, *J. Molec. Biol.*, **23**, 47 (1967).
6. R. M. Meighan and R. H. Cole, *J. Phys. Chem.*, **68**, 503 (1964).
7. H. A. Scheraga, R. A. Scott, G. Vanderkooi, S. J. Leach, K. D. Gibson, T. Ooi, and G. Nemethy. In *Conformation of Biopolymers*, Vol. 1, p. 43. Ed. G. N. Ramachandran, Academic Press, London (1967).
8. M. Davies, *Some Electrical and Optical Aspects of Molecular Behaviour*, Pergamon, Oxford (1965).
9. R. J. Kurland and E. B. Wilson, *J. Chem. Phys.*, **27**, 585 (1957).
10. J. E. Grabenstetter and M. A. Whitehead, *J.C.S. Faraday II*, **69**, 962 (1973).

11. G. Del Re, *J. Chem. Soc.*, **1958**, 4031.
12. A. Streitwieser, *Molecular Orbital Theory for Organic Chemists*, Wiley, New York (1961).
13. J. D. Roberts, *Molecular Orbital Calculations*, Benjamin, New York (1962).
14. L. Pauling, *The Nature of the Chemical Bond*, 3rd Edition, p. 235. Cornell University Press (1960).
15. E. G. Cox and G. A. Jeffrey, *Proc. Roy. Soc. (London)*, **A207**, 110 (1951).
16. E. Hirota, R. Sugisaki, C. J. Nielsen, and G. O. Sorensen, *J. Molec. Spectrosc.*, **49**, 251 (1974).
17. H. Eyring, *Phys. Rev.*, **39**, 746 (1932).
18. C. P. Smyth, *Dielectric Behaviour and Structure*, p. 234. McGraw-Hill, New York (1955).
19. P. Debye and F. Bueche, *J. Chem. Phys.*, **19**, 589 (1951).
20. A. Wada, *J. Chem. Phys.*, **30**, 328 (1959).
21. C. A. Coulson and U. Danielson, *Arkiv. Fysik.*, **8**, 239 (1954).
22. H. Tsubomura, *Bull. Chem. Soc. Japan*, **27**, 445 (1954).
23. A. Wada, in *Poly-α- Amino Acids*, pp. 369–90. Ed. G. D. Fasman, Marcel Dekker, New York (1967).
24. J. Applequist and T. G. Mahr, *J. Am. Chem. Soc.*, **88**, 5419 (1966).
25. A. D. Buckingham, *Aust. J. Chem.*, **6**, 93, 323 (1953).
26. P. Doty, J. H. Bradbury, and A. M. Holtzer, *J. Am. Chem. Soc.*, **78**, 947 (1956).
27. M. Sharp, *J. Chem. Soc. A*, **1970**, 1596.
28. E. H. Erenrich and H. A. Scheraga, *Macromolecules*, **5**, 746 (1972).
29. T. Matsumoto, N. Nishioka, A. Teramoto, and H. Fujita, *Macromolecules*, **7**, 824 (1974).
30. H. Nakamura, Y. Husimi, G. P. Jones, and A. Wada, *J.C.S. Faraday I*, **73**, 1178 (1977).
31. A. Wada and H. Kihara, *Polymer J.*, **3**, 82 (1972).
32. F. Perrin, *J. Phys. Radium*, **5**, 497 (1934).
33. J. D. Watson and F. H. C. Crick, *Nature*, **171**, 737, 964 (1953).
34. G. P. Jones, M. Gregson, and M. Davies, *Chem. Phys. Letters*, **4**, 33 (1969).
35. M. Gregson, G. P. Jones, and M. Davies, *Trans. Faraday Soc.*, **67**, 1630 (1971).
36. H. Block and A. M. North, *Advan. Molec. Relax. Proc.*, **1**, 309 (1970).
37. K. Deutsch, E. A. W. Hoff, and W. Reddish, *J. Polym. Sci.*, **13**, 565 (1954).
38. C. H. Bamford, W. E. Hanby, and F. Happey, *Proc. Roy. Soc. (London)*, **A205**, 30 (1951).
39. C. H. Bamford, L. Brown, A. Elliott, W. E. Hanby, and I. F. Trotter, *Proc. Roy. Soc. (London)*, **B141**, 49 (1953).
40. A. Tanaka and Y. Ishida, *J. Polym. Sci. Polym. Phys.*, **11**, 1117 (1973).
41. K. Hikichi, K. Saito, M. Kaneko, and J. Furuichi, *J. Phys. Soc. Japan*, **19**, 577 (1964).
42. A. J. McKinnon and A. V. Tobolsky, *J. Phys. Chem.*, **72**, 1157 (1968).
43. T. Takahashi, A. Tsutsumi, K. Hikichi, and M. Kaneko, *Macromolecules*, **7**, 806 (1974).
44. J. A. E. Kail, J. A. Sauer, and A. E. Woodward, *J. Phys. Chem.*, **66**, 1292 (1962).
45. Y. Ishida, *J. Polym. Sci.*, **A-2**, 1835 (1969).
46. D. A. Brant and P. J. Flory, *J. Am. Chem. Soc.*, **87**, 663, 2791 (1965).
47. R. A. Scott and H. A. Scheraga, *J. Chem. Phys.*, **45**, 2091 (1966).
48. I. D. Kuntz, *J. Am. Chem. Soc.*, **94**, 8568 (1972).
49. J. T. Edsall, *The Proteins*, Vol. 1B, Academic Press, New York (1953).

Note added in proof:

Since this Chapter was written a valuable article by Wada has been published entitled, 'The α-helix as an electric macrodipole'. The dielectric properties of PBLA and PBLG in solution are described, and also included is an appendix by G. Parry Jones on high field effect measurements on PBLG.

A. Wada, *Adv. Biophys.* (M. Kotani, Ed), Univ. Tokyo Press, Vol. 9, 1–63 (1977).

Amino-acids in chains

Are the cause, so the X-ray explains

Of the stretching of wool

And its strength when you pull

And show why it shrinks when it rains.

A. L. PATTERSON

as quoted by W. T. Astbury,
Transactions Faraday Soc., 34, 378 (1938).
(*Reproduced by kind permission of The Chemical Society, London.*)

Chapter 3

Dielectric Properties of Solvated Biomolecules

We will now extend the description of the intrinsic dielectric properties of biological materials to include effects associated with solvation. As in the previous chapter, attention will be given primarily to amino acids, polypeptides, and protein molecules. Of fundamental relevance to such dielectric studies are considerations of the ionic properties of amino-acid molecules, and of the so-called 'oil-drop' model to describe the structural feature of typical protein molecules.

IONIC PROPERTIES OF AMINO ACIDS

Amino acids, containing as they do both carboxylic and amino groups, are amphoteric in that they are capable of acting as acids as well as bases. This property is fundamental to the understanding of many of the dielectric and electrochemical properties of solvated biopolymers, and is therefore worthy of some detailed consideration.

According to the Brönsted formulation an acid is defined as any substance which can give rise to protons. Furthermore, for every acid there is a conjugate base, which results from the acid as a consequence of dissociation of the proton. Also, a base may be defined as any compound or ion which can bind protons, so we can also say that to every base there is a conjugate acid. This can be illustrated by the schematic reaction

$$AH^+ \rightleftharpoons B + H^+ \tag{3.1}$$

in which AH^+ is an acid and B its conjugate base. Thus all undissociated carboxylic acids are acids in the Brönsted scheme and their anions are bases, as we see from the reaction

$$RCOOH \rightleftharpoons RCOO^- + H^+$$

The acidic dissociation constant of the acid AH^+ according to equation (3.1) is given by

$$K_a = \frac{[B][H^+]}{[AH^+]}$$

where the terms in brackets refer to activities of the species in question. The activities for ideal solutions represent the concentration values, but in practice they refer to *effective* concentrations. For titration and buffering behaviour it is of interest to solve for $[H^+]$, the activity of hydrogen ions. This is given by

$$[H^+] = \frac{K_a [AH^+]}{[B]} \qquad (3.2)$$

By taking the common logarithm of both sides of equation (3.2)

$$\log[H^+] = \log K_a + \log \frac{[AH^+]}{[B]}$$

We now make use of the well-known symbol p to denote 'negative logarithm of' and obtain

$$pH = pK_a + \log \frac{[B]}{[AH^+]} \qquad (3.3)$$

Thus in the buffer system consisting of acid A and its conjugate base B, the pH is related through equation (3.3) to the pK_a of the acid, and the ratio of the concentrations of form B to form AH^+ existing in the solution. We remind ourselves also, that in the reaction

$$HOH \rightleftharpoons H^+ + OH^-$$

water is an acid, and the hydroxyl ion its conjugate base. In this way, for water

$$K_a = \frac{[H^+][OH^-]}{[H_2O]}$$

In aqueous solution, water usually occurs in such great excess that its activity can be considered constant, which by convention is taken to have the value unity. The acidic dissociation constant of water is therefore usually written

$$K_a = [H^+][OH^-]$$

and has a value of about 10^{-14} at 25 °C increasing to about 10^{-12} at its boiling point. The room temperature pK_a value for water can therefore be taken as 14, and its pH as 7, since in equation (3.3), $\log([B]/[AH^+]) = \log([OH^-]/[H_2O]) = -7$.

Typical α-amino acids are solids that do not melt until they decompose at almost 300 °C, and they are appreciably soluble only in water. This behaviour suggests that they are salt-like molecules and that they should be more correctly represented by the dipolar ion (often called *zwitterion*) form

$$R{-}\overset{\displaystyle H}{\underset{\displaystyle NH_3^+}{C}}{-}COO^-$$

rather than as the uncharged species described in Chapter 2. Further support for such a suggestion is that the process of solvation of isoelectric amino acids in water is accompanied by a strong negative volume change, best explained in terms of the aggregation of water molecules around charged groups. Such neutral solutions of amino acids have very high permittivities, as would be expected if the amino acids exist in the dipolar—ionic form. Also, amino acids in the solid state or in neutral solution do not absorb infrared radiation at 1720 cm^{-1}, a wavenumber character-istic of a non-ionized carboxyl group. Instead, they exhibit absorptions near 1400 and 1600 cm^{-1}, which is characteristic of the carboxylate ion.

Amino acids can be titrated in aqueous solution, and such titration reveals the two expected ionizable groups, having ionization constants of about 10^{-2} ($pK_a \simeq 2$) and 10^{-9} ($pK_a \simeq 9$) mole/litre. If the dipolar ion, or zwitterion, does in fact represent the structure of a neutral amino acid, then these observed ionization reactions correspond to

$$RCH(NH_3^+)COOH \rightleftharpoons RCH(NH_3^+)COO^- + H^+$$

and

$$RCH(NH_3^+)COO^- \rightleftharpoons RCH(NH_2)COO^- + H^+$$

where the $pK_a \simeq 2$ ionization is that of a carboxyl group, and the $pK_a \simeq 9$ ionization is that of an aliphatic ammonium ion. These pK_a assignments are very reasonable when compared with the properties of related amines and carboxylic acids. On the other hand, if the non-polar form of the amino acid were involved in the major ionization equilibria, as in

$$RCH(NH_3^+)COOH \rightleftharpoons RCH(NH_2)COOH + H^+$$

$$RCH(NH_2)COOH \rightleftharpoons RCH(NH_2)COO^- + H^+$$

then we would have to assign the unlikely values of $pK_a \simeq 2$ ionization to an aliphatic ammonium ion, and that of $pK_a \simeq 9$ to a carboxylic acid. In fact one can estimate that there are of the order 10^5 times more dipolar amino-acid molecules than non-ionized ones present in an aqueous amino-acid solution.

Table 3.1 summarizes the dissociation or acid ionization pK_a constants for various amino acids, and is based on the Brönsted formulation.[1] Alanine and glycine are given as examples of simple amino acids having only two ionizable groups, namely the α-carboxyl and α-amino group. But as we have seen in Chapter 2, the amino acids in a protein molecule are linked through these groups and hence they will be unfunctional. Of prime importance for dielectric studies of proteins, therefore, will be consideration of those amino acids having ionizable groups residing in the side chain, and these are listed in Table 3.1. Also included are the pH values, denoted by pI, at which the amino acid carries no net (positive or negative) electric charge. This situation in which the amino acid is electrically neutral is called the isoelectric point, but strictly speaking the term isoionic should be used, since ions other than protons may be bound to amino acids or proteins. For the simple amino acids the pI value is given by the average of the two pK_a values. For the more complex ones, the pI value is given, to a very good approximation, by the average of two of the ionization constants, the choice as to which pair of pK_a's to average being determined simply by considering the overall charge situation. Two

Table 3.1   Ionization constants and pH values at the isoelectric point of α-amino acids in water at 25 °C

Amino acid	pK_a Values			Isoelectric point pI
	carboxyl group $(-COOH)$	ammonium ion $(-NH_3^+)$	Side chain	
Ala	2.35	9.87	—	6.11
Gly	2.35	9.78	—	6.06
Arg	2.01	9.04	12.48	10.76
Asp	2.10	9.82	3.86	2.98
Glu	2.10	9.47	4.07	3.08
His	1.77	9.18	6.10	7.64
Lys	2.18	8.95	10.53	9.47
Tyr	2.20	9.11	10.07	5.63
Cys	1.04	2.05	8.0, 10.25	5.02

pK_a values are listed in Table 3.1 for the sulphydryl group of cysteine. This is because ionization of this group and the adjacent α-amino group occur in the same pH region. The lower pK_a corresponds to ionization of sulphydryl adjacent to the ammonium ion (NH_3^+), and the higher to sulphydryl adjacent to the neutral amino NH_2.

The significance of amino-acid dissociation and ionization, for the understanding of some of the dielectric properties of amino acids and proteins in solution, is given schematically in Figure 3.1. The scheme of Figure 3.1, taken with the data of Table 3.1, can be used as the basis for the isolating of amino acids from solution. The pI value for the simple amino acids, having only two ionizable groups, is of the order 5.5. So, if an aqueous solution of various amino acids is maintained at pH 5.5 in an electrical conductivity cell, the acidic and basic species will become separated on passing an electrical current through the cell. The negatively charged aspartic acid and glutamic acid will be attracted towards the anode, and arginine, histidine and lysine will migrate towards the cathode, and their concentrated solutions collected. One's next thought might be to subject the concentrated basic solution again to electrical transport at pH 9.5 and to collect arginine at the cathode, histidine at the

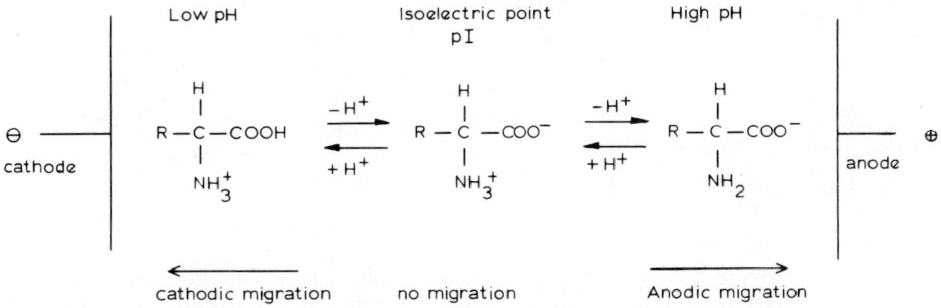

Figure 3.1 A schematic summary of the ionic properties of simple amino acids

anode, leaving lysine behind (the pI values for histidine, lysine and arginine being 7.64, 9.47, and 10.76 respectively). But we should first remind ourselves of the logarithmic nature of equation 3.3, relating the degree of ionization with the pH and pK_a values. For example, at pH values one unit either side of the isoelectric point, for any particular amino acid, 90 per cent of the molecules will still possess their neutral dipolar–ionic or zwitterionic form, whereas at a pH two whole units away from the pI value, only 1 per cent of them will be in this form. This demonstrates that the method just suggested for separating the basic amino acids would not be very efficient. A better method would be to conduct electrical transport at pH 7.7, whereupon lysine and arginine migrate towards the cathode, leaving histidine behind.

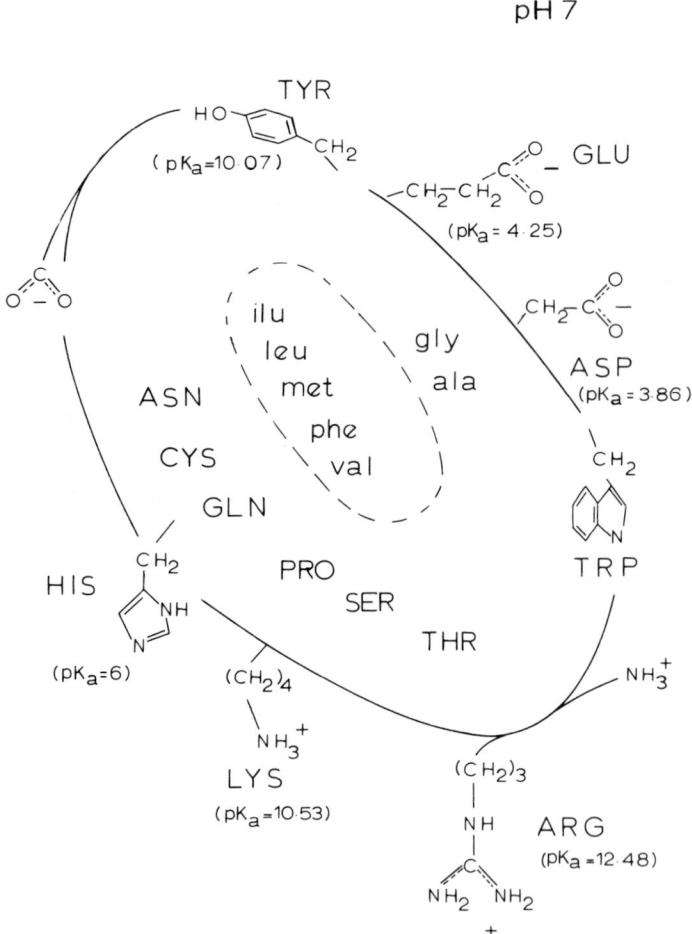

Figure 3.2 A schematic summary of the structural and ionic character, and polarity, of the 20 α-amino-acid residues in a protein molecule at pH 7

If, instead of a steady electric field, an a.c. field was applied to the electrode system of Figure 3.1, and if the cell contained a pure amino-acid solution, then we will expect the measured permittivity to have a high, relatively constant, value for a pH range extending about one unit either side of the pI value. Outside this pH range, where the concentration of dipoles diminishes, the permittivity will fall and only ionic conduction will contribute to the measured dielectric loss, this contribution decreasing with increasing frequency.

A schematic summary of some of the structural and ionic features of the 20 α-amino-acid residues, commonly used as the building blocks for protein molecules, is given in Figure 3.2. The purpose of this figure is to characterize the residues in terms of their polarity and relative locations in a protein molecule. The polar residues are written in upper case, the non-polar ones in lower case. The larger non-polar residues Ilu, Leu, Met, Phe, and Val are indicated to be those invariably located in the protein interior, whereas alanine and glycine are so small that they can sometimes be located at the surface. The neutral polar residues are most often on the outside, but can also be inside when their polar groups are effectively neutralized by hydrogen bonding or, in the case of Cys, by the formation of

Cytochrome-c.

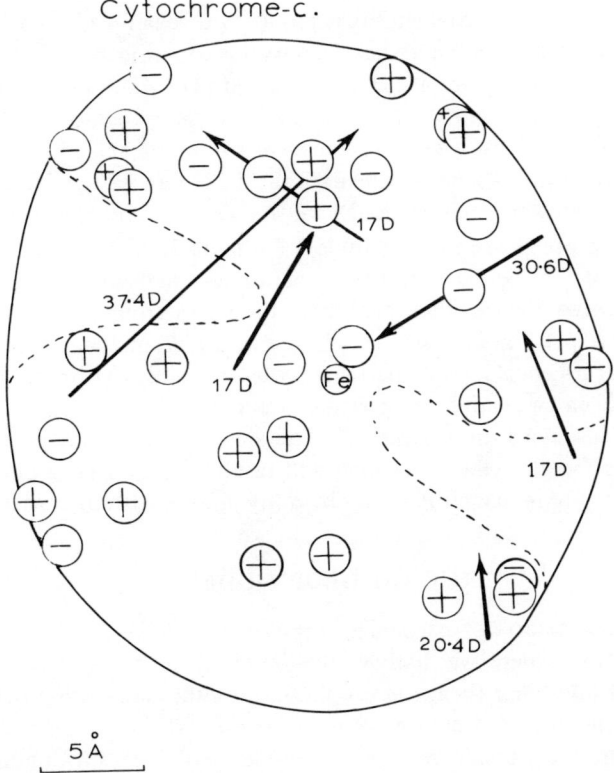

5 Å

Figure 3.3 The cytochrome-c molecule showing the location of the charged groups at pH 7, and the α-helical permanent dipole moments

Table 3.2   The isoelectric points for some proteins

Protein	pI	Protein	pI
α-Casein	4.1	Lysozyme	11.1
β-Casein	4.5	Myoglobin	7.0
γ-Casein	5.9	Myosin-A	~5.3
Collagen	~6.7	Ovalbumin	4.6
Cytochrome-c	~10.0	Pepsin	~1.0
Fibrinogen	~5.7	Rhodopsin	4.5
Haemoglobin (human)	7.1	Salmine	12.1
Insulin	5.4	Serum albumin	4.8

disulphide bridges. By contrast, the acidic and basic ionizable polar groups practically always appear at the surface. The situation in Figure 3.2 shows these various ionizable groups in the charged state prevalent at pH 7. Histidine, which will only be 8 per cent protonated at pH 7, and tyrosine, with its very weakly acidic hydroxyl group, have both been shown as uncharged.

Obviously, depending on their amino-acid composition, different proteins will have differing numbers of charged polar groups. For example, from Table 2.2 we can see that bovine serum albumin has a net excess of acidic groups. As a result this protein has its isoelectric point at pH 4.7, whilst at pH 7.4 the macromolecule has a net negative charge equivalent to 19 electrons.[2] Horse cytochrome-c, on the other hand, has an excess of basic groups, its isoelectric point being near pH 10.[3] The cytochrome-c molecule is shown schematically in Figure 3.3, and is based on the 2.8 Å resolution structure given by Dickerson et al.,[3] and the isoelectric point values for some proteins are given in Table 3.2. Figure 3.3 shows what Dickerson et al. term to be the back view of the cytochrome-c molecule, and indicates the locations of the two crevices that lead up to the haem group. The locations of the charged groups at pH 7 are shown, together with dipole moments arising from the α-helical, or near α-helical, conformational structures in the molecule. These dipole moments have been calculated using Wada's value of 3.4 debye per peptide unit in an α-helix, as discussed in Chapter 2. The net vectorial sum of these dipole moments and surface distributed charges will result in quite a large dipole moment for the molecule, whose magnitude will obviously change with pH variations.

THE 'OIL-DROP' MODEL

Another structural feature of particular importance to us is the so-called 'oil-drop' model for proteins, where we discover the general rule 'non-polar in, polar out'. This means that non-polar (hence hydrophobic) residues are mainly to be found in the interior of the protein structure, whereas the polar (hence hydrophilic) ones are to be found on the outside surface in contact with the surrounding aqueous medium. The fundamental significance of this is best illustrated by considering the hypothetical situation where the inverse of this rule has been applied. If a protein molecule having non-polar hydrophobic side chains protruding from its surface is placed in water, then water molecules surrounding each non-polar side chain will be forced to form a cage-like structure having lower entropy than the normal liquid water structure, since the number of possible water molecule arrangements and rotations will be more restricted. If these non-polar side chains, distributed over the protein molecule surface, were now all collected tightly together in one place, then the liberated water molecules would be able to adopt a more random configuration and the entropy of the solution would therefore increase. This is the fundamental physical reason why oil droplets separate spontaneously in water. Furthermore, it can be calculated[4] that, chiefly as a result of this entropy effect, a protein molecule gains an additional 4 kcal of free energy of stabilization for every non-polar side chain that is removed from an aqueous to a non-polar hydrophobic environment. Another factor of relevance here is the formation of hydrogen bonds. A hydrogen bond (e.g. A—H⋯B) is an attraction between two closely spaced electronegative atoms with a proton shared between them, or in other words the result of an attraction between two dipoles, one formed by the A—H bond and the other produced by atom B and a pair of its unshared valence electrons. As an example we shall consider the following interaction for N-methylacetamide:

In solvents such as benzene or carbon tetrachloride, the formation of a single amide—amide H-bond results in a favourable heat of formation ΔH, of -3 or -4 kcal/mole. In a solvent such as water, which is itself highly associated through H-bonds, the extent of association between amide molecules is small and is influenced by interactions between solvent and the less polar parts of the molecule. (For N-methylacetamide the equilibrium constant in water has been estimated at 0.005.) Intermolecular H-bonds between amide molecules are not favoured in aqueous solution, because if water has unrestricted access to the amide unit, H-bonds of at least equal strength may be formed between water and the amide. In other words, the ΔH for formation of an interamide H-bond in water appears to be close to zero. This then is another reason why the interior regions of a protein should essentially be of a hydrophobic nature, otherwise the formation of structure stabilizing hydrogen bonds would be less favourable.

There is still another, quite subtle and possibly profound, consequence of this 'oil-drop' model, in that it allows for the possibility of some non-aqueous electrochemistry taking place in the interior of the protein molecule. Water, being a polar liquid, has a high static permittivity value of the order 80 at room temperature. The force of attraction F_A between two counter-charges q^+ and q^- will be given according to Coulomb's law as

$$F_A = \frac{q^+ q^-}{80r^2} \cdot \frac{1}{4\pi\epsilon_0}$$

where r is the counter-charge separation in such an aqueous medium of permittivity $\epsilon = 80$. This attractive force is relatively weak, and as such we can say that an aqueous medium tends to keep counter-charged particles apart, resulting in a relatively slow rate of reaction for many chemical processes. In the non-polar hydrophobic interior of proteins, however, we have seen that the effective molecular permittivity will have a value of the order $\epsilon = 2$, so that the attractive force between counter-charges will be some 40 times greater than in an aqueous medium. This effect, coupled with the fact that strong electric forces can now be used to influence any reactants, must surely be of relevance to the catalytic process of many enzymes. We shall return to this theme in later chapters when dealing with the possible existence of electronically conducting pathways in protein structures, and their relevance to some biological functions.

AQUEOUS SOLUTIONS OF BIOMOLECULES

Mention has already been made of the fact that amino acids are solids that do not melt until they decompose at high temperatures, and that they are appreciably soluble only in water. The difficulty of dissolving amino acids and proteins in non-polar liquids arises from the polar nature of these biomolecules, and the ability to dissolve in water has for some time led to considerable importance being attached to the accurate analysis of the dielectric properties of aqueous solutions of biomolecules.

From Figure 3.1 it can be seen that amino acids near their isoelectric point will have quite large dipole moments. Taking the simplest amino acid, glycine, then from a straightforward structural model of this molecule, the distance between the positive charge on the nitrogen atom of the amino group and that of the negative charge lying between the oxygen atoms of the carbonyl group should be about 3.3 Å. As a result of internal polarization, the effective charge separation should be a little less than this, giving the effective dipole moment as a little under 16 debye units ($\mu \simeq 1.6 \times 10^{-19} \times 3.3 \times 10^{-10} = 5.3 \times 10^{-29}$ SI units = 15.9 D). This value compares favourably with the early estimate by Wyman[5] of 20 Debye units from dielectric measurements on aqueous solutions of glycine and α-aminobutyric acid. At about the same time, a more detailed estimate was given by Kirkwood[6] based on the theory of solutions of dipolar ions, and a value of 15 debye units was derived for glycine. From consideration of the factors leading to the ionization equilibrium scheme of Figure 3.1, we will expect the dielectric properties of amino-acid and protein solutions to vary markedly with changes of pH, and indeed such effects can be observed. The early measurements of Shutt[7] showed that egg albumin exhibits two sharp minima in permittivity value in the pH range 4.1–5.9, and Dunning and Shutt[8] found that the permittivity of glycine solution is constant from pH = 4.5 to pH = 7.5, but falls sharply on both sides of this pH range. These early dielectric measurements were taken as convincing evidence for the existence of the dipolar ionic (zwitterionic) form for α-amino acids, first proposed by Küster[9] in 1897, and emphasized by Bjerrum some 26 years later.[10]

The dipole moment for the glycine dipolar ion is much greater than that of ordinary polar molecules, and therefore strong electrostatic interactions between these ions and the solvent polar molecules is to be expected. Indeed, Kirkwood, who did much to develop the dielectric theory for polar solutions,[11] described such an amino-acid dipolar ion as 'a superpolar molecule, surrounded by an intense electrostatic field'.[12] Also, the large dipole moments of such biomolecules should, according to the theory of Debye as expressed by equation 1.18, result in large permittivity values for their solutions. This was confirmed by the early workers[13,14] who found that the aliphatic amino acids and peptides produce remarkably large permittivity increments in aqueous solution and mixed polar solvents. However, not surprisingly considering the strong electrostatic interactions and non-spherical form of many biomolecules, attempts to interpret the measurements on the basis of the Lorentz field and the Clausius—Mossotti formula failed completely.

As a result of these difficulties of interpretation, the molecular polarizabilities of amino-acid and protein solutions have often been compared using the following equation

$$\epsilon = \epsilon_1 + \delta c \tag{3.4}$$

where ϵ and ϵ_1 are the permittivities of the solution and pure solvent, respectively, and c is the concentration of the solute (moles/litre). If the molecular weight of the solute is not accurately known, then the concentration can be expressed in grams of solute per litre, with the corresponding dielectric increment $\Delta\epsilon/c = (\epsilon - \epsilon_1)/c = \delta$ being replaced by a weight increment $\Delta\epsilon/g = \delta/M$, where M is the molecular weight and g is the grams of solute per litre. For low concentrations (typically up to 0.01 mole fraction of solute) δ in equation (3.4) is usually found to be constant, so that the measured permittivity varies linearly with concentration. For aqueous solutions of glycine at 25 °C, δ has a constant value of around 22.6 for concentrations up to 2.5 moles/litre.[5]

Kirkwood was the first to attempt an accurate description of this problem. As an extension of Onsager's theory of the dielectric polarization of polar liquids, Kirkwood[15] developed a general statistical theory applicable to polar liquids and polar mixtures. In this theory it becomes apparent that the dipole moment μ of an individual molecule cannot be obtained from permittivity measurements alone. However, a moment $\bar{\mu}$, equal to $(\mu\bar{\mu})^{1/2}$, may be calculated in a straightforward manner, the moment $\bar{\mu}$ being the sum of the dipole moment of a molecule and the moment which it induces in its neighbouring molecules by hindering their rotation relative to itself. For the dipolar ions of the aliphatic amino acids in dilute aqueous solution, it appears that in fact $\bar{\mu}$ does not differ much from μ, the molecular dipole moment. From Kirkwood's theory,[15] then to a fair approximation at 25 °C, we may write

$$\bar{\mu} = 3.30 \, \delta_0^{1/2} \tag{3.5}$$

where δ_0 is the infinite dilution value for the permittivity increment δ of equation (3.4). As a result of the linear dependence of the permittivities of the very dilute solutions upon concentration, it is normally unnecessary to distinguish between δ_0 and δ.

Wyman[5] obtained a value of 22.6 for δ_0 for glycine in water, which from equation (3.5) leads to a moment $\bar{\mu}$ of 15.7 debye units. From salting-in data obtained from the solubility relations of glycine in alcohol—water mixtures, Kirkwood[6] calculated the value for μ for glycine to be 15.0 debye units. From the relation $\bar{\mu} = (\mu \bar{\mu})^{1/2}$, this gives $\bar{\mu}$ to be 16.4 D. In this way, $\bar{\mu}$ differs from the true dipolar ion moment μ by only 10 per cent. The difference of 1.4 D between μ and $\bar{\mu}$ represents the vector sum of the moments induced in the solvent environment through orientation of adjacent water molecules by the positive and negative groups of the dipolar ion. Since this additional moment is only a relatively small fraction of the total moment of the dipolar ion, we may conclude that $\bar{\mu}$ is a rather good approximation to μ, the dipolar ion moment. The effective dipole separation $\bar{\mu}/e$ is 3.27 Å, in good agreement with the observed structural value of 3.17 Å.

Of all the materials that can be dissolved in water, with few exceptions such as solutions of urea, amino acids and proteins are the only ones to give rise to positive values for δ in equation (3.4). This results from water having such a high permittivity value of the order 80, that replacement of part of it by a substance of lower permittivity will tend to result in a lower permittivity value for the mixture. For these other materials δ has a negative value and is usually termed as a dielectric decrement. A list of dielectric increments δ for some amino acids and peptide chains in aqueous solution at 25 °C is given in Table 3.3. Included in this table are values derived using equation (3.5) for the moment $\bar{\mu}$, equal to $(\mu \bar{\mu})^{1/2}$. As we have seen for the case of glycine, the value for $\bar{\mu}$ will overestimate the true molecular

Table 3.3   Dielectric increments and derived dipole moment $\bar{\mu}$ for some amino acids and glycine peptide chains in aqueous solution at 25 °C

Sample	Dielectric increment δ	Debye units (from eqn. (3.5))	References
α-Alanine	23.2	15.9	13
	27.7	17.4	14
l-Asparagine	28.4	17.6	14
	20.4	14.9	16
l-Aspartic acid	27.8	17.4	14
d-Glutamic acid	26	16.8	14
l-Glutamine	20.8	15.1	16
Glycine	22.6	15.7	13
	26.4	17.0	14
l-α-Leucine	25	16.5	14
Glycine dipeptide	70	27.6	14
	70.5	27.7	17
Glycine tripeptide	113	35.1	13
	128	37.3	17
Glycine tetrapeptide	159	41.6	13
Glycine pentapeptide	215	48.4	13
Glycine hexapeptide	234	50.5	13
Glycine heptapeptide*	290	56.2	5

* In 5.14 molar urea

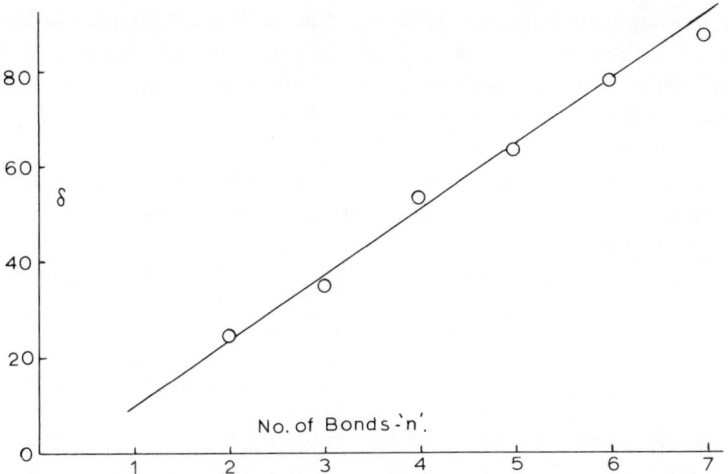

Figure 3.4 The variation of the dielectric increment δ with the number n of bonds between the terminal charged groups of the various amino acids in aqueous solution

moment μ, but for the simple amino-acid molecules at least, this overestimate will be no greater than about 10 per cent. A more extensive list of dielectric increments has been given by Wyman,[18] which has recently been extended by Hasted.[19] Oncley[20] has given the weight increment $\Delta\epsilon/g$ values for sixteen protein molecules.

From Table 3.3 it can be seen that the values of δ for the α-amino acids are not only positive and large, but are of nearly the same magnitude. Also, there is a satisfactory agreement between the results of Wyman ($\delta = 20.4-23.2$) obtained in the frequency range 40–120 MHz, those of Devoto ($\delta = 25-28.4$) obtained at 330 MHz, and those of Hedestrand[21] ($\delta = 23.0-23.6$) obtained at 1 MHz. This indicates that for aqueous solutions of α-amino acids at 25 °C there is unlikely to be a sizeable dielectric dispersion in the frequency range 1–330 MHz. From the list of dielectric increments given by Wyman,[18] it is found that not only the α-amino acids form a class having nearly similar δ values, but so also do the β- and γ-amino acids, for example. Furthermore, there is a regular increase of δ with the number n of chemical bonds between the amino and carboxyl groups ($n = 2$ for an α-amino acid, 3 for a β-amino acid, and so on). This is demonstrated in Figure 3.4, where the average values of δ, derived from Wymann's list[18] for the amino acids, have been plotted against n, the number of chemical bonds between the amino and carboxyl groups. This shows δ to increase linearly with n, by a factor of about 14 for each additional chemical bond length.

The least-squares fit of the data of Figure 3.4 to the form of a linear equation $y = ax + b$ gives the linear relationship between the dielectric increment δ and number of bond lengths n as

$$\delta = 14.42n - 8.72$$

This behaviour is understandable in terms of these amino acids existing in aqueous solution predominantly in the form of dipolar ions. With increasing distance (number of bonds) between the charged amino and carboxyl end groups, the molecular dipole moment, and hence δ, will increase in magnitude. The linear dependency of the dielectric increment on the number of atomic bonds n implies that in aqueous solution amino acids have almost complete freedom of rotation about their atomic bonds. This follows from the proportionality between δ and $\bar{\mu}^2$ as given by equation (3.5), and the fact that $\bar{\mu}^2$ is proportional to $\langle r \rangle^2$, the square of the mean distance between the charged end groups. For linear atomic chains, assuming complete freedom of rotation about bonds, this mean distance is given by[22]

$$\langle r \rangle^2 = \frac{1 - \cos \theta}{1 + \cos \theta} \cdot nb^2 \tag{3.6}$$

where n is the number of bonds of length b, and θ is the covalent bond angle, assumed here to be $110°$. This formula only gives accurate values when n is large. Eyring[23] has given the more complex equation

$$\langle r \rangle^2 = b^2 [n + 2(n - 1)\cos \phi + 2(n - 2)\cos^2 \phi + \cdots + 2 \cos^{n-1}\phi] \tag{3.7}$$

which holds for any chain length, including when $n = 1$, where $\phi = (180° - \theta)$. For a freely rotatable dipolar ion, then we have

$$\delta \propto \bar{\mu}^2 \propto \langle r \rangle^2 \propto n$$

as shown by Figure 3.4. For a rigid linear carbon chain, the distance between the end chain groups is given by

$$r = nb \sin(\theta/2)$$

and the dielectric increment δ would vary as n^2. Ignoring polarization effects due to the surrounding polar water molecules, electrostatic attraction between the amino and carboxyl terminal groups, the space occupied by the chain atoms themselves, and the question of the exact location of the negative charge at the carboxyl group, then from equations (3.5) and (3.6) we would expect $\delta \gtrsim 10.1n$, which gives a reasonable prediction of the observed behaviour for amino acids.

The dipolar ionic nature of aqueous solutions of amino acids is further emphasized by their anhydride derivatives having negative δ values. The formation of dipolar ions is not possible for anhydrides, so we have the situation, for example, of γ-aminobutyric acid in water exhibiting a dielectric increment[14] $\delta = 51$ whereas its anhydride, pyrrolidone exhibits a dielectric decrement[14] with $\delta = -1$. Another example is glycine anhydride,[14] which has a large dielectric decrement in water of $\delta = -10$.

The amino-acid polypeptides can be expected to form a homologous series similar to the aliphatic amino acids, although they will contain an additional source of polarity associated with the peptide residues as described in Chapter 2. From Table 3.2, the first seven peptides of glycine can be seen to have values for δ increasing linearly with the number of constituent chemical bonds n between the

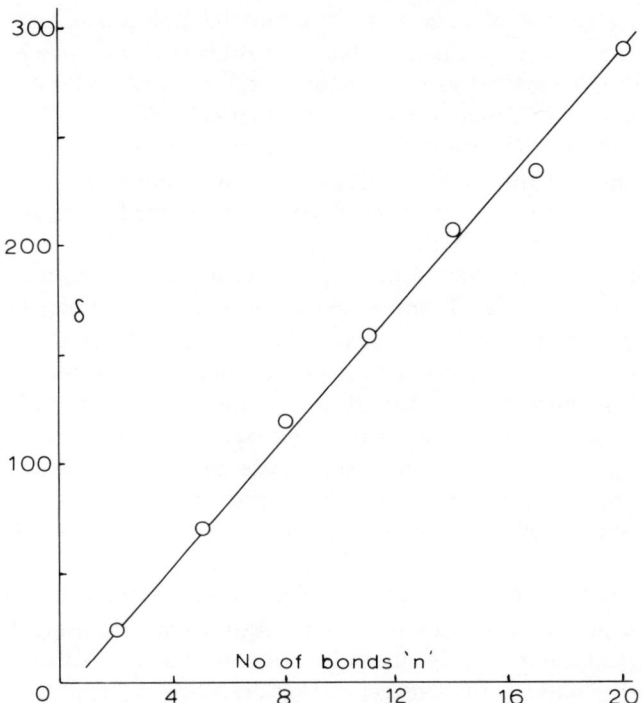

Figure 3.5 The variation of the dielectric increment δ with the number n of bonds between the terminal groups of the glycine polypeptides in aqueous solution

terminal NH_3^+ and COO^- groups, as shown in Figure 3.5. The least-squares fit of the data of Figure 3.5 gives the linear relationship between δ and the number of bond lengths n for the glycine polypeptides as

$$\delta = 14.51n - 5.87$$

in close agreement with the results of Figure 3.4 for the simple aliphatic amino-acid molecules. From Chapter 2, we would expect the peptide moments μ_p to oppose the dominant moment resulting from the charged amino and carboxyl terminal groups. The fact that such a reduction of the dependence of δ with bond number n for the polypeptides, as compared with the amino-acid molecules, is not observed, would indicate a completely random configuration for the polypeptide structures in water. The slight difference in dielectric behaviour of the amino acids and their polypeptides could result primarily from differences in solvent polarization effects. These effects are likely to be complex for the polypeptides, demonstrated to some extent by the fact that glycine heptapeptide is almost completely insoluble in water,[18] the dielectric results for this polypeptide being obtained using urea as the solvent. Ignoring the dielectric increment measurement for glycine heptapeptide, then the least-squares fit for the remaining glycine peptide data of Figure 3.5 gives

$$\delta = 14.49n - 3.87$$

indicating that the use of urea as the solvent did not produce any significant difference to the results. In a more detailed investigation by Conner *et al.*[24] of the 30 MHz dielectric increment values of aqueous solutions of twelve polypeptides, it was concluded that sufficient rotational freedom existed within the molecules to permit an almost random distribution of the orientations about the atomic bonds. In the various di- and tripeptides investigated by these authors, the principal effect of side chains upon the molecular moments was considered to be associated with modification of solvent polarization effects.

Wyman[18] has also investigated the dielectric properties of amino acids dissolved in solvents other than water. From his measurements it can be concluded that for solvents having permittivity values greater than around 20–25 the values of the dielectric increment δ remain fairly constant and similar to those derived for water. In solvents of permittivity less than 20 or 25, the dielectric increments decrease markedly with decreasing permittivity, and specific differences between the solvents become larger. The fundamental processes involved are obviously complex, and this area of work in investigating the dielectric properties of amino acids dissolved in solvents having low permittivities is one that could usefully be further developed.

A more accurate determination of the molecular dipole moment, than that given by Kirkwood's formula of equation (3.5), is possibly obtained using Buckingham's theory[25] for polar mixtures, which has been reviewed by Hasted (reference 19, p. 191). This theory extends the Onsager model, described in Chapter 1, to consider the situation for ellipsoidal molecules and cavity fields, and gives a method for evaluating the root mean-square value of the molecular moment of polar molecules dissolved in a polar solvent. The calculations are quite involved, in that use has to be made of tabulated geometry factors, and a value is required for the refractive index of the polar mixture. A comparison of the dipole moments derived using equation (3.5) against values quoted by Hasted (reference 19, p. 197) and derived using Buckingham's theory, is given below for three amino-acid molecules.

Molecule	$\bar{\mu}(D)$ (eqn. 3.5)	$\mu(D)$ (Hasted)
Glycine	16.3	13.3
β-Alanine	19.4	17.5
ϵ-Aminocaproic acid	29.1	28.3

The dipole moments obtained using Buckingham's theory are seen to be of the correct order of magnitude, in that they have values slightly lower than those derived using equation (3.5), which we know overestimates the dipole moment values.

From the consistency of the dielectric increment values obtained at various frequencies, we have already seen that aliphatic amino acids in aqueous solution are unlikely to exhibit significant dielectric dispersions in the frequency range up to about 10^8 Hz. At frequencies above this, however, we will expect such a solution to exhibit a dielectric dispersion associated with the inability of the solute molecules to contribute to the orientational polarization component of the total permittivity. At such high frequencies, assumed to be below the dispersion region for water itself, we can treat such biomolecules as behaving as a non-polar solute in water. The dielectric property of such a mixture can then be given, using one of many mixture theories, as for example[26,27]

$$\frac{\epsilon - \epsilon_\omega}{\epsilon_s - \epsilon} = \frac{\beta v}{1 - v}$$

where ϵ, ϵ_ω and ϵ_s are the permittivities of the mixture solution, of the water solvent, and of the solute, respectively. The parameter β is a shape factor and v is the volume fraction of the solute. For low solute concentrations

$$\epsilon = \epsilon_\omega - \beta v(\epsilon_\omega - \epsilon_s)$$

and as now, at frequencies above the solute orientational dispersion, we have $\epsilon_s < \epsilon_\omega$, then the mixture solution will exhibit a dielectric *decrement*. This behaviour is demonstrated by the 20 °C results obtained by Shepherd and Grant[28] for an isoelectric 2 M ϵ-aminocaproic acid solution, as shown in Figures 3.6 and 3.7.

Included in Figures 3.6 and 3.7 are the variations of ϵ' and ϵ'' for water, and it can be seen that above 10^8 Hz the dielectric increment reduces steadily in value, until at around 10^9 Hz the solution begins to exhibit a dielectric decrement. The dielectric loss is seen to be much greater than that for water, and the maximum loss occurs at a frequency of 4×10^8 Hz, nearly two decades below that for the

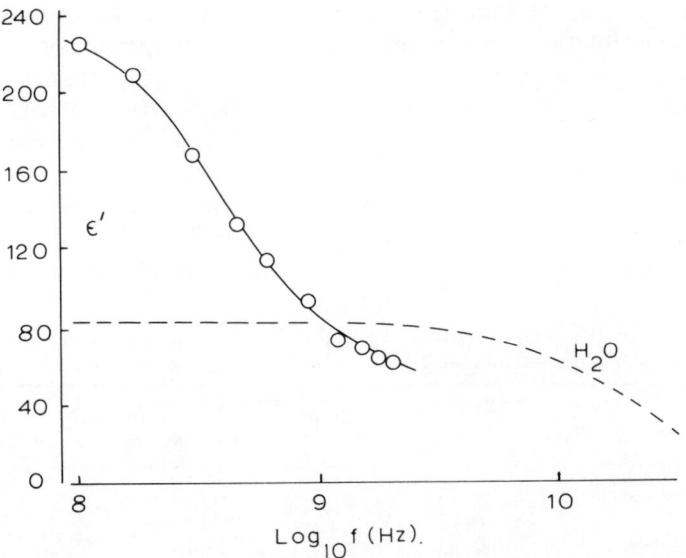

Figure 3.6 The frequency dependence of ϵ' for an isoelectric 2 M ϵ-aminocaproic acid solution, compared with that for water[28]. (Reproduced from J. C. W. Shepherd and E. H. Grant, *Proc. Roy. Soc. London A,* **307**, 335, 345 (1968), by permission of the Royal Society)

86

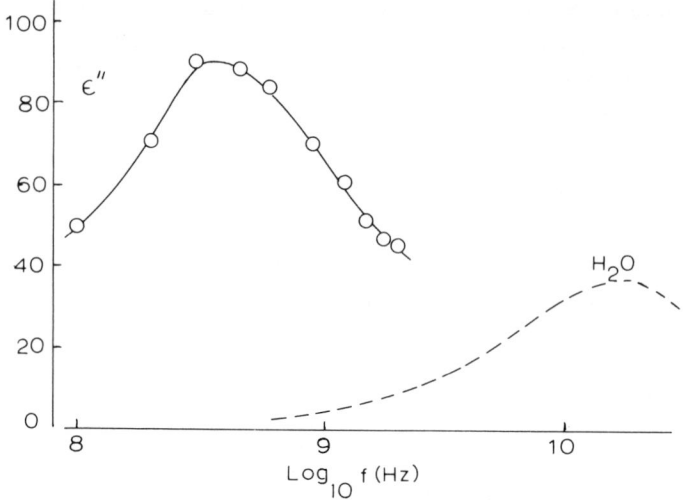

Figure 3.7 The frequency dependence of ϵ'' for the same
ϵ-aminocaproic acid solution of Figure 3.6. (Reproduced from
J. C. W. Shepherd and E. H. Grant, *Proc. Roy. Soc. London A,*
307, 335, 345 (1968), by permission of the Royal Society)

maximum loss of normal water. The derived dipole moment value of 28.3 D is in
good agreement with the value of 29.1 D given earlier here for ϵ-aminocaproic acid
using equation (3.5). Measurements were also obtained for aqueous proline and
hydroxyproline solutions, and Shepherd and Grant interpreted the fairly broad
dielectric loss curves in terms of three separate contributions, comprising water of
normal structure, the amino-acid solute, and thirdly, to a first approximation, all
water with structure modified by the presence of the solute molecules. The
structure of normal water involves a complicated, and as yet not completely
understood, association of hydrogen bonds and dipolar interactions. The presence
of the large dipole of the amino-acid molecules could influence the surrounding
water molecules by inducing a larger than normal dipole moment in them, or

Table 3.4   Dielectric weight increments and derived dipole moment values for
some aqueous protein solutions at 25 °C

Protein	Dielectric increment (m^3/kg)	Dipole moment (Debye)	Reference
Horse carboxyhaemoglobin	0.33	480	20
Myoglobin	0.15	170	20
Whale myoglobin	0.12	146	30
Horse myoglobin	0.15	163	30
Egg albumin	0.10	250	20
Horse serum albumin	0.17	380	20
Horse serum γ-pseudo- globulin	1.08	1100	20

causing two or more to rotate together as one unit, or even giving them an effective density greater than that of normal water.[29] Water molecules adjacent to the charged amino and carboxyl groups will have short-range structures differing from those near to the hydrophobic groups. Matters will be further complicated by the likelihood of there being significant dipole–dipole interactions between the amino-acid solute molecules. The measurements by Shepherd and Grant were made at various solute concentrations and temperatures between 0 and 50 °C, and their observations and critical assessments, together with the comments of their work by Hasted (reference 19, p. 215), make an important contribution to the development of an understanding of the dielectric behaviour of amino-acid solutions.

Finally, the dielectric weight increment values and derived dipole moments for some aqueous protein solutions are given in Table 3.4. The dipole moment values given in this table were derived using the equation first derived and used by Oncley,[20] namely

$$\mu^2 = \frac{2\epsilon_0 MkT\delta}{N_A}$$

where here δ is the dielectric weight increment and M is the protein molecular weight. In Table 3.4, some 30 years spans the time between the values given by Oncley[20] and those derived from the work of South and Grant[30] for the myoglobin molecule, and the very close agreement in dielectric increment values indicates the very high experimental standards set by the early pioneers of this field of study. With the passing of time, more accurate values are available for protein molecular weights, so that small changes may need to be made to some of the earlier derived dipole moment values.

DIELECTRIC DISPERSION – RELATION TO MOLECULAR GEOMETRY

If the polar solute molecules are spherical, and large by comparison with the solvent molecules, then the orientational relaxation of the solute molecules can usefully be described using Debye's rotational model, outlined in Chapter 1. In this model the dipolar solute molecules are considered as spheres whose rotation is opposed by the viscosity of the surrounding solvent medium. The orientational relaxation time τ, as given by equation (1.45), is

$$\tau = 4\pi\eta a^3 / kT$$

where η is the solvent viscosity and a is the radius of the solute molecule. In this way, we expect dielectric dispersion measurements of aqueous solutions of bio-molecules to give us information about the size of the biomolecule. Because of the proportionality of the relaxation time with molecular volume given by equation (1.45), we will expect the frequency of maximum loss for protein solutions to be lower than that for simple polypeptides, which in turn will exhibit dispersions at lower frequencies than that for amino-acid solutions. Such a trend is observed, with aqueous protein solutions exhibiting dispersions typically in the range 20 KHz ~ 20 MHz, but unfortunately an interpretation of the results is complicated by the fact that very few biomolecules are spherical, their shapes often being more appropriately described as ellipsoids of revolution.

One protein molecule that can almost be considered as spherical is oxyhaemoglobin, whose three mutually perpendicular diameters, as obtained from X-ray diffraction measurements,[31] have lengths of 50, 55, and 60 Å. Grant et al.[32] have measured the relaxation times τ for aqueous solutions of oxyhaemoglobin as a function of temperature T and solution viscosity η, and as predicted by equation (1.45), there is a linear dependency of τT upon η, such that $\tau T/\eta$ has a mean value of 3.25×10^{-2} m s^2/kg. From equation (1.45) this gives the effective molecular radius as 33 Å, which is rather larger than the mean X-ray value. This increase in effective volume of oxyhaemoglobin most likely arises from water molecules being bound to the polar hydrophilic groups at the surface of the molecule (a consequence of the 'oil-drop' model!). Such a bound coverage of the water of hydration will relax with the protein molecule and so increase its effective volume. A uniform coverage of water, two water molecules thick, is required for the oxyhaemoglobin molecule to have the effective volume indicated by the dielectric results. The mechanisms and physical characteristics of the binding of water to proteins will form the subject matter of the next chapter. At room temperature, the maximum dielectric loss for oxyhaemoglobin solution occurs at a frequency of around 10 MHz and the derived molecular dipole moment is 400 D.[32]

By considering biomolecules as ellipsoids of revolution, having semiaxes a,b,b, we will expect there to be two characteristic relaxation times; τ_a for orientation involving rotation of the a-axis around the b-axis, and τ_b for rotation of the b- about the a-axis. The early work of Perrin,[33] who considered the diffusion constants of an ellipsoid in a viscous medium, is very relevant here, and has been extensively related to the case of protein solutions by Edsall (reference 20, pp. 506–542). Oncley,[20] using Perrin's equations has obtained the relaxation times τ_a and τ_b in terms of the ratio a/b of the semiaxes of the ellipsoid and the relaxation time τ_0 of a sphere of the same volume, with τ_0 being given by

$$\tau_0 = 4\pi\eta ab^2/kT \tag{3.8}$$

In summarizing the relaxation times of nineteen amino acids and peptides, Oncley[20] found that the dielectric relaxation times for glycine and α-alanine were somewhat smaller than that predicted by equation (3.8), but with increasing molecular size the relaxation times became increasingly larger than expected, but by never more than 50 per cent. In their measurement for glycine polypeptides and the simpler di- and tripeptides of alanine and leucine, Conner and Smyth[34] interpreted the lower relaxation time values for glycine and alanine as attributable to the inner or microscopic solvent viscosity being lower than the experimentally determined macroscopic viscosity. The increased values, obtained by Conner and Smyth, for the relaxation times of the peptides were used to calculate the axial ratio a/b of the molecular ellipsoids of revolution, the values increasing from 1.3 for diglycine to 2.1 for pentaglycine. Triglycine, alanyldiglycine and leucyldiglycine all had values for a/b of around 1.57, indicating that their molecular shapes were similar, whereas alanylleucylglycine, with a value for a/b of 1.33 appeared to be more nearly spherical as a result of the large isobutyl side group midway between its charged end groups. Also, the values of a/b for the various glycine poly-peptides[34] show a linear dependence upon the square root of n, the number of constituent chemical bonds between the end groups, which is to be expected if these molecules are randomly distributed in all possible configurations.

Whereas it might be expected that the amino-acid and peptide molecules, being not much larger than the solvent molecules, will not conform very closely to the Debye rotational model, this will not be the case for the much larger protein molecules. Assuming that protein molecules in solution behave hydrodynamically as rigid ellipsoids of revolution generated by rotating an ellipse with semiaxes of length 'a' and 'b' about the a-axis, and that the dipole moment is directed in a set direction with respect to these axes, then from Perrin's work[33] the characteristic relaxation times are given by

$$\tau_i = \frac{4\pi\eta ab^2}{kT} P_i(a/b)$$

$$= \frac{3V\eta}{kT} P_i(a/b) \tag{3.9}$$

where the subscript i refers to the relaxation of either the a- or the b-axis about the other. In equation (3.9), V represents the effective volume of the protein molecule in solution, and the two functions $P_i(a/b)$, whose values were calculated by Perrin, depend only on the axial ratio a/b. From the experimentally determined relaxation time value, two curves can be drawn of the possible combinations of axial ratio and the effective molecular volume V, as shown in Figure 3.8. These two curves of Figure 3.8 correspond to the assumption that the measured relaxation time is

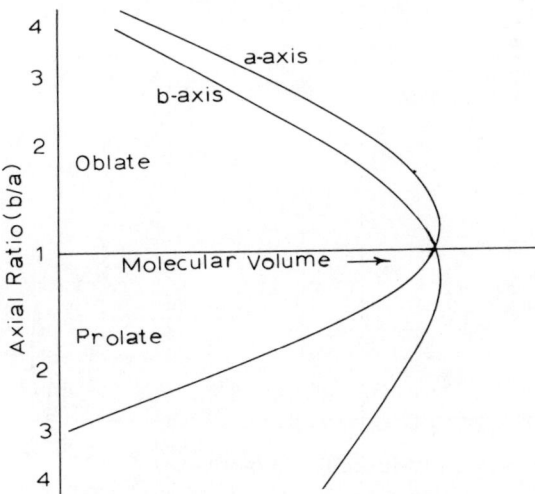

Figure 3.8 Plots of the Perrin functions[33] of axial ratio against effective molecular volume for ellipsoids with rotation of either the a- or b-axis

90

associated with the rotation of one of the principal axes, with the molecular dipole moment pointing along the rotating axis. In general the dipole moment orientation will not coincide with an axis direction, and a curve lying between the two shown in Figure 3.8 will be more appropriate, since the dipole moment will have a component along each axis. Depending upon whether ellipses are rotated about their minor or major axis, the resulting surfaces are either oblate or prolate spheroids, respectively. In Figure 3.8, the term 'oblate' corresponds to $b > a$ and the molecule is disc-shaped. The term 'prolate' refers to the case $a > b$ and the molecule will be cigar-shaped.

An example of the application of Perrin's functions is shown in Figure 3.9 using the dielectric data obtained by South and Grant[30] for whale myoglobin. From the X-ray diffraction data of Kendrew et al.[35] myoglobin has the form of an oblate spheroid, so only the Perrin functions applicable to an oblate spheroid are shown in Figure 3.9. From the X-ray data[35] the molecular volume of the myoglobin molecule is about 15 nm^3 (1.5 x 10^4 Å^3), which from Figure 3.9, if the molecule retains its conformational shape in going into solution, implies an axial ratio b/a of about 5. The X-ray data suggests an axial ratio of 2, which from Figure 3.9 gives the effective molecular volume to be around 35 nm^3. Such an increase in effective volume is compatible with there being a hydration layer of about two water molecules thickness which relaxes with the myoglobin molecule. A similar applica-

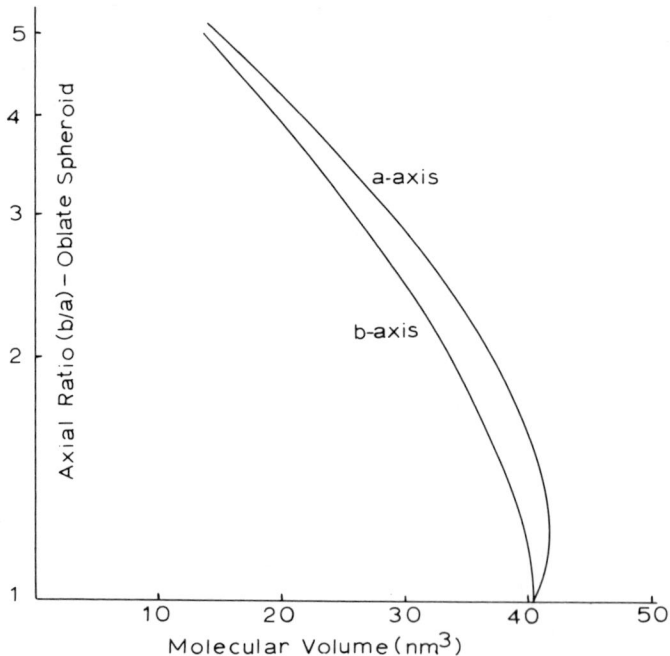

Figure 3.9 The Perrin plot of axial ratio against effective molecular volume for whale myoglobin. (Reproduced from G. P. South and E. H. Grant, *Proc. Roy. Soc. London A,* **328**, 371 (1972), by permission of the Royal Society)

tion of the Perrin functions for the oxyhaemoglobin molecule[32] indicates a similar hydration coverage, equivalent to about 0.45 g of bound water per g of protein.

This additional information regarding bound water content is both a valuable bonus and at the same time a complication regarding interpretation of dielectric results in terms of the size and shape of protein molecules. Hendrickx *et al.*[36] have derived and proposed a set of parameters to check the interpretation of the dielectric behaviour of protein solutions as a rigid dipole relaxation of ellipsoids of revolution, but unless there is a *uniform* thickness of bound water covering the protein molecule (which from the next chapter we shall see is probably unlikely) their derived axial ratio for transferrin of 4.5 for a prolate ellipsoid of revolution, may not accurately reflect the actual molecular geometry.

OTHER INTERPRETATIONS OF PROTEIN DIELECTRIC DISPERSION

So far we have considered the dielectric dispersion of aqueous protein solutions to arise essentially from the partial orientation of the protein dipole moment under the action of an externally applied electric field. It has also been assumed that this dipole reorientation corresponds to the Debye model of rotation of the macromolecule as a rigid unit. However, other mechanisms and theories have in the past been proposed to describe the dielectric behaviour of protein solutions.

The first alternative to Debye rotation comes from Debye himself, who with Falkenhagen[37] proposed the *ion atmosphere* model, where the protein molecule is considered to exist within a cloud of counter-charge ions. The dielectric dispersion is assumed to be associated with the polarization and resulting displacement of these ions from the centre of the protein molecule charge about which the ion atmosphere is formed. This model does not appear to be very relevant for protein solutions, since the model predicts that the dielectric relaxation time τ should vary inversely with the ionic strength of the protein solution, whereas for myoglobin,[30] for example, it is found that τ is only weakly dependent on the solution conductivity (ionic strength). Also, through the process of electrodialysis, small ions can be removed from the solution, leaving on average only two or three ions per protein molecule, which can hardly be described as an 'ionic atmosphere'. Furthermore, such electrodialysis treatment has no effect on the protein or its dielectric properties.[30]

Kirkwood and Shumaker,[38] however, have proposed a *proton fluctuation* model which cannot be dismissed so readily. This model proposes that the dielectric dispersion of protein molecules, containing as they do a number of neutral and negatively charged basic sites to which protons may be bound, arises from the fluctuations and redistributions of surface charge due to proton migration on the surface. From our discussion, at the beginning of this chapter, of the ionic properties of amino acids, we know that many of the residue side chains along the polypeptide chains of protein molecules are acidic or basic, and thus able to lose or gain a proton to become either negatively or positively charged. The carboxyl and amino groups terminating polypeptide chains are also ionizable. For such haem-containing proteins as myoglobin, haemoglobin, and the cytochromes, the iron atom and the two propionic acids of the haem group are also ionizable. The probability, f_j of such an ionizable group j having a bound proton depends on the solvent proton concentration, and hence solvent pH, and also on the ionization constant pK_a of that group, and is given by

$$f_j = (1 + 10^{pH - pK_a})^{-1}$$

The average charge $q\langle x_j \rangle$ on group j is then given by

$$q\langle x_j \rangle = q(f_j - \gamma_j)$$

where q is the protonic charge and γ_j is 0 for basic groups and 1 for acidic groups. Except in very acid solutions, the number of basic sites usually exceeds the average number of bound protons, so that many possible proton distributions exist, each one differing only slightly in their free energy values. Kirkwood and Shumaker[38] show by a statistical method that fluctuations in charge and configuration of mobile protons on the protein surface can give rise to a non-vanishing mean-square dipole moment $\Delta\mu^2$, even if the mean permanent molecular dipole moment is zero. They consider that the process of redistribution of protons possessing a translatory diffusion constant D on the surface of a sphere of radius 'a' should have a relaxation time of the order $\tau = b^2/D$. For the myoglobin molecule we have $a \simeq 3.5$ nm (35 Å) and with $D \simeq 10^{-10}$ m^2/s, a not unreasonable value, we have a relaxation frequency of 8 MHz, as observed experimentally. Examples of their theoretical calculations for $\Delta\mu$, as compared with experimentally observed values μ_{obs}, are given below for four protein molecules in aqueous solution:

Protein	μ_{obs}(D)	$\Delta\mu$(D)
β-Lactoglobulin	700	270
Ovalbumin	260	440
Horse haemoglobin	480	620
Horse serum albumin	700	680

Of these examples, only β-lactoglobulin has an observed dipole moment exceeding that which could arise solely from proton fluctuations.

Since only those ionizable groups with pK_a values nearly equal to the pH of the solution will contribute significantly to $\Delta\mu^2$, the *proton fluctuation* model predicts a pH-dependent dielectric dispersion. Assuming no electrostatic interaction between the protons, then $\Delta\mu^2$ will have maximum values at pH values coincident with the pK_a of the conjugate acid groups, and minima at intermediate values. Kirkwood and Shumaker consider that electrostatic interaction between the protons will cause suppression of the maxima and minima in $\Delta\mu^2$ away from the isoelectric (isoionic) point, and $\Delta\mu^2$ becomes a monotonically decreasing function of pH. For ovalbumin, the derived dipole moment is calculated to be 449 D at pH 2, decreasing steadily with increasing pH to the value 5 D at pH 10.

Contrary to the treatment by Kirkwood and Shumaker, Scheider[39] shows that charge fluctuations on the surface of a protein molecule do not introduce independent components into the dielectric relaxation spectrum. If the molecular relaxation time τ_m is much greater than the ionized charge fluctuation time τ_q, then there will be practically no detectable fluctuation dielectric dispersion at any frequency, even if the fluctuation mean-square moments are large. This corresponds to the situation where the state of ionization for the charge-binding sites on a molecule is in such rapid fluctuation, the molecule acts as if the sites were only partially ionized. Conversely, if $\tau_q > \tau_m$ then Scheider shows that this corresponds to the situation where a given instantaneous charge configuration remains fixed on a molecule for a period of many relaxations. So, although the magnitude of the total dielectric dispersion of a protein solution may be strongly dependent on fluctuation moments, in all cases the measured relaxation times are primarily related to τ_m, the molecular relaxation time.

Other theories for protein dispersion include O'Konski's[40] *surface conductivity* model, which extends the ion atmosphere model of Debye and Falkenhagen to include effects associated with surface ionic conductivity at the interface between the protein molecule and the counter-charge ion atmosphere. Schwan[41] has developed equations to show that Maxwell–Wagner, or *interfacial polarization*, effects can account for the observed dielectric dispersions. However, as Hasted indicates (reference 19, p. 211) such interfacial polarization theories require what appears to be unrealistically large values for the protein conductivity in order for the derived relaxation times to be in agreement with experimental findings. Jacobson[42] has suggested that the *structured water* of hydration is solely responsible for the observed dielectric dispersion and that the protein molecule itself contributes nothing; and finally, Schwarz[43] has developed a *surface ionic mobility* model to account for the dielectric behaviour of protein solutions. South and Grant[30] have considered the relevance of these various models for the specific case of the myoglobin molecule in water. They conclude that only the *Debye rotation* and the *proton fluctuation* model contribute significantly to the dielectric properties of this protein in aqueous solution.

In general, then, we can consider the dipole moment of a protein molecule in aqueous solution to arise from two main components, namely the permanent dipole moment component μ and the fluctuating proton component $(\Delta\mu^2)^{1/2}$. The permanent dipole moment component of the protein molecule can be represented in terms of the average volume charge density $\rho(r)$ at any point r in the molecule of volume V, as

$$\mu = \int_V \rho(r) r \, dV$$

$$= q \sum_j \langle x_j \rangle r_j + \int_v \rho(r) r \, dv$$

where the first term on the right of the above equation represents that component of the permanent moment arising from the charged ionizable groups, and the second term represents the contribution from the remainder of the molecular volume, v. This second term contains contributions from the peptide residue moments μ_p and side-chain moments μ_s. Ignoring electrostatic interactions, the

94

fluctuating proton dipole moment $(\Delta\mu^2)^{1/2}$ is given by[38]

$$\Delta\mu^2 = q^2 \sum_j (f_j - f_j^2) r_j^2$$

The total dipole moment, equal to $(\langle\mu^2\rangle + \Delta\mu^2)^{1/2}$ for the whale and horse myoglobin molecules in aqueous solution has been calculated by South and Grant[30] using the above formulae, together with the known X-ray structure data for these molecules. The contribution of the non-ionizable portions of the molecule was approximated to just that arising from the peptide residues of the α-helical portions of the protein molecule, using Wada's value (Chapter 2, reference 20) of 3.4 D per residue. The experimental and theoretical results obtained for horse myoglobin are shown in Figure 3.10. In Figure 3.10 it is seen, particularly at the higher pH values, that the contribution from the fluctuating proton dipole component is not sufficient to account for the observed dipole moment values, whereas that due to bound surface charges appears to be significant at all the pH values. The peptide moment component in the core of the protein molecule appears to have very little effect on the overall moment value. Owing to the vectorial nature of the various components, for example, the protein core dipole tends to oppose the surface charge dipole, the total dipole moment, $(\langle\mu^2\rangle + \Delta\mu^2)^{1/2}$, is sometimes less than some of the individual components.

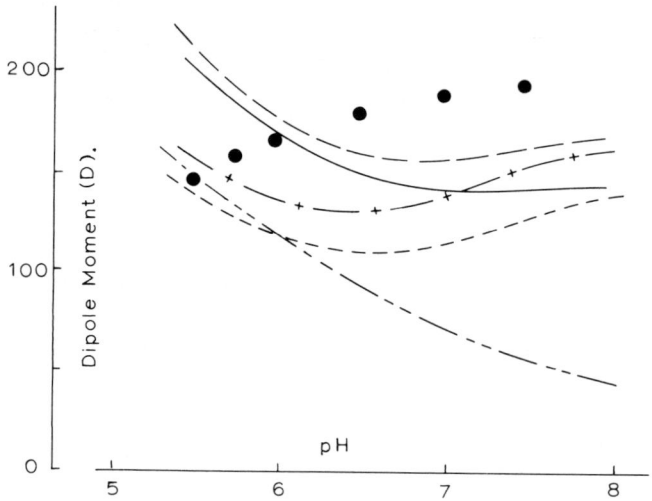

Figure 3.10 The experimental and calculated dipole moment for horse myoglobin as a function of pH.
● experimental values; — + — surface charges only; — — — — fluctuation dipole only; — — — — surface charges + core dipole; — — — — surface charges + fluctuation dipole; ———— vectorial sum of all dipole components. (Reproduced from G. P. South and E. H. Grant, *Proc. Roy. Soc. London A*, **328**, 371 (1972), by permission of the Royal Society)

In comparing the theoretical and practical moment values, it is to be remembered that in deriving the value from the experimentally determined dielectric increment data, several approximations have to be made regarding any dipole—dipole correlation effects and the effective internal permittivity of the protein molecule. Also, regarding the theoretical dipole moment values, it may not be valid to assume that the pK_a's for similar chemical groups are identical, since local environment effects may invalidate such an assumption, especially at pH values away from the isoionic point. Other complications include the possibility that some of the long ionizable side chains may not have the same locations in solution as those suggested by the X-ray data from protein crystals, and also the fact that the bound surface charge field may induce a significant polarization within the core of the molecule. Taking all these factors into account, we can conclude that the results of Figure 3.10 show a most acceptable agreement between experiment and theory, and make an invaluable contribution to our understanding of the dielectric properties of aqueous protein solutions.

HELIX—RANDOM COIL TRANSITION EFFECTS

In Chapter 2, the α-helix configuration for polypeptide chains was described. A feature of the α-helix is that the individual peptide residue moments μ_p are totally additive, giving even a modest chain length a sizeable permanent dipole moment. By contrast the residue moments in a randomly coiled polypeptide chain are not

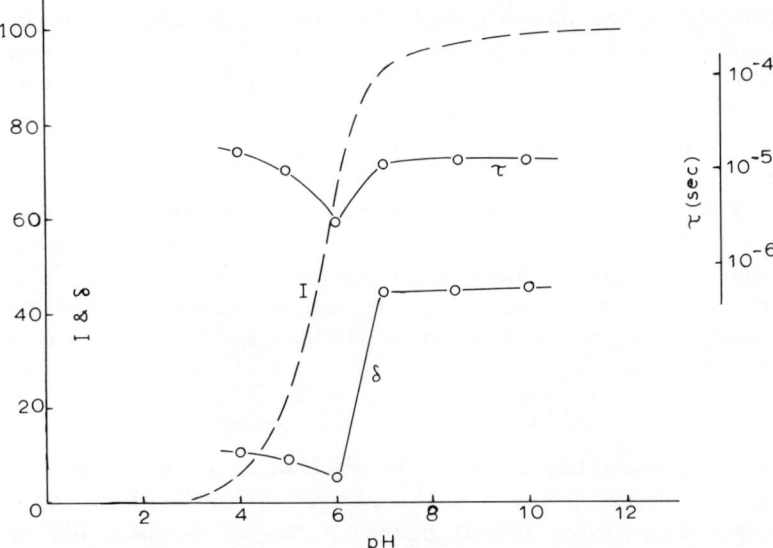

Figure 3.11 The dielectric increment δ and relaxation time τ of poly-L-glutamic acid (PGA) at various pH values in aqueous solution,[45] together with the degree of ionization I in 0.2 M NaCl/dioxane (2:1) solvent[46]. (Reproduced from A. Wada, *J. Chem. Phys.*, **30**, 328 (1959), by permission of the American Institute of Physics)

totally additive, and the resultant dipole moment may even approach a value of zero, if complete vectorial cancellation of the various side-chain and peptide unit moments occurs. We will expect solvated regular α-helix and random coil polypeptide chains to have different dielectric properties.

The most thoroughly investigated polypeptide, regarding studies of the helix to random coil transition, is poly-L-glutamic acid (PGA), and practically every possible physical technique available has been used for such studies.[44] In aqueous solution at pH values below pH 5, PGA assumes a regular α-helix conformation, but as the pH of the solution is increased, a transition to the random coil configuration takes place. PGA contains ionizable carboxyl groups in the side chains, and with increasing pH these side chains will become ionized to $-COO^-$, and it is reasonable to assume that charge repulsion between these ionized groups is sufficient to destabilize the α-helix. It is also possible, that apart from the interpeptide hydrogen bonds, the PGA α-helix is also stabilized through H-bonds between adjacent uncharged carboxyl groups. With increasing pH and ionization of the carboxyl groups, these extra stabilizing bonds would disappear. In aqueous solution, the helix—coil transition in PGA can be detected as a change in the dielectric increment δ and relaxation time τ, as shown by the results of Wada[45] in Figure 3.11. Also included in Figure 3.11 are the results of Doty *et al.*[46] for the degree of ionization of PGA as a function of pH, obtained in a 0.2 M NaCl/dioxane (2:1) solvent. At an ionization level of around 40 per cent, it can be seen that the dielectric properties of PGA are already significantly different from those at the low pH values.

From our previous discussion of the way the peptide residue moments are additive in the helical form, we might have expected the dielectric increment δ to decrease steadily as the transition to the random coil conformation proceeded. In fact δ in Figure 3.11 does decrease in the pH range 4—6, but then increases rapidly to its final value of 45. The rapid increase in δ in the pH range 6—7 is most likely to be associated with the increase in the polarizability of PGA as a result of the ionized carboxyl groups and the surrounding atmosphere of counter-charge ions. The relaxation time τ variation may involve effects associated with combinations of molecular rotations, together with induced polarizations arising from the movement of counter-ions along the lines of Schwarz's ion mobility model,[43] and a chemical relaxation associated with the helix conformational changes. According to Schwarz and Seelig,[47] poly-γ-benzyl-L-glutamate (PBLG) exhibits a chemical relaxation effect, with τ_{ch} given by

$$\tau_{ch} = N^2 x(1-x)/K_h \qquad (3.10)$$

where x is the molar fraction of peptide bonds which are in the helical form, N is the number of peptide bonds in the average polypeptide chain, and K_h is the effective rate coefficient of helix growth, of value of the order 10^{10} s^{-1}. If we assume that at pH 3, 50 per cent of the peptide bonds are involved in helix formation, and that at pH 6 this proportion reduces to 5 per cent, then with N ($\simeq 10^3$) and K_h ($\simeq 10^{10}$ s^{-1}) remaining constant, the variation of τ for PGA in the range pH 4—6, as shown in Figure 3.11, is adequately described by equation (3.10). Whether the rapid increase of τ in the pH range 6—7 results from either an increasing contribution from a rotational relaxation or a surface ionic conductivity effect, is not clear.

The surface ionic conductivity effect appears to be relevant for DNA in aqueous solution. DNA is a double-stranded helical nucleic acid, which we have seen from Figure 2.9 does not possess a permanent dipole moment arising from the peptide units, although a rotational orientation of such a permanent dipole has in the past been used to describe the dielectric dispersion of aqueous solutions of DNA.[48,49] Rather than considering the existence of a permanent dipole moment for DNA, it is more realistic to consider polymers such as DNA and RNA in aqueous solution as cylindrical polyelectrolyte macromolecules, bathed in an electrolyte consisting of counter-ions and co-ions. Under the influence of an applied electric field, these counter-ions will tend to be displaced along the polymer, giving rise to an induced dipole moment. The resulting dielectric dispersion will depend on the frequency dependence of the ionic surface conductivity, and for rod-shaped molecules the corresponding relaxation time is given by[50]

$$\tau = \pi \epsilon L^2 / 2uzq^2$$

where ϵ is the effective permittivity of the surrounding ionic atmosphere of z ions per unit length, u is the counter-ion surface mobility, and L is the molecular length. Such an L^2 dependence for τ has been found for solvated DNA,[51] giving support for the surface ionic conductivity model. Vasilescu[52] has described dielectric measurements investigating the role of the ionic atmosphere that surrounds DNA in electrolyte solutions such as NaCl solution, with particular emphasis being placed on effects associated with conformational changes. The addition of ionic salts into water causes large changes in its physical and dielectric properties, which we will not consider here. Hasted, in his book,[19] has given an extensive account of the dielectric properties of electrolytic solutions.

To summarize, we have seen in this chapter how dielectric measurements on solvated biomolecules can lead to an increased understanding of the ionic, dipolar, and atomic configuration characteristics, of their size, geometry, and conformational changes, and also an insight into the extent of water that is strongly bound to protein molecules when in aqueous solution. Before proceeding with the dielectric properties of living tissue and cells, the next chapters will be concerned with a more detailed account of the hydration process, with the dielectric properties of water bound to biomolecules, and dielectric effects associated with the heterogeneity of hydrated biological systems.

References

1. *CRC Handbook of Chemistry and Physics*, **53**, C741 (1972/73).
2. T. Peters, in *Advances in Clinical Chemistry*. Eds. O. Bodansky and C. P. Stewart, **13**, 37 (1970).
3. R. E. Dickerson, T. Takano, D. Eisenberg, O. B. Kallai, L. Samson, A. Cooper, and E. Margoliash, *J. Biol. Chem.*, **246**, 1511 (1971).
4. W. Kauzmann, *Advan. Protein Chem.*, **14**, 1 (1959).
5. J. Wyman, *J. Am. Chem. Soc.*, **56**, 536 (1934).
6. J. G. Kirkwood, *J. Chem. Phys.*, **2**, 351 (1934).

7. W. J. Shutt, *Trans. Faraday Soc.*, **30**, 893 (1934).
8. W. J. Dunning and W. J. Shutt, *Trans. Faraday Soc.*, **34**, 479 (1938).
9. W. Küster, *Z. Anorg. Chem.*, **13**, 135 (1897).
10. N. Bjerrum, *Z. Physik. Chem.*, **104**, 147 (1923).
11. *John Gamble Kirkwood; Collected Works. Proteins.* Ed. G. Scatchard. Gordon & Breach, New York (1967).
12. J. G. Kirkwood, *Chem. Rev.*, **24**, 233 (1939).
13. J. Wyman and T. L. McMeekin, *J. Am. Chem. Soc.*, **55**, 908 (1933).
14. P. Devoto, *Gazz. Chim. Ital.*, **60**, 520 (1930); **61**, 897 (1931); **64**, 76 (1934).
15. J. G. Kirkwood, *J. Chem. Phys.*, **7**, 911 (1939).
16. H. Greenstein and J. Wyman, *J. Am. Chem. Soc.*, **58**, 463 (1936).
17. A. Cavallaro, *Arch. Sci. Biol.*, **20**, 567 (1934).
18. J. Wyman, *Chem. Rev.*, **19**, 213 (1936).
19. J. B. Hasted, *Aqueous Dielectrics*, p. 193. Chapman & Hall, London (1973).
20. J. L. Oncley in *Proteins, Amino Acids and Peptides*, Chapter 22. Eds. E. J. Cohn and J. T. Edsall. Reinhold, New York (1943).
21. G. Hedestrand, *Z. Physik. Chem.*, **135**, 36 (1928).
22. W. Kuhn, *Z. Physik. Chem.*, **175A**, 1 (1935).
23. H. Eyring, *Phys. Rev.*, **39**, 746 (1932).
24. W. P. Conner, R. P. Clarke, and C. P. Smyth, *J. Am. Chem. Soc.*, **64**, 1379 (1942).
25. A. D. Buckingham, *Aust. J. Chem.*, **6**, 93, 323 (1953).
26. H. Fricke, *J. Phys. Chem.*, **57**, 934 (1924).
27. L. Lewin, *J. Instn Elect. Engrs*, **94**, 65 (1947).
28. J. C. W. Shepherd and E. H. Grant, *Proc. Roy. Soc. (London) A*, **307**, 335, 345 (1968).
29. J. C. W. Shepherd, Ph.D. Thesis, University of London (1967).
30. G. P. South and E. H. Grant, *Proc. Roy. Soc. (London) A*, **328**, 371 (1972).
31. M. F. Perutz, N. G. Rossman, A. F. Cullis, H. Muirhead, and G. Will, *Nature*, **185**, 416 (1960).
32. E. H. Grant, G. P. South, S. Takashima, and H. Ichimura, *Biochem. J.*, **122**, 691 (1971).
33. F. Perrin, *J. Phys. Radium*, **5**, 497 (1934).
34. W. P. Conner and C. P. Smyth, *J. Am. Chem. Soc.*, **64**, 1870 (1942).
35. J. C. Kendrew, H. C. Watson, B. E. Standberg, R. E. Dickerson, D. C. Phillips, and V. C. Shore, *Nature*, **190**, 666 (1961).
36. H. Hendrickx, R. Verbruggen, M. Y. Rosseneu-Motreff, V. Blaton, and H. Peeters, *Biochem. J.*, **110**, 419 (1968).
37. P. Debye and H. Falkenhagen, *Phys. Z.*, **29**, 121, 401 (1928).
38. J. G. Kirkwood and J. B. Shumaker, *Proc. Nat. Acad. Sci. USA*, **38**, 855 (1952).
39. W. Scheider, *Biophys. J.*, **5**, 617 (1965).
40. C. T. O'Konski, *J. Phys. Chem.*, **64**, 605 (1960).
41. H. P. Schwan, *Advan. Biol. Med. Phys.*, **5**, 147 (1957).
42. B. Jacobson, *J. Am. Chem. Soc.*, **77**, 2919 (1955).
43. G. Schwarz, *J. Phys. Chem.*, **66**, 2636 (1962).
44. G. D. Fasman, in *Poly-α-Amino Acids*, pp. 499–604. Ed. G. D. Fasman. Marcel Dekker, New York (1967).
45. A. Wada, *J. Chem. Phys.*, **30**, 328 (1959).
46. P. Doty, A. Wada, J. T. Yang, and E. R. Blout, *J. Polym. Sci.*, **23**, 851 (1957).
47. G. Schwarz and J. Seelig, *Biopolymers*, **6**, 1263 (1968).
48. L. G. Allgen, *Acta Physiol. Scand.*, **22**, 76 (1950).
49. S. Takashima, *J. Molec. Biol.*, **7**, 455 (1963).

50. G. Schwarz, *Z. Phys. Chem. Frankfurt*, **19**, 286 (1959).
51. N. Ise, M. Eigen, and G. Schwarz, *Biopolymers*, **1**, 343 (1963).
52. D. Vasilescu, in *Physico-Chemical Properties of Nucleic Acids* Vol. 1, pp. 31–66. Ed. J. Duchesne. Academic Press, New York (1973).

Note added in proof:

Two publications of significant interest have recently appeared. The first is the proceedings of a conference which covered the theoretical and experimental aspects of biomolecular dielectric relaxations, and also electrical and transport phenomena in membranes.

Electrical Properties of Biological Polymers, Water, and Membranes (S. Takashima and H. M. Fishman, Eds.) *Annals New York Acad. Sci.*, 303 (1977).

The second is a book which can be considered as complimentary to the subject content of this chapter since it contains valuable details of experimental procedures not included here.

E. H. Grant, R. J. Sheppard, and G. P. South, *Dielectric Behaviour of Biological Molecules in Solution*, Oxford University Press (1978).

Chapter 4

Water in Biological Systems

Die Welt rundet sich im Tautropfen

J. W. Goethe

Life as we know it could neither have been created, nor continue to exist now, without water. Water is its *mater* and *matrix*.[1] The removal of water from all living systems results in death, or at best dormancy, a fact that led the earliest of biologists to recognize the importance of water in biological processes. The great Victorian biologist, T. H. Huxley, went as far as to write, 'so we may as well make a beginning of science by studying water'.[2] We know in considerable detail the distribution and rate of consumption of water in biological systems.[3,4] For example, some marine invertebrates are composed of 97 per cent water, whilst at the other extreme some bacterial spores contain 'only' around 50 per cent water. An adult human has a water content of from 65 to 70 per cent, it being most concentrated in the nervous tissue (84 per cent H_2O) and least concentrated in the fatty adipose tissue (30 per cent H_2O). On average, adults consume more than 900 litres of water a year in the form of drink and that held in solid food, and the annual turn-over of water in the photosynthetic green plants and marine organisms is 6.5×10^{11} tons. But despite the extent and detail of such data, our knowledge of the contribution water actually makes to the life processes at the molecular level is practically zero.

The fact is that until comparatively recently very few biologists thought of water as anything other than a mere space filler, or suspending medium for the active molecules, in living organisms. Because it is so familiar and ubiquitous, an apparently inert even dull liquid, it is not really surprising that water should become so taken for granted and its fundamental importance overlooked. As Sir Oliver Lodge is attributed[1] to have remarked, 'The last thing a deep sea fish could discover is water'!

Water has largely been ignored by the other scientific disciplines too, with active interest on a significant scale only beginning in the 1960s. With this recent work, much of which is detailed in a useful series of volumes edited by Franks,[5] it is clear that the water molecule, which after all is composed of only three atoms, provides some very severe problems. We shall see that as yet we do not even have a clear idea of the structure of liquid water. Our ignorance of the basic molecular properties of this familiar liquid is well demonstrated by the response that followed the observations by Derjaguin[6] of the condensation of a liquid from an atmosphere of unsaturated water vapour in narrow glass capillaries. This liquid was found to have a density of 1.3 g/cm^3, a high viscosity, and a freezing point below 0 °C, and soon came to be described by some as 'polywater'. Many laboratories began to study this exciting new form of water, much to the alarm of others[7] who regarded it as 'the most dangerous material on earth' in that if allowed to trickle into the oceans it could turn them solid, 'converting the earth into a reasonable facsimile of Venus'! We now know that impurities dissolved by the water in passing through the narrow capillaries caused all this excitement,[8] although this should not detract from the fact that such studies by Derjaguin of the physical properties of water constrained within macromolecular sized channels are of great fundamental importance, especially for biological studies.

From what we do know, water is anything but dull or inert. The fact that all living processes, involving as they do organic macromolecules, should be so dependent on the only naturally occurring inorganic liquid, is by itself remarkable. There are very few chemicals that do not dissolve to some extent in water, and it is also one of the most reactive and corrosive of known chemicals, readily reacting with ions and molecules. But despite this, and unlike other hydrides, it is not physiologically harmful! The physical properties of water are also extraordinary. According to its molecular size, the melting and boiling points should be about 100 K lower than they in fact are, and its heat of vaporization, heat of fusion, and surface tension is higher than that of the comparable hydrides H_2S or NH_3, or even than that of most other common liquids. These physical properties can only arise from there being strong forces of attraction between the molecules in liquid water.

These strong intermolecular forces arise from the specific distribution of electrons in the water molecule. Each of the two hydrogen atoms share a pair of electrons with the oxygen atom through overlap of the 1s electron orbitals of the hydrogen atom and sp hybrid orbitals of oxygen. The conclusion from quantum mechanical calculations is that the orbitals of the oxygen atom overlapping the hydrogens are almost of pure 2p character, and that the lone-pair oxygen electrons approximate to sp hybrids.[9] The lone-pair electron orbital lobes project above and below the atomic plane of the water molecule, and their exact contribution to the molecule's dipole moment of 1.84 debye is not known. From spectroscopic and X-ray analyses the HOH bond angle is 104.5° (see Figure 1.9(a)) and the average hydrogen–oxygen interatomic distance is 0.0957 nm. The highly electronegative oxygen atom tends to withdraw the single electrons from the hydrogen atoms, leading to its electrical asymmetry (dipole moment) and the ability of the molecule to act as proton donors to form two hydrogen bonds with other water molecules. The lone-pair electron orbitals are able to act as proton acceptors to form hydrogen bonds with two more water molecules. Apart from forming hydrogen bonds with other water molecules, water can also hydrogen-bond with other proton donor or acceptor chemical groups, such as the amino and carboxyl groups. In Chapter 2 we

102

saw the importance of intramolecular hydrogen bonds in stabilizing the structure of proteins. The hydrogen bonds in liquid water have a bond energy of about 4.5 kcal/mole each, and are made and broken very rapidly, with the half-life of each hydrogen bond being about 10^{-11} seconds.

The formation of hydrogen bonds in water is accompanied by a redistribution of electronic charges, with the hydrogen atom in the bond losing electronic charge. The oxygen atoms gain charge, with the oxygen on the proton-donating molecule gaining the most. The greatest electron loss occurs for the hydrogens immediately attached to the proton-acceptor molecule, whilst the other hydrogen attached to the proton-donor molecule gains electronic charge upon the formation of the hydrogen bond. The charge redistributions are shown in Figure 4.1(a). As a result

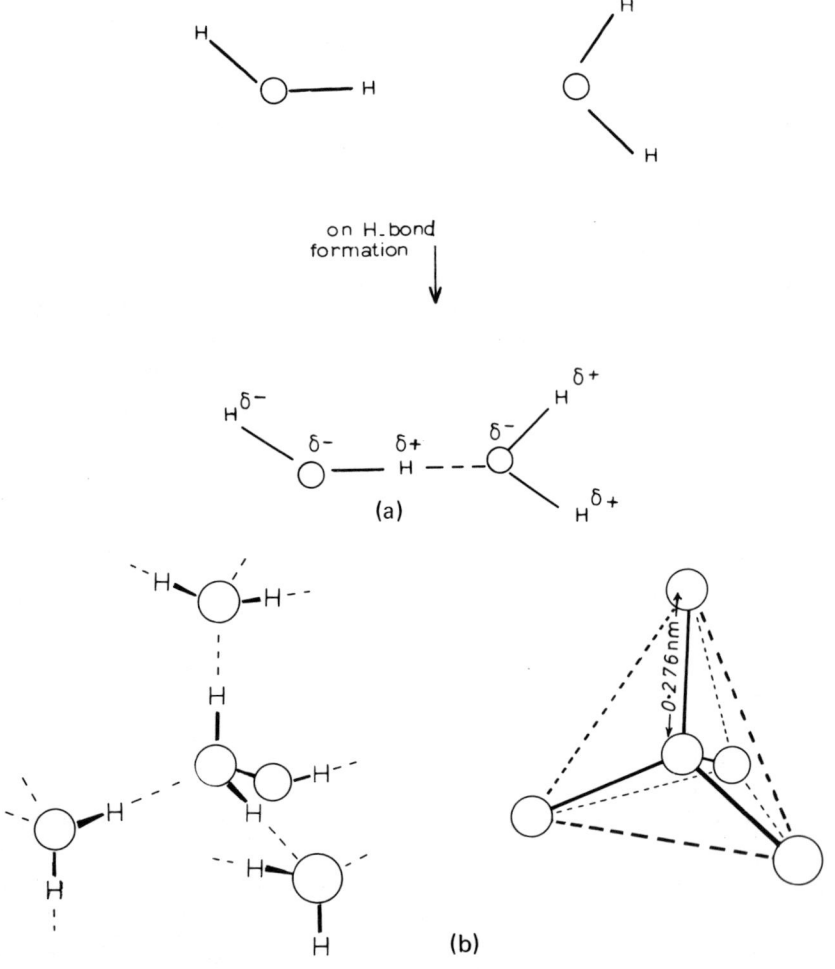

Figure 4.1 (a) Electronic charge distribution involved in the hydrogen bonding of two water molecules. (b) The tetrahedral hydrogen-bonding structure of ice I

of such charge redistribution, the OH···OHOH hydrogen-bonding structure is
H   H   H
more stable than the HO···HOH···OH trimeric structure. In this latter structure,
H   H
the third water molecule is having to donate charge to a hydrogen atom which
has already gained negative charge through formation of the first hydrogen
bond. An extensive account of the theory of hydrogen bonding in water has been
given by Rao.[10] Dipolar—dipolar interactions between water molecules lead to the
formation of an extensive hydrogen-bond network. In the most common crystalline
form of ice, namely the ice I structure, each water molecule is tetrahedrally
hydrogen-bonded to its four nearest neighbours, forming a regular lattice having an
average oxygen interatomic distance of 0.276 nm, as shown in Figure 4.1(b). In
liquid water at 0 °C, each water molecule is hydrogen-bonded on average with 3.6
other water molecules, and the inter-oxygen distance is increased slightly. From the
value of the heat of fusion of ice at 1.43 kcal/mole then it would appear that only a
relatively small fraction of the total hydrogen-bond network in ice is broken. Even
so this is sufficient for the ice structure to lose its rigidity and become liquid. The
hydrogen bonds (being made and broken about 10^{11} times a second) in liquid
water provide the strong intermolecular attractive forces that give rise to the
extraordinary physical properties described earlier.

With its small mass and the fact that its single electron is tightly held by the
oxygen atom, there is a tendency for a hydrogen ion (proton) to dissociate from
the oxygen atom to which it is covalently bound. Provided the internal energy of
each molecule is favourable, such a proton can jump to the oxygen atom of the
adjacent water molecule to which it is hydrogen-bonded, as shown below:

In this dissociation reaction two ions are produced, the hydronium ion H_3O^+ and
the hydroxide ion OH^-, and this dissociation is an equilibrium process described by

$$2H_2O \rightleftharpoons H_3O^+ + OH^-$$

for which we can write the equilibrium constant

$$K_a = \frac{[H_3O^+][OH^-]}{[H_2O]^2}$$

where the brackets indicate concentration in moles per litre. In a litre of pure water
at 24 °C at any given time, there is only 10^{-7} mol of H_3O^+ ions and OH^- ions.
From Chapter 3, we see that this leads us to a value of 7.0 for the pH of pure water.
We should also note, that in Chapter 3 the usual convention of using the symbol
H^+ to designate the hydronium ion was used. In fact H^+ ions do not exist as such
to any significant extent, they rapidly become hydrated to the H_3O^+ ion, which in
turn is further hydrated through additional hydrogen bonding to the $H_9O_4^+$ ion, as
shown in Figure 4.2(a). The hydroxide ion OH^- is also hydrated in liquid water.

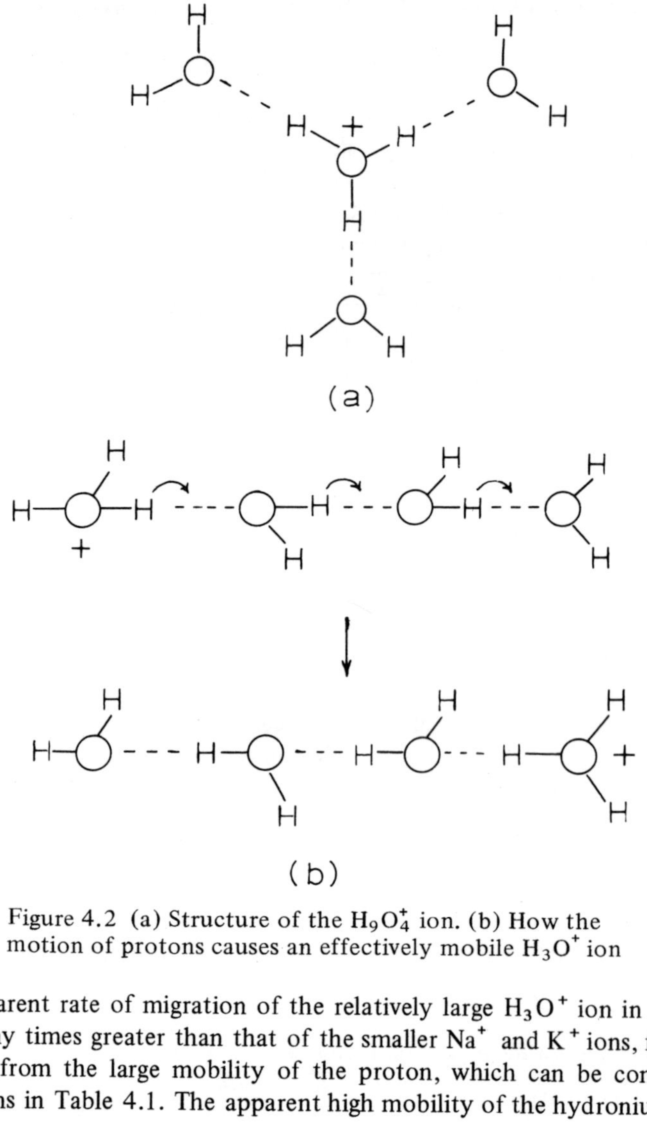

Figure 4.2 (a) Structure of the $H_9O_4^+$ ion. (b) How the
motion of protons causes an effectively mobile H_3O^+ ion

The apparent rate of migration of the relatively large H_3O^+ ion in an electrical
field is many times greater than that of the smaller Na^+ and K^+ ions, for example.
This arises from the large mobility of the proton, which can be compared with
other cations in Table 4.1. The apparent high mobility of the hydronium ion arises

Table 4.1   The electrical mobility of some ions at 25 °C in
dilute aqueous solution

Cation	Mobility $(m^2/V\ s)$	Anion	Mobility $(m^2/V\ s)$
H^+, H_3O^+	36.3×10^{-8}	OH^-	20×10^{-8}
Rb^+, Cs^+	7.7×10^{-8}	Br^-	7.8×10^{-8}
K^+, NH_4^+	7.6×10^{-8}	Cl^-, I^-	7.7×10^{-8}
Na^+	5.0×10^{-8}	F^-	5.4×10^{-8}

from the rapid transference of protons from hydronium ions to neighbouring hydrogen-bonded water molecules, as shown in Figure 4.2(b). Effectively, a positive charge can move a considerable distance with little movement of the water molecules themselves, at a rate much faster than that of the H_3O^+ ion itself. It is probably with such a mechanism in mind that led Szent-Györgyi to remark once 'that water was the only molecule he knew that could turn around without turning around'. Needless to say, the use of the pH scale in designating the concentration of H_3O^+ and OH^- ions in fluids is of great importance in biological studies, and Table 4.2 lists the pH of some biological fluids.

Table 4.2   The pH of some biological fluids

Fluid	pH
Pancreatic juice	7.8—8.0
Blood plasma	7.4
Interstitial fluid	7.4
Intracellular fluids	
liver	6.9
muscle	6.1
Saliva	6.35—6.85
Cow's milk	6.6
Gastric juice	1.2—3.0

Concerning the structure of liquid water, the early X-ray diffraction measurements in the 1930s[11,12] were interpreted to be characteristic of tetrahedral water coordination, strongly suggestive of the structure of liquid water being similar to that of ice. Starting from these observations, a number of theories of water structure have been developed, and comprehensive accounts of these theories have been given by Eisenberg and Kauzmann,[13] Kavanau,[14] Nemethy and Scherage,[15] and Ben-Naim.[16] Basically, the fundamental proposals for the models of water structure may be classified into two categories: (1) continuum models in which liquid water is treated as an uninterrupted three-dimensional lattice of tetrahedrally coordinated hydrogen-bonded molecules, where thermal, electrostatic, and steric perturbations are thought to produce hydrogen-bond stretching and bending deformations rather than bond breakage; and (2) mixture of 'flickering cluster' models in which water is viewed as a collection of differently hydrogen-bonded species in which each water molecule can fluctuate through states where neither, one, or both of its hydrogens are engaged in hydrogen bonding. Even after the application of an extensive range of physicochemical techniques to the problem of liquid water structure, no clear-cut discrimination between these two basic models, let alone between the numerous hybrids and variants of each, has been forthcoming. As a typical example of the confusion confronting us, we find that estimates of the percentage of broken hydrogen bonds in water at 0 °C range from 2.5 per cent[17] to 71.5 per cent.[18]

One hopeful prospect for future theoretical work will possibly lie in the development of computer procedures to simulate the dynamic behaviour of water molecules. Rahman and Stillinger[19,20] have successfully applied such a technique to simulate the time-dependent behaviour of a large number of water molecules in the liquid state. Further developments could well lead to a closer understanding of the relationships between the structure and the dynamic and electronic properties of water associated with macromolecular systems. Molecular orbital calculations[21,22] on aggregates of some water molecules have shown that at certain intermolecular geometries the strength of the hydrogen bonds increases as the number of associated molecules increases. This shows that the formation of clusters of water molecules can be regarded as a cooperative process. More recent molecular orbital calculations[23] for one- and two-dimensional $(HF)_n$ and $(H_2O)_n$ chains confirm this, where in the one-dimensional $(H_2O)_n$ chains for example, the average hydrogen-bond energy increases from a value of 8.7 kcal/mole for $n = 2$, to 10.9 kcal/mole for $n = 8$. This increase in hydrogen-bond energy is considered to be brought about by charge transfer along the chain, which leads to an increase of both the basicity of the electron lone pair and the acidity of the proton involved in the hydrogen bond. Energy band structures were also calculated[23] and for an infinite one-dimensional H_2O chain the width of the forbidden band between the valence and conduction bands was found to be of the order of 20 eV ($\sim$460 kcal/mole). More recently Bell and Salt[24] using a statistical treatment have derived Helmholtz free energy values and phase diagrams for a three-dimensional lattice model of the water/ice system. The water molecule was regarded as having four 'bonding arms', two of positive and two of negative polarity. A bond is formed between two nearest neighbour molecules when a positive arm links with a negative arm. The lattice considered was of the body-centred-cubic type, where the lattice sites are either vacant or occupied by molecules with four hydrogen-bonded 'arms', two positive and two negative, directed in the tetrahedral structure. Interaction energies also occur for unbonded nearest neighbours and second-nearest neighbours. As well as disordered states, two long-range ordered structures were found to be possible; one an open bonded network, like ice I(c), in which half the b.c.c. lattice sites are vacant, and the other a close-packed structure with intertwined bonded networks, like ice VII. It was found for suitable energy parameters, that four phases exist; the vapour, liquid, open ice, and close-packed ice phases. Two critical points were found, an open ice/liquid/vapour triple point and at a low temperature a close-packed ice/open ice/vapour triple point. Between the triple points an increase of pressure causes the open ice to transform into a liquid of higher density than the ice, while below the lower critical point it causes an open ice to close-packed ice transformation. This model is able to represent some of the most important characteristic phenomena displayed by water, including that of 'ice floating on water'.

BIOLOGICAL WATER

If there are uncertainties regarding the molecular characteristics of bulk liquid water, then the situation for the aqueous state in biological systems poses even greater problems. The ways in which the water in living organisms can differ from

Table 4.3   Average concentration of ions in body fluids $(mM/litre)$[25]

	Na^+	K^+	Ca^+	Mg^+	Cl^-
Marine invertebrates	370−550	7−24	8−21	6−58	430−590
Terrestrial invertebrates	3−262	1−46	2−47	6−188	15−270
Reptiles and birds (plasma)	130−180	3−6	2−6	1−2	103−148
Mammals (plasma)	145−166	3−6	2−10	1−2	100−116

normal water include effects associated with dissolved ionic salts and other chemicals, and the influence of macromolecular surfaces.

The concentration of ions in biological fluids can be quite high, as shown in Table 4.3, where it can be seen that Cl^- ions can reach concentrations of 0.59 M in marine invertebrates, whilst in mammalian plasma the concentration of Na^+ ions is of the order 0.15 M.[25] In the cytoplasm of the algae *Nitella translucens* the three major ions are K^+ (0.119 M), Cl^- (0.065 M) and Na^+ (0.014 M),[26] which can be taken to be typical of the ionic concentrations for the higher plant cells in general. Robinson and Stokes[27] have formulated an equation for estimating the average separation of ions in solution as a function of solute concentration. Based on a geometric model consisting of a time-average cubic lattice of ions, the average interionic distance r may be calculated from the equation

$$r = 0.95 \, c^{-1/3} \text{ nm} \tag{4.1}$$

where c is the concentration of salt in moles per litre. Figure 4.3 shows the variation of the average interionic distance in aqueous NaCl solutions as a function of the concentration of NaCl. We see that at 0.5 M NaCl, the approximate concentration of NaCl in marine invertebrate body fluids, the Na^+ and Cl^- ions are separated by just 1.20 nm, which taking into account the diameter of these ions leaves sufficient space for only three water molecules between ions. Estimates for the average size of H_2O clusters in pure liquid water range from about 6 to 10^5 molecules[15,28,29] so we see that the molecular structure of water must be significantly altered when salts are dissolved in it in concentrations of the order of that occurring in biological fluids. Edzes and Berendsen[30] have given a recent review of the localization, mobility, and thermodynamics of ions in biological cells, and useful texts on the physicochemical properties of aqueous electrolyte solutions have been given by Harned and Owen[31] and Samoilov.[32]

There is little doubt that ions in aqueous solution modify the liquid water structure. X-ray scattering data[33] indicate that in concentrated potassium chloride solutions each ion is surrounded by approximately seven water molecules and hence must disrupt the normal tetrahedral water structure. Brady[34] has demonstrated that the effect of LiCl in solution appears to result in the complete breakdown of water structure, since the radial distribution curve indicated the absence of the first water−water neighbours usually found around 0.29 nm, and there was nearly complete absence of the second water−water neighbours at 0.475 nm. The most obvious ways ions can influence the orientation of water molecules are those associated with just the physical disruption involved in substituting a foreign

108

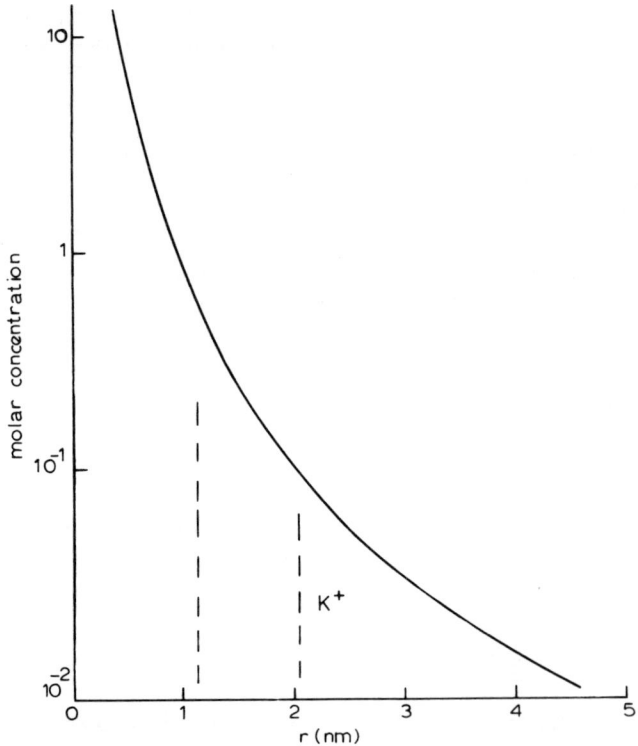

Figure 4.3 Variation of the average interionic distance
r in electrolytes as a function of the concentration of
the solute ion. The dashed lines indicate the average
concentration of Na^+ and Cl^- ions in animal body fluids
and K^+ in plant cytoplasm

chemical structure into a pure water lattice, and electrostatic interactions between
the point electric charge (monopole) of the ion and the electric dipole moments of
the surrounding water molecules. Gurney[35] has calculated the mutual electrostatic
potential energy between a potassium cation and a water dipole in a vacuum
separated by the sum of their van der Waals' radii, to be −16.1 kcal/mole. The
magnitude of this interaction alone leads us to expect on purely electrostatic
reasonings, that a drastic reorientation of water structure should be observed in the
immediate vicinity of an ion. Consequently, there must exist a structural mismatch
region of water molecules between such an inner hydration sphere and the distant
unperturbed water. This concept has led to a postulated model[36,37] for aqueous
solutions in which there are envisaged three concentric regions about an ion: (1) an
innermost region of polarized, relatively immobilized, and electrorestricted water
molecules; (2) an outer region containing water having the normal liquid structure;
and (3) an intermediate structure-broken region in which the normal tetrahedral
structure-orientating influence of the bulk water is in competition with the radially
orienting electric field of the ion. This concept leads to a classification of ions based

on their capabilities of either promoting structure by reinforcing either the inner or outer regions, or breaking structure by making the intermediate region about the ion most important. For low charge density ions such as K^+ and Cl^-, there are about 12 water molecules in the inner cosphere and about 37 water molecules in the cosphere representing the intermediate, structure-broken region. According to the model of Frank and Evans,[36] the total entropy change for the water in a mole of the inner cosphere is about -12 cals/deg[38] around K^+ and Cl^- ions, whilst in the disrupted cosphere the entropy changes are $+12$ and $+10$ cals/deg for the K^+ and Cl^- ion, respectively. From consideration of various ionic viscosity coefficients, von Hippel and Schleich[39] conclude that ions of high charge density increase the order of the water structure, presumably by electrostatic reorientation and immobilization of neighbouring water molecules. Ions in the high charge density category include Ba^{2+}, Ca^{2+}, Li^+, SO_4^{2-}, and Sr^{2+}. On the other hand monovalent ions with large radii, and hence low charge density, seem to function as structure breakers primarily because of the steric stresses involved in accommodating these ions in the water lattice. Such structure breakers include Br^-, Cl^-, I^-, K^+, NH_4^+, NO_3^-, and SCN^-. However, as pointed out by von Hippel and Schleich, such an oversimplification can be misleading. For example, the F^- ion, which has a similar ionic radius to the K^+ ion, and is therefore of fairly low charge density, is in fact an effective water structure maker, unlike the K^+ ion.

Even ignoring the influence of dissolved ions and other chemicals, we can expect the structure of pure water to be influenced by interfaces. For example, in a compact spherical volume of water containing 10^3 water molecules at least half of them will be at the surface, and it is very unlikely that the resulting molecular configurations will resemble free bulk water. We can imagine that it will require perhaps a minimum of 5×10^5 water molecules, occupying a volume of about 10^4 nm^3 (effective diameter 16.8 nm) before any significant amount of the molecules within this volume will behave as normal bulk water. It is now generally accepted that the physicochemical properties of pure bulk liquid water do not exhibit thermal anomalies, in that the property being studied varies smoothly with temperature.[40-43] It is therefore of interest to note that there is considerable evidence to indicate that water associated with interfaces does exhibit temperature anomalies. For example, the surface tension of capillary water has been observed to exhibit an inflection in its temperature variation at around 15 °C,[44] an effect also observed much later by Pethica et al.[45] The disjoining pressure and viscosity of a 10 nm layer of water between quartz plates has been shown to exhibit abrupt changes around 15, 31 and 47 °C.[46] (The human body temperature fits neatly within the range 31–47 °C!) . An anomaly near 31 °C has also been observed in the infrared spectrum of water in a 12 μm calcium fluoride cell,[47] and in nuclear magnetic resonance (n.m.r.) studies of relaxation times of water in colloidal suspensions.[48] It is reasonable to assume that these thermal anomalies arise as a result of phase changes in the interface-influenced structures of the water, and that these molecular structures do not resemble those of normal bulk water. Thermal anomalies have also been observed in biological materials, many examples of which have been cited by Drost-Hansen.[49] Such temperature effects, observed for example in studies of enzymatic reactions,[50] ionic conduction in membranes,[51] haemolysis of erythrocytes,[52] permeability of liposomes,[53] and transmission along nerve cells,[54] can be taken as reasonable evidence for the existence of structured water in biological systems, whose structural phases vary, sometimes abruptly, with temperature. The conclusion by Clifford *et al.*,[55] that thermal anomalies in colloidal suspensions are associated with the existence of pores and extending long-chain chemical groups on the particle surfaces, is obviously of pertinence to biological systems. Estimates of the content of 'structured' water in biological systems seem to depend on the physical technique employed. The results of infrared studies of water in membranes and other systems[4] suggest that most of the water in biological structures is indistinguishable from bulk water. N.m.r. measurements can be divided into those, such as by Cope,[56] that indicate the existence of structured water to a considerable extent, and the many more (see, for example, references 57–60) that give the upper limit of structured water in biological systems as no more than about 20 per cent of the total water content. Regarding Cope's measurements,[56] an alternative explanation has been presented[61] in terms of quadrupole effects, which if correct, would significantly reduce the estimate of the content of immobilized water. Of the two techniques mentioned, the n.m.r. results can possibly be taken to be more indicative of the state of water content, if only because the infrared measurements may be insensitive to structural changes as a result of the presence of broad, overlapping, absorption bands that occur in the spectra of aqueous systems.

We have considered the effects that ionic salts and macromolecular surfaces have on the molecular properties of water as separate factors, but in fact they are not independent effects. We can imagine that, in general, water located at a protein molecule surface in a typical biological fluid will also be within the sphere of influence of at least one ion as well. The interdependence of the thermodynamic potentials of the components of such a system can be represented in terms of the Gibbs–Duhem equation.[62] For the protein–water–salt system this equation may be written in the form

$$N_p \, d\mu_p + N_\omega \, d\mu_\omega + N_s \, d\mu_s = 0$$

or

$$d\mu_\omega = -\left(\frac{N_p}{N_\omega} \, d\mu_p + \frac{N_s}{N_\omega} \, d\mu_s \right)$$

where N represents the number of moles of each component, μ the thermodynamic chemical potential, and the subscripts p, ω, and s refer to protein, water, and salt respectively. Thus, for example, the change of a salt component in an aqueous protein solution can change μ_ω through either a direct interaction between the salt and the water, through an indirect mechanism involving the protein and the salt, or as a result of a combination of both processes.

The most commonly known and practically useful interactions between protein solutions and salts are those termed the salting-in and the salting-out effects. If a neutral salt is progressively added to an aqueous protein solution, two macroscopic effects are observed. First the protein solubility increases (salting-in) and then, after passing through a maximum, it starts to decrease again (salting-out). The salting-in process can be reasonably well understood in terms of non-specific electrostatic interactions between the electrically charged protein molecules and the ionic environment.[63] These interactions bring about a net decrease in the activity coefficient of the protein, which is reflected as an increase in the net solubility. To a first approximation the effectiveness of various electrolytes in salting-in is independent of the ion type and depends only on ionic strength. By contrast, the effectiveness in inducing the salting-out effect depends very strongly on salt type. Thus sulphates, phosphates, and citrates are characterized by a high molar effectiveness in salting-out, whereas chlorides and acetates are quite ineffective. Beyond the initial salting-in region, the solubility S of the protein (g per litre) in a given salt at concentration c (moles per litre) is given by

$$\text{Log } S = \beta - Kc$$

where K is the so-called salting-out coefficient and β is effectively the hypothetical solubility of the protein in pure water in the absence of a salting-in effect. Whereas the protein solubility can vary widely as a function of pH, temperature and protein type, for example, K is quite independent of all these parameters and depends primarily on salt type.[64] The relative effectiveness of ions in salting-out proteins can be ranked as first described by Hofmeister in 1888 from his studies of the salting-out of euglobulins from aqueous solution.[65] This characteristic ranking of ionic effectiveness in processes such as salting-out is called the Hofmeister or lyotropic series, and is discussed in detail by von Hippel and Schleich.[39] These authors have made an interesting suggestion regarding the interactions between ions, water, and protein molecules. Basically their idea is that the extent to which local water is perturbed by ions will determine the extent to which it can be organized by the exposed non-polar groups of the protein molecule into a cage-like or clathrate-type molecular structure. This then will determine the free energy of transfer of these groups from the hydrophobic region inside the protein molecule into the aqueous environment, and so influence the protein structure in terms of the 'oil-drop model' as described in Chapter 3. Essentially there will be competition between three organizing forces all of which are attempting to impose a particular, and different, type of order on the local water; namely the organizing forces of the non-polar groups, the ions, and the unperturbed water itself. The water structure destabilizing ions can be viewed to make less water 'available' to form cage-like structures around the exposed non-polar groups than is available in the unperturbed water lattice, while structure stabilizers must have the property of somehow loosening the unperturbed water lattice and making more water available for organization about the non-polar groups. A prediction of this model would be that the type of water structure reorganization induced by the ions is immaterial regarding the structural stability of the protein molecule, only the *extent* of reorganization would be important. This hypothesis appears to have some foundation based on the observed effects of ionic water structure makers and breakers on the stability of ribonuclease.[39]

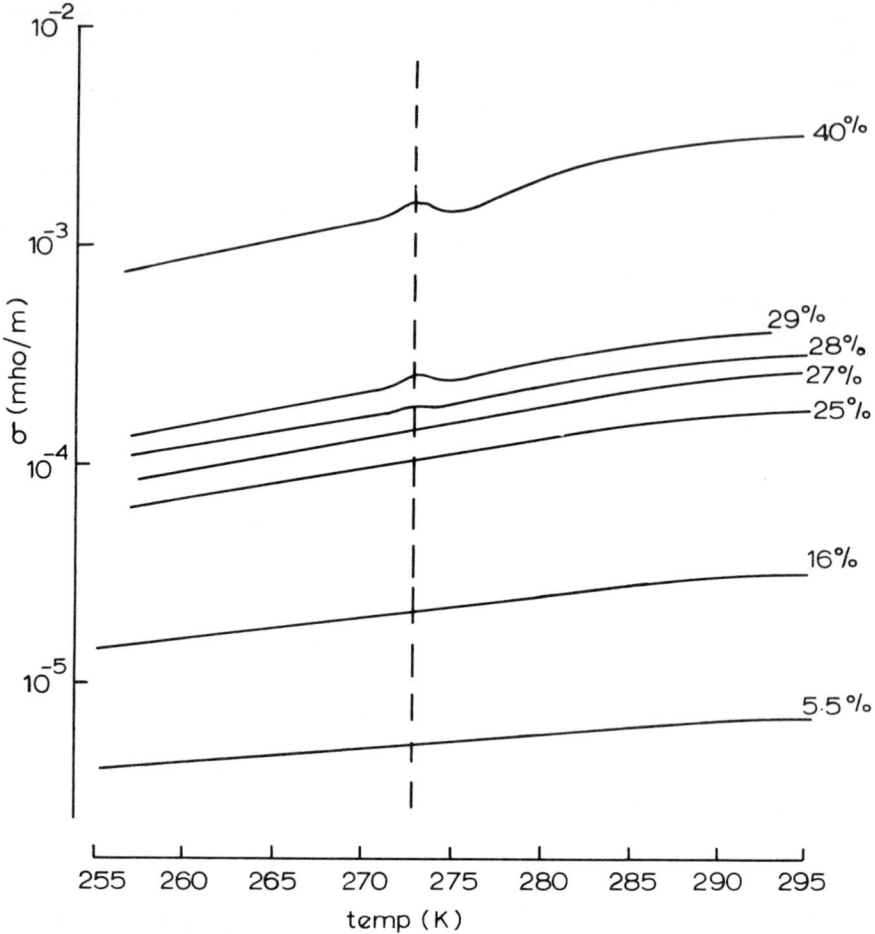

Figure 4.4 The conductivity at 1 MHz of a compressed powder sample of bovine serum albumin as a function of temperature and percentage weight content of water. (Reproduced from P. Batho and R. Pethig, unpublished work, cited in *Dielectric and Related Molecular Processes*, Vol. 3, The Chemical Society, London (1977), by permission of The Chemical Society)

Finally, we may note that the existence and extent of an ordered form of water differing from that of normal bulk water can be inferred from freezing experiments aimed at detecting the presence, or otherwise, of phase transitions around 0 °C. The first application of this idea for dielectric studies appears to have been that of Lovell *et al.*[66] who made measurements of the temperature dependence of the 3.33 cm wavelength ($\sim 10^{10}$ Hz) permittivity of water-loaded brick and hardened cement paste. A discontinuity around 0 °C was found only for the brick, indicating that the water absorbed in the cement paste was so 'bound', or chemically adsorbed, as to be unable to undergo a phase change at the freezing temperature of normal bulk water. A.c. conductivity measurements[67] at 1 MHz for compressed powder samples of bovine serum albumin (BSA) as a function of temperature and water content are shown in Figure 4.4. For water contents less than about 28 per cent by weight, the a.c. conductivity of the BSA samples exhibited no discontinuity around 0 °C. This supports the viewpoint that a considerable proportion of the water content is bound to the protein in such a way as to inhibit its 'crystallization' into ice, as would occur for normal bulk liquid water at 0 °C. From differential scanning calorimetry measurements of the heat of fusion of solvent water for BSA, casein, collagen, and β-lactoglobulin, Berlin et al.[68] found that about 50 per cent by weight of the water content does not participate in any phase transitions down to −70 °C. These workers considered that this came as a result of the water already existing in a frozen ice-like structure at room temperature. Whereas from n.m.r. measurements[69,70] for partially hydrated collagen, the rotational mobility of the water molecules was observed simply to slow down with decreasing temperature, so that even at −50 °C the water molecules retained a considerable degree of rotational freedom. Also, from infrared absorption measurements[71] of water in DNA samples, using a difference technique which clearly distinguishes ice from liquid water, it is found that an inner layer of about ten water molecules per nucleotide is incapable of crystallization even when surrounded by frozen water. This inner hydration sheath exhibited stretching mode spectra identical to that for normal liquid water. We can conclude, contrary to Berlin et al., that the strongly bound water in biomacromolecular systems is not of an ice-like structure, but should be more correctly viewed in terms of it being bound in such a manner as to be unable to assume the ice structure without considerable, thermodynamically unfavourable, molecular rearrangements.

X-RAY STUDIES OF WATER IN PROTEIN CRYSTALS

In the electron density maps calculated from X-ray diffraction data for protein crystals, regions of very low density can be observed which correspond with the space occupied by water of crystallization. The boundary between these regions and the protein molecules themselves is not sharp, as can be seen from Figure 4.5, which shows the electron density map for carboxypeptidase A.[72] It is often found that amino-acid side chains, usually hydrophylic ones such as tyrosine or lysine, projecting into the surrounding solvent are less well defined than those in the body of the protein molecule as if they were waving about in the water. Lysine side chains are sometimes shown by a low forked distribution corresponding to two alternative chain positions.[73] Electron density peaks representing water molecules, which may be as well defined in position as the atoms of the protein molecule, can be seen attached to the polar groups, often linked by hydrogen bonds to peptide oxygen and nitrogen atoms. Other water molecules can be traced which are bonded to only one polar group and which are weaker in definition, and beyond these the water pattern becomes progressively less distinct. An example of such a water molecule electron density peak is given in Figure 4.6 and is based on the 0.28 nm resolution electron density maps of carboxypeptidase A in the region of the Zn atom as given by Lipscomb et al.[72] and by Dickerson and Geis.[74] In fact, it has

114

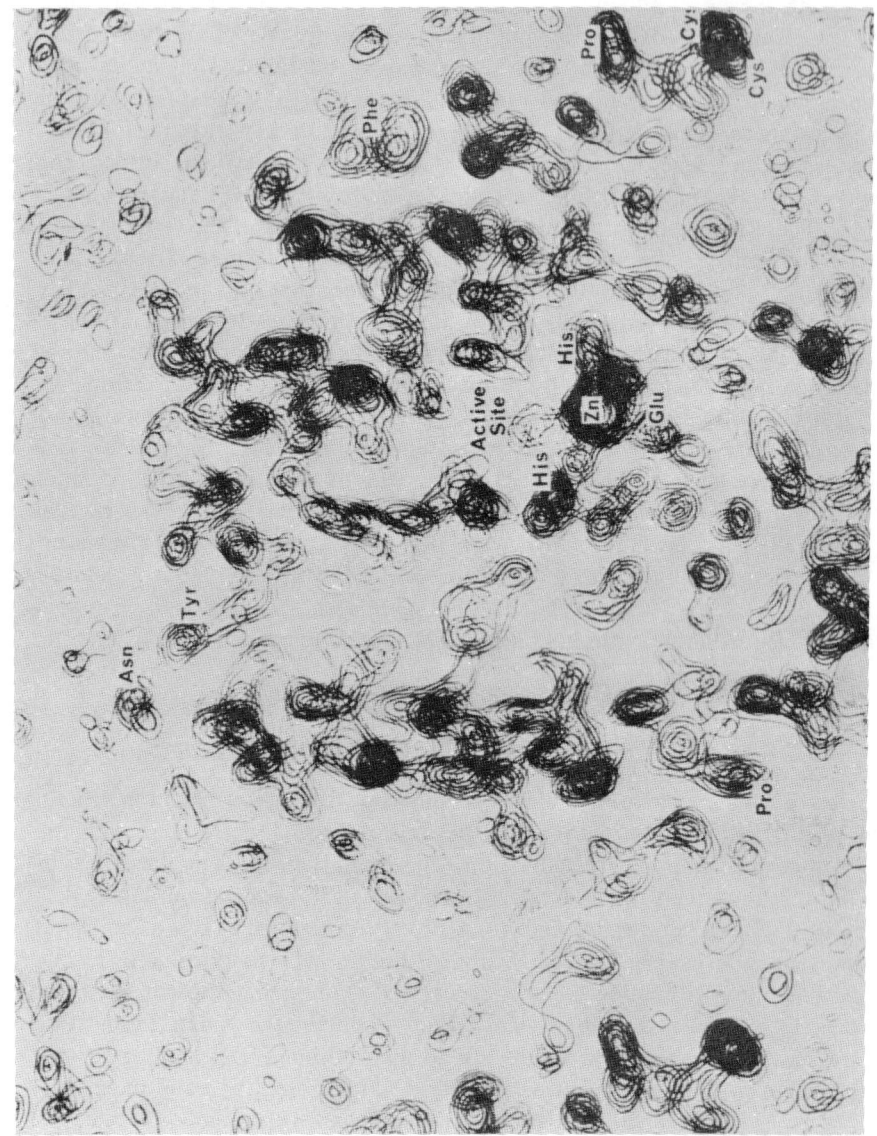

Figure 4.5 The electron density map of carboxypeptidase A. (Reproduced from reference 72 by courtesy of the senior authors)

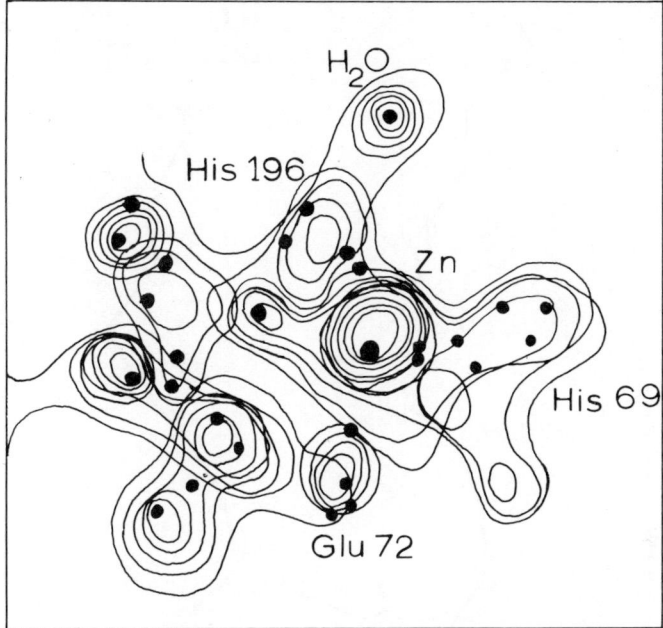

Figure 4.6 Part of the electron density map of carboxy-peptidase A in the region of the Zn atom, showing its three protein ligands and a water molecule which is hydrogen-bonded to histidine 196. Residues 288−300 form an α-helix running vertically along the left side. (Based on reference 72 by courtesy of the senior authors)

been suggested[75] for carboxypeptidase that some hundreds of water molecules are identifiable, although only eleven are found[76] to be localized actually within the protein structure as for the water molecule of Figure 4.6. Of these eleven internal water molecules in carboxypeptidase, two of them are bound by four hydrogen bonds, seven by three H-bonds and one by only two H-bonds. The water molecule associated with the zinc atom, which forms the active site of this enzyme, is displaced when binding of the substrate occurs in the 'pocket' around the active site.[77] Other examples can be found where water molecules appear to be involved in the catalytic activity of an enzyme. As for carboxypeptidase A, a water molecule in carbonic anhydrase C is found to be associated with the zinc atom at the active site and becomes displaced on activation of the catalytic process.[78] More water molecules, a total of nine altogether, are also found in the 'pocket' or 'cleft' at the active site of carbonic anhydrase C, and the electron density pattern can be interpreted in terms of these water molecules being in the ice I lattice configuration.[78] One of the two water molecules found in the enzyme ribonuclease S also appears to be involved in the catalytic process.[79]

Some of the strongly bound water can definitely be associated with structure stabilizing rôles for their macromolecular substrates. From X-ray diffraction studies

116

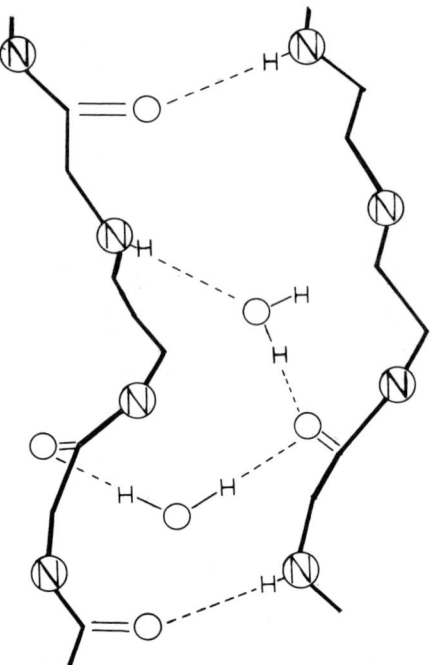

Figure 4.7 The binding of water molecules in collagen. (Reproduced from G. N. Ramachandran and R. Chandrasekharan, *Biopolymers*, **6**, 1649 (1968), by permission of John Wiley & Sons, Inc.)

of collagen,[80] a very specific arrangement for the bound water molecules has been deduced, where two water molecules are hydrogen bonded to the collagen triple helix for every three amino-acid residues, as depicted in Figure 4.7. By bridging across two different polypeptide chains, these bound water molecules can be seen to stabilize the triple helix structure of the collagen macromolecule. A more extensive configuration of water molecules has been found for the hemicellulose, β,D-xylan hydrate.[81] In this structure each water molecule participates in four hydrogen bonds, with the water molecules forming a helical chain to stabilize the three-fold polysaccharide helical conformation by forming hydrogen bonds to oxygen atoms and hydroxyl groups in these helices.

In their description of the atomic structure of tosyl-α-chymotrypsin, Birktoft and Blow[82] made a careful study of the content of solvent molecules. No strong evidence could be found to suggest that the bound solvent molecules were anions or cations such as SO_4^{2-} or NH_3^+, and based on the finding[83] that few, if any, cations bind to chymotrypsin around pH 4, these bound molecules were assumed to be water. Account was taken of density peaks that appeared clearly in the 'averaged' electron density map, indicating those water molecules that in general were bound in identical manner in each of the crystallographically independent protein molecules. Fifty such water molecules were identified, with 13 of them being bound internally. Most of these internal water molecules participate in three or four hydrogen bonds, either with the polypeptide chain or other water molecules, and one such water molecule is shown hydrogen-bonded to Ser 45, Val 53, and Gly 196 in Figure 4.8. Each of the regions associated with the buried charged amino-acid side groups of Ile 16, Asp 102, and Asp 194 contains an extensive hydrogen-bonded network including bound water molecules. It is possible that these water molecules act so as to increase the local dielectric permittivity and help reduce the area of influence of the charged groups which would otherwise tend to destabilize the protein structure. Most of the external water molecules were bound by only one hydrogen bond to the protein surface, and no extensive clusters of ordered, cage-like structured, water molecules were to be found around exposed hydrophobic amino-acid side groups. Some of the external water molecules form more

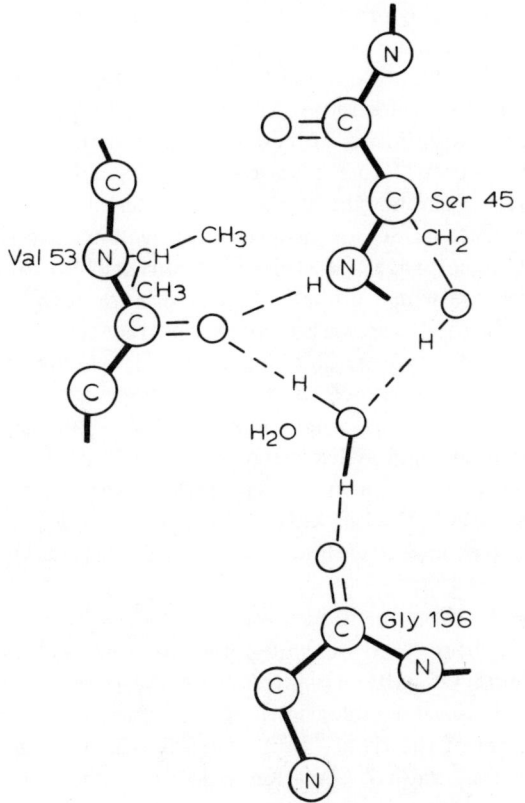

Figure 4.8 A hydrogen-bonded water molecule in tosyl-α-chymotrypsin. (Reproduced with permission from J. J. Birktoft and D. M. Blow, *J. Molec. Biol.*, **68**, 187 (1972). Copyright by Academic Press Inc. (London) Ltd.)

than one hydrogen bond to the surface of the same macromolecule, whilst a few others bridge neighbouring chymotrypsin molecules, at their regions of closest approach, by hydrogen bonds to each macromolecule.

In the past considerable significance has been placed by some workers on the frequently observed repeat distances of 0.47 nm, or multiples thereof, in biological systems.[84] This distance is of the order of that expected for the second neighbour distance in ice-like water structures, and could be taken to indicate extensive structuring of water molecules between the macromolecular constituents of living organisms. High resolution X-ray and neutron diffraction studies have revealed the existence of water molecules bound to protein surfaces; e.g. 105 for the rubredoxin molecule,[85] 250 for subtilisin,[86] and more than 100 water molecules for each myoglobin molecule.[87] No apparent structuring of these surface water molecules appears to be present, and the hundreds more water molecules that make up the remaining 30–50 per cent weight content of protein crystals give weak or no electron density peaks suggesting they have a quite random distribution of inter-molecular spacings. Using protein crystals as models for more complex systems such as membranes, viruses, and even cellular complexes, then we would not expect an extensive content of highly ordered ('ice-like') water. A review[88] of the n.m.r. and infrared absorption data for biological systems also supports this viewpoint. In this way no particular significance should be placed, perhaps, on the water structures proposed by Berendsen[84] to explain the intriguing 0.47 nm 'biological dimension'.

To counteract this thought there is the interesting fact that when peptides, amides, or monosaccharides are dissolved into water, there are no large changes in such thermodynamic properties as the heat capacity, for example. This is in sharp contrast to the case where non-polar molecules are introduced into water, which appears to result in the formation of a clathrate-type cage structure of water molecules around the hydrophobic molecules.[89,90] Quite often, the interaction of polar solutes in water is known to involve hydrogen-bond interactions, and in this respect it is of interest to note that for many polar organic molecules the spacing of carbonyl, ester, ether, and hydroxy oxygen atoms is about 0.48 nm.[91] This is close to the interoxygen spacing in the ice lattice, and suggests that if there does exist an 'ice-like' structure in biological water, then many biologically important molecules can be incorporated into it without disturbing the basic hydrogen-bonded network of the water.

As has already been mentioned various nuclear magnetic resonance (n.m.r.) techniques have been used to study the structure and properties of water in biological systems. Of particular interest are the observations that the relaxation times of water in several biological tissues are sensitive to the morphological or physiological state of the tissue.[92-94] Recently, Chang et al.[95] found that there is more than one longitudinal relaxation time for water in rat skeletal muscle, and that one of these relaxation times is quite sensitive to the freshness of the muscle. These studies clearly indicate that the spin-lattice relaxation of water protons within a muscle varies as the tissue changes from the normal living state to a non-functional or dead state. However the changes in the n.m.r. parameters were not too dramatic, as might be expected since the difference between life and death may only involve quite subtle changes in the cellular water structure. Damadian[92] was the first to observe that tumorous rat tissue had a larger tissue water proton spin-lattice relaxation value than the corresponding normal tissue. This basic observation has now been confirmed in many laboratories for various tumours in animals and humans. Floyd *et al.*[96] have reported that significant water proton spin-lattice relaxation value changes occur in the liver, spleen, and body serum of rats experiencing the very early stages of chemically induced cancer development. The use of n.m.r. techniques restricts the observation of relaxation times to within the range from about 10^{-13} to 10^{-5} s. Dielectric measurements also cover this range and can extend it to times in excess of 10^5 s, and it will be of great interest and value to explore the differences in dielectric properties of the water in cancerous and healthy tissues. With respect to water and cancer it is of interest to note that the water content and sodium concentration of tumour cells is higher than for normal cells.[97] The potassium concentrations remain relatively unchanged. This increase in tumour sodium concentration and the consequent decrease in membrane potassium selectivity could account for the low membrane electric resting potentials observed for cancer cells.[98,99]

Finally, the unique role that water must play in the functioning of biological systems is dramatically demonstrated when H_2O is replaced by D_2O in the growth media of cell cultures. D_2O acts as a poison and its presence affects the characteristics of almost every function of a cell.[100] For concentrations of D_2O of the order 99 per cent, very few cells survive and only a few species of algae, yeasts, and bacteria have been able to be grown.[101]

SORPTION OF WATER BY BIOLOGICAL MATERIALS

One method of obtaining more information regarding the interaction of water with biological systems is through the measurement of the sorption of water vapour. Sorption may be defined as the process in which vapour molecules are attached to sites in solids.[102] The vapour may be water or any other gas, and the sites may be distributed throughout the bulk phase of an amorphous or partially crystalline solid. The terms 'adsorption' and 'absorption' have been used to describe sorption processes. Adsorption is the uptake of vapour at the surface of an insoluble solid phase, and absorption is generally taken to be the dissolving of the vapour molecules within a non-volatile material phase. Although the term 'adsorption' is commonly used to describe the uptake of water vapour by proteins and polypeptides, Kuntz and Kauzmann[103] have pointed out that this is in fact a poor description of the process, which essentially is one of solvation or absorption. We can avoid having to make such fine distinctions by referring simply to 'sorption' processes, or more specifically, to the uptake or binding of water molecules to biological materials.

From a general point of view, the sorption of water by biological materials is only a special case of the problem of the binding of any small molecule to macromolecular materials. The phenomena of sorption processes are so diverse that an overall quantitative treatment is difficult, if not impossible. Not only are the pressure, temperature, and concentration of the material to be taken up by a surface important, but so also is the actual surface area and the chemical and physical properties of this surface. We should, therefore, expect that only an equation based on purely thermodynamic considerations could be developed to describe all sorption processes.

Gibbs[104] deduced a general equation based on thermodynamic considerations for the adsorption of a gas by a solid surface. Basically, this equation can be written as

$$h = -f(c, T, \ldots) \left(\frac{\partial \gamma}{\partial c} \right)_a \tag{4.2}$$

where the amount of the gas adsorbed h is a function of the gas concentration c, of the absolute temperature T, and other variables. The adsorption is positive when the adsorbate is concentrated on the surface of the adsorbent, and this situation arises if the surface tension γ decreases with increasing concentration. For negative adsorption, when the adsorbate tends to leave the interfacial layer, the surface tension increases with the concentration. It is assumed that the surface area 'a' of the adsorbent surface remains constant during the concentration changes. Equation (4.2) does not allow numerical calculations to be made, since it implies only the existence of an adsorption process without describing the definite relationships between the various variables. This equation was also developed later by Thomson[105] and Milner[106] and hence is frequently called the Gibbs–Thomson adsorption rule. If it is assumed that the solute obeys the ideal gas laws or Van't Hoff's laws in dilute solutions, then this equation can be reduced to the more useful form[107,108]

$$h = \frac{-c}{RT} \cdot \frac{d\gamma}{dc} \tag{4.3}$$

where R is the standard gas constant. This equation requires an empirical determination for $d\gamma/dc$ and as such does not give any insight into the mechanism of adsorption. This disadvantage was avoided by Langmuir[109] who developed a quantitative theory of adsorption based on molecular kinetic considerations.

Langmuir assumed that the atoms or molecules of the gas are bound at discrete points or active centres on the surface of the solid, to form a monomolecular film. Each active centre is assumed to hold just one atom or molecule by adsorptive forces similar in nature to chemical forces. These forces are taken to be of short effective range and hence are independent of whether adjacent centres are empty or occupied. With these physical conditions we can deduce the so called Langmuir adsorption isotherm equation as follows.[110] Let the number of active centres per unit surface area be N_t, and let only n be occupied by atoms or molecules. The rate r_1 of adsorption of new atoms will then be proportional to the number of free sites $(N_t - n)$, and to the pressure p, so that

$$r_1 = \beta(N_t - n)p$$

The rate r_2 of evaporation or desorption of the atoms already bound will be proportional to n, so that

$$r_2 = \alpha n$$

at equilibrium $r_1 = r_2$ and so

$$\beta(N_t - n)p = \alpha n$$

where β and α are proportionality constants. Hence

$$\frac{n}{N_t} = \frac{\beta p}{\alpha + \beta p}$$

Replacing β/α by b and n/N_t by h/h_m we obtain

$$h = \frac{h_m bp}{1 + bp} \tag{4.4}$$

The ratio n/N_t is equal to h/h_m because the amount h in grams of an adsorbate is proportional to the number of molecules n per unit surface area. The factor h_m is the amount adsorbed when all sites N_t are occupied, and hence represents the monolayer capacity. Equation (4.4) represents the Langmuir adsorption isotherm and its form is shown in Figure 4.9(a). The equation can also be written in the linear form

$$p/h = p/h_m + 1/h_m b \tag{4.5}$$

and produces a linear plot as shown in Figure 4.9(b). Such a plot can be taken as a test as to whether or not the observed adsorption process follows the Langmuir

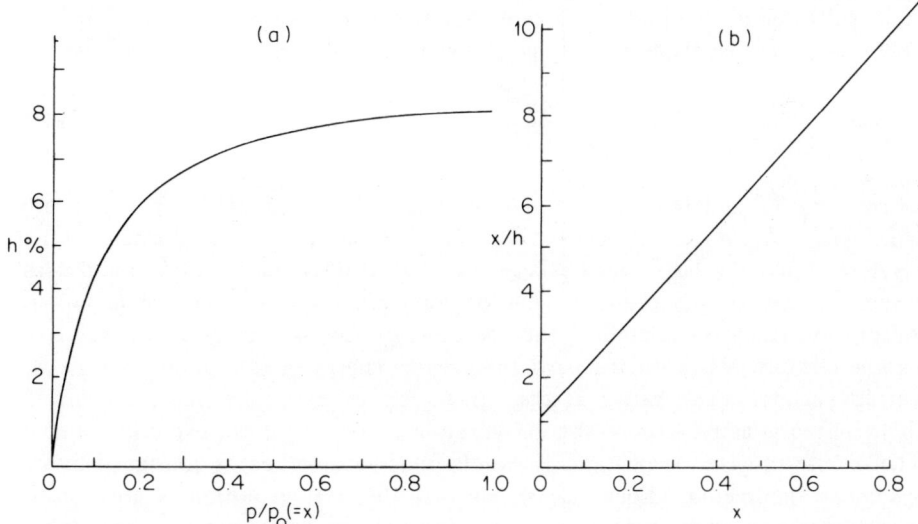

Figure 4.9 (a) The Langmuir adsorption isotherm. (b) The Langmuir adsorption isotherm plotted in the form of equation (4.5)

isotherm. The monolayer capacity h_m can be calculated from the slope of this linear plot, and the factor b is obtained from the extrapolated intercept on the p/h axis. From equation (4.4) we see that $1/b$ is equal to the pressure at which half the available monolayer sites are occupied. In plots such as those of Figure 4.9 the pressure is often conveniently expressed in terms of the relative vapour pressure p/p_0 where p_0 is the saturated vapour pressure of the adsorbate. In practice it is not common to have adsorption processes limited to the formation of only an adsorbate monolayer, it being more often the case that multilayers will be formed. The first monolayer coverage will often be formed at a low pressure value, corresponding to a partial pressure p/p_0 value of 0.15 or less, and it is of greater value to rewrite equation (4.5) in the form

$$1/h = 1/h_m + 1/ph_m b \tag{4.6}$$

A plot of h^{-1} against p^{-1} will now produce a straight line over the adsorption isotherm region where Langmuir's theory is valid, with the extrapolated intercept on the h^{-1} axis leading to the value for h_m. If accurate adsorption measurements can be made in the partial pressure range 0.02–0.1, then the plot of equation (4.6) will be more useful than that of equation (4.5) for determining an accurate assessment of the monomolecular coverage h_m.

Refinements of the Langmuir theory to include lateral interactions between the surface active sites[111] can lead to changes in the form of the adsorption isotherm, but these modified isotherms do not have the sigmoidal characteristic shape commonly found experimentally for the uptake of water by biological materials. An extension of the Langmuir theory, to take into account the formation of adsorbed layers more than one molecule thick, was made by Brunauer, Emmett, and Teller.[112] As for the Langmuir theory, the atoms or molecules in a layer are assumed to be in thermodynamic equilibrium. For an adsorption process limited to just n successive layers, the theory gives the amount of adsorbed material h as

$$h = \frac{h_m Cx[1 - (n + 1)x^n + nx^{n+1}]}{(1 - x)[1 + (C - 1)x - Cx^{n+1}]} \tag{4.7}$$

where $x = p/p_0$ and the factor C is approximately equal to $\exp(-(E_1 - E_f)/RT)$, where $(E_1 - E_f)$ is the difference in the heats of adsorption obtained when the first layer (E_1) and the last layer (E_f) are adsorbed. It is assumed that E_f for the last layers is equal to the ordinary heat of condensation of the adsorbate vapour. Adsorption is always related to the liberation of thermal energy, and hence is an exothermic process. Both the partial molal free energy of the vapour and the net entropy change are negative as the sorbed vapour molecules become relatively immobilized at active sites on the sorbing surface, resulting in the evolution of heat. The reverse process, desorption, is endothermic because heat is consumed by the system. Experimental experience shows that the heat evolution is greatest for adsorption on a clean surface, i.e. when the first molecules are adsorbed. Subsequently the heat released gradually decreases with further adsorption, and finally does not differ appreciably from the latent heat of condensation. The total heat of adsorption is always nearly twice as high as the heat of liquefaction, which indicates that there are at least two heat releasing processes: the adsorption due to the attractive van der Waals forces, and the condensation of the vapour on the monolayer already formed. The heat effect in the first process is larger than in the latter.

In the case of the formation of a monolayer film ($n = 1$), then equation (4.7) reduces to the form

$$h = \frac{h_m Cx}{1 + Cx}$$

which is identical to the Langmuir equation (4.4).

With the simplifying assumption that the heats of adsorption of all layers except the first are nearly equal and do not differ from the heat of condensation of the vapour adsorbate, then equation (4.7) reduces to the so-called BET equation:

$$\frac{x}{h(1 - x)} = \frac{1}{h_m C} + \frac{(C - 1)x}{h_m C} \qquad (4.8)$$

The form of the adsorption isotherm satisfying the BET equation (4.8) is shown in Figure 4.10(a). A straight line should be obtained by plotting $x/h(1 - x)$ against x, as shown in Figure 4.10(b), and the applicability of the BET isotherm can be checked from such a plot, with values for h_m and C being obtained from the slope and intercept values.

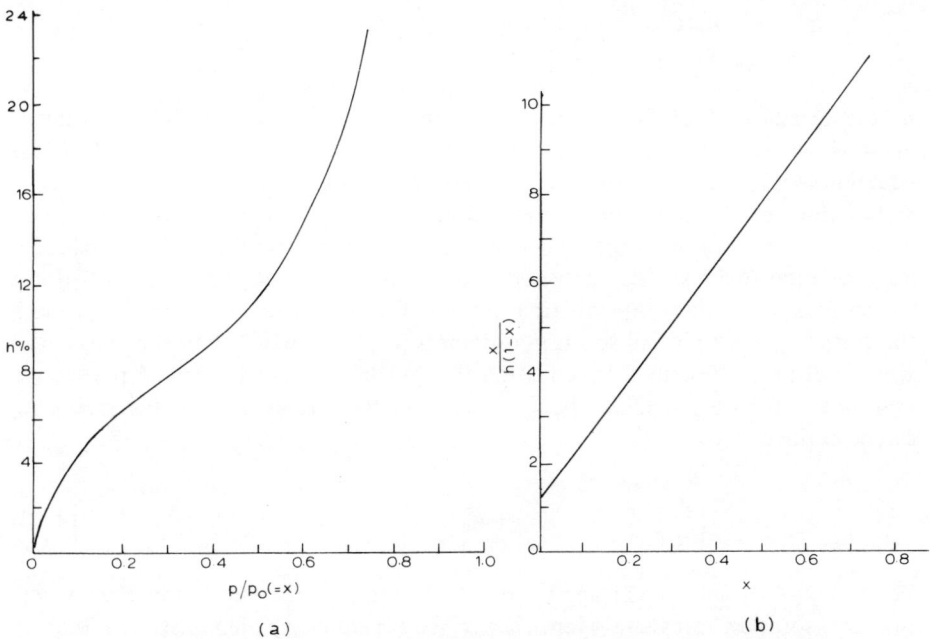

Figure 4.10 The BET adsorption isotherm of equation (4.8)

Gascoyne and Pethig[113] have shown that a completely general formula describing sorption isotherms can be derived from a general statistical mechanical treatment using Grand Partition Functions. This formula has the form

$$h = h_m x \frac{\partial}{\partial x} \ln f(x)$$

(4.9)

where $f(x)$ is the partition function. Substitution for $f(x)$ in equation (4.9) allows the corresponding sorption isotherm to be established. For example, setting the activities of sorption sites in the second and subsequent hydration layers to zero leads to the partition function $f(x) = 1 + Cx$. Substitution into equation (4.9) yields the isotherm equation

$$h = \frac{h_m Cx}{1 + Cx}$$

(4.10)

which is the Langmuir equation (4.4) for monolayer adsorption. If the activities of the second and subsequent layers are assumed to be equal, then the resulting partition function is a geometrical progression

$$f(x) = 1 + Cbx + C(bx)^2 + C(bx)^3 + \cdots$$

where (Cb) is the activity of molecules sorbed in the first layer, and b is the activity in all subsequent layers. Substitution of $f(x)$ into equation (4.9) yields the isotherm equation

$$h = \frac{h_m Cbx}{(1 - bx)[1 + b(C - 1)x]}$$

(4.11)

In the special case of the activity in the second and subsequent sorption layers being equal to that of the bulk condensed liquid, b becomes unity. In this case equation (4.11) reduces to the BET isotherm of equation (4.8).

Equation (4.9) allows the partition function for the sorption sites to be determined directly from the sorption isotherm data. To complete the analysis a value for h_m, the monolayer site capacity, is required. At sufficiently small values of the partial pressure x, then any partition function for multilayer sorption will approach the form $f(x) = 1 + Cx$. All sorption isotherms for the case of identical primary sites will therefore be described by equation (4.10) when x is very small. A plot of x/h against x will have a gradient $1/h_m$ at very low hydrations, so that the factor h_m can be defined as

$$\frac{1}{h_m} = \lim_{x \to 0} \frac{\partial}{\partial x} \left(\frac{x}{h} \right)$$

(4.12)

Equations (4.9) and (4.12) apply only to sorption isotherms of materials whose primary sorption sites are identical. It may readily be demonstrated that an extended form of equation (4.9) which describes sorption by materials with N

different types of primary sites is

$$\prod_{j=1}^{N} (f_j(x))^{h_{mj}} = \exp\left\{\int_0^x \frac{h}{x}\,dx\right\}$$

where h_{mj} are the primary site capacities of the N different types of sites whose respective partition functions are $f_j(x)$.

In Table 4.4 values are presented for the percentage hydration h_m and activity (Cb) for the first bound monolayers, together with values for the parameter b, obtained from a computer-assisted analysis of the hydration isotherms for various biomacromolecules.[113] Values derived from the conventional BET graphical analysis are also included in Table 4.4, from which it can be seen that the BET theory can lead to significant errors.

Table 4.4   Values for h_m, (Cb), and b derived from a computer fit of equation (4.11) for the hydration isotherms of various biomacromolecules. Values derived from the BET theory are included for comparison

Material	Equation (4.11)			BET theory $(b = 1)$	
	h_m	(Cb)	b	h_m	(Cb)
Bovine serum albumin	7.87	9.63	0.81	6.5	12.2
Cytochrome-c	8.27	13.4	0.88	7.5	9.8
Lysozyme	8.15	10.5	0.82	7.4	14.9
Lecithin	7.18	3.79	0.88	5.2	8.5
DNA	11.5	17.3	1.05	12.1	13.3

From Table 4.4 it can be seen that the activity parameter b of equation (4.11) is less than unity for lecithin and the proteins studied, but has a value greater than unity for DNA. This can be regarded to be of significance in terms of their behaviour in aqueous solution. The value of b less than unity for the proteins implies that their outermost hydration layer is formed by water molecules of lower activity than bulk water, so that extra layers of water are required to be associated with the hydrated macromolecule before it can become fully accommodated into normal bulk water. In other words these globular proteins can be considered to have slight hydrophobic qualities, and the value of b can have direct relevance to such phenomena as their salting-in and salting-out behaviour, and to other macromolecular interactions in solution. For materials such as DNA, having a value b greater than unity, the outermost hydration shell is already indistinguishable from bulk water and no additional modification of the surrounding bulk water is required for such macromolecules to become fully solvated. Further studies of such aspects of hydration isotherms could lead to an increased insight into the various biomacromolecular—water interactions that take place in biological systems.

SORPTION EXPERIMENTS AND OTHER THEORIES

The various methods of directly measuring the sorption of water vapour by proteins and other biological materials have been outlined by McLaren and Rowen.[102] The two variables which have to be measured to obtain sorption isotherms are h, the amount of vapour sorbed, and p, the vapour pressure. Fixed vapour pressures may be obtained by using salt solutions or sulphuric acid solutions, by varying the temperature of the water in equilibrium with its vapour, or by adjusting the volume of the water vapour. Some tables for obtaining fixed vapour pressures using salt solutions and sulphuric acid solutions are given in Tables 4.5 and 4.6. The oldest and perhaps most widely used method is the so-called 'weighing bottle and salt solution' method. Its main advantage is that it requires no expensive or specialized equipment, but it lacks sensitivity in the low vapour pressure region, and the presence of

Table 4.5   Equilibrium relative humidity of some sulphuric acid solutions at 25 °C*

% Sulphuric acid and density (g/ml)	R.H. %	% Sulphuric acid and density (g/ml)	R.H. %
80%, 1.720	1.0	40%, 1.300	57.0
75%, 1.660	2.2	35%, 1.258	66.8
70%, 1.605	5.1	30%, 1.217	75.9
65%, 1.549	9.9	25%, 1.176	82.9
60%, 1.495	17.0	20%, 1.138	88.5
55%, 1.421	26.8	15%, 1.101	92.9
50%, 1.392	36.9	10%, 1.066	96.2
45%, 1.344	46.8	5%, 1.030	98.6

*From: R. E. Wilson, *J. Indust. Eng. Chem.*, **13**, 326 (1921)

Table 4.6   Equilibrium relative humidity of some saturated salt solutions at 25 °C*

Salt	R.H. %	Salt	R.H. %
Cesium fluoride	3.39 ± 0.94	Sodium bromide	57.57 ± 0.40
Lithium bromide	6.37 ± 0.52	Cobalt chloride	64.92 ± 3.50
Zinc bromide	7.75 ± 0.39	Potassium iodide	68.86 ± 0.24
Potassium hydroxide	8.23 ± 0.72	Strontium chloride	70.85 ± 0.04
Sodium hydroxide	8.24 ± 2.1	Sodium nitrate	74.25 ± 0.32
Lithium chloride	11.30 ± 0.27	Sodium chloride	75.29 ± 0.12
Calcium bromide	16.50 ± 0.20	Ammonium chloride	78.57 ± 0.40
Lithium iodide	17.56 ± 0.13	Potassium bromide	80.89 ± 0.21
Potassium acetate	22.51 ± 0.32	Ammonium sulphate	80.99 ± 0.28
Potassium fluoride	30.85 ± 1.3	Potassium chloride	84.34 ± 0.26
Magnesium chloride	32.78 ± 0.16	Strontium nitrate	85.06 ± 0.38
Sodium iodide	38.17 ± 0.50	Potassium nitrate	93.58 ± 0.55
Potassium carbonate	43.16 ± 0.39	Potassium sulphate	97.30 ± 0.45
Magnesium nitrate	52.89 ± 0.22	Potassium chromate	97.88 ± 0.49

*From L. Greenspan, *J. Res. Nat. Bur. Standards,* **81A**, 89 (1977)

other permanent gases results in long time intervals being required for the attainment of equilibrium conditions. Greater sensitivity at low vapour pressures is achieved using a gravimetric method, where weight determinations are made using a calibrated silica or tungsten helical spring suspended in an evacuable glass thermostatted jacket. An electrical sorption balance, first described by Gregg,[114] has a sensitivity of the order 3×10^{-4} g for a test adsorbate sample of around 15 grams, and can be used to study the dynamics of sorption and desorption processes. Volumetric methods can provide reproducible and sensitive measurements, but only after tedious calculations using the gas laws and various correction factors, and they also have the disadvantage of not being particularly useful for the study of the dynamics of sorption processes. A new technique uses a resonating quartz crystal which is coated with the biological material to be investigated.[115] The adsorption of water vapour onto the deposited biological material effectively increases the mass of the resonant crystal, and the frequency of resonance is lowered. Using modern frequency counter techniques, frequency changes to within 1 part in 10^8 are readily measured, which for a typical quartz crystal oscillator corresponds to a mass change of around 5×10^{-10} g. If AT-cut crystals are used, frequency changes

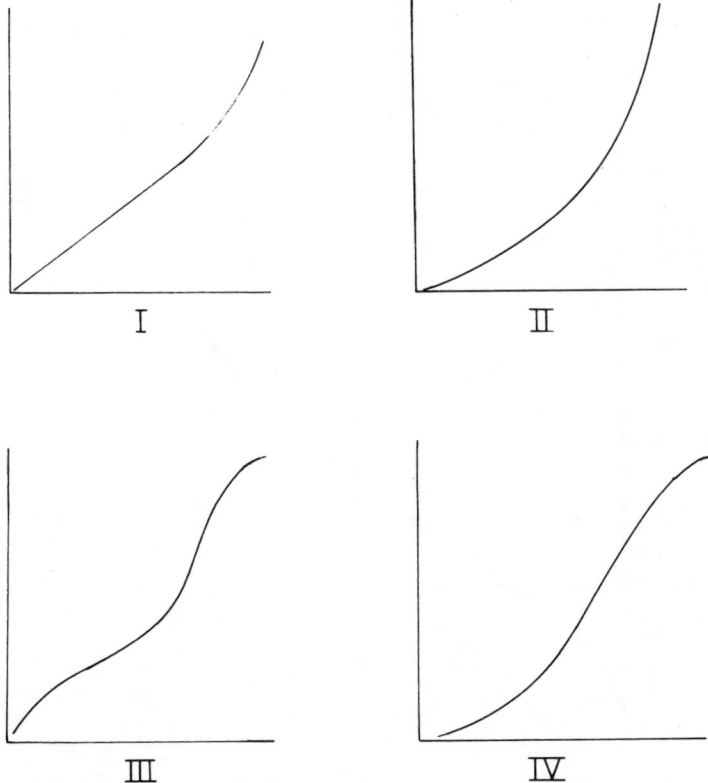

Figure 4.11 Various forms of sorption isotherms. See text for explanation

due to temperature and pressure variations are very small, and the sensitivity is such that the sorption of just one water molecule on any one of 10^4 protein molecules (molec. wt. $\simeq 5 \times 10^4$) can be detected. This makes the technique very useful for aiding the study of the dielectric properties of biological materials with very low water content.[113] Useful experimental outlines for various techniques have also been given by Adamson[111] and Clark,[116] and Ling[117] describes much of the early work carried out on the measurement of water sorption isotherms for proteins.

Several types of experimentally determined sorption isotherms may be distinguished.[112,118] Apart from isotherms which may be described in terms of the Langmuir and BET theories as shown by Figures 4.9 and 4.10, other types may be obtained, as shown in Figure 4.11.

Apart from what can be termed the van der Waals' type adsorption processes described by the Langmuir and BET theories, two other processes can be considered; namely solution sorption, represented as $h = \alpha x$ (Henry's Law) where α is a constant, and also what can be termed a kind of capillary condensation which gives rise to a great increase in the rate of sorption dh/dx with increasing relative vapour

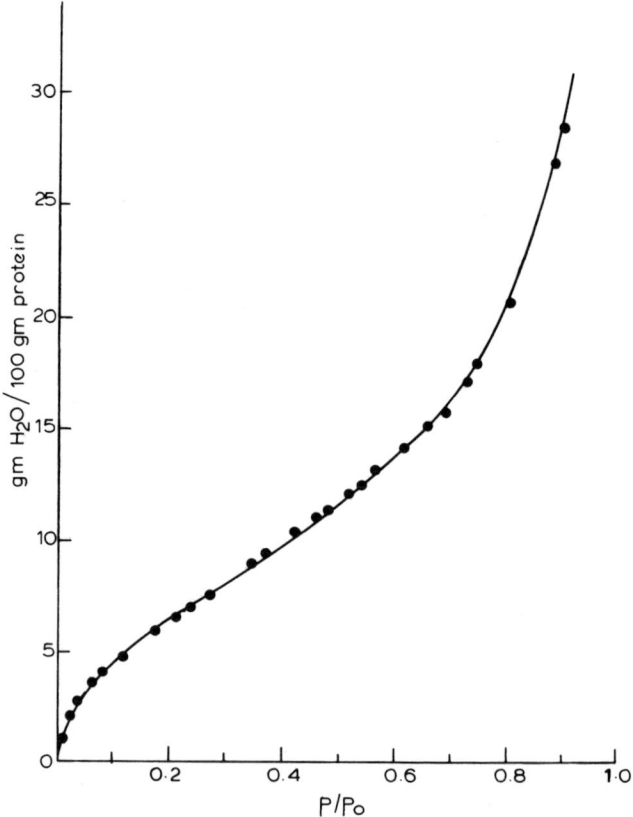

Figure 4.12 The water sorption isotherm for bovine serum albumin at 25 °C

pressure.[119] The isotherm type I of Figure 4.11 appears to be composed of a combination of these two additional processes, and the sorption of water by many vinyl polymers is of this form.[103] According to Gregg[120] the reproduction of isotherms of the type II, III, and IV is limited, with there being only moderate agreement between experiment and theory. For example, isotherms of type II may be described in terms of the BET equation (4.8) with $C < 1$, and types III and IV may be described by a somewhat complicated modification of equation (4.7).

The water sorption isotherms for proteins, polypeptides, polyamides and many biological systems show the common feature of being sigmoids (S-shaped), as represented by the type of isotherm of Figure 4.10(a). A typical example of such an isotherm is that for the sorption of water by bovine serum albumin at 25 °C as shown in Figure 4.12. The results were obtained using the quartz crystal resonator technique.[113] Other examples are given for 25 °C water sorption on the poly-glycine-DL-alanine copolymer[102] and the tobacco mosaic virus[121] as shown in Figure 4.13. The common feature for such isotherms for biological materials is that at low relative vapour pressures, of the order $p/p_0 \lesssim 0.05$, the amount of water sorbed increases rapidly. A distinct rate of change of sorption with vapour pressure, appearing as a 'knee' in the isotherm characteristic, occurs at a relative vapour pressure of around 0.1, and up to $p/p_0 \simeq 0.8$ the water content increases fairly slowly but steadily. At relative vapour pressures greater than about 0.9, the water sorption again begins to increase rapidly and indeed the rate of sorption dh/dx must approach an infinitely high value at relative vapour pressures of unity (100 per cent relative humidity) if the material exhibits any detectable water solubility. The isotherms obtained at different temperatures are similar in shape, with the amount of sorption decreasing with rising temperature. This follows from equation (4.3), since with rising temperature there are decreases not only in the term $d\gamma/dc$ (the surface tension falls with temperature), but also in the term c/RT. Considering the kinetic processes involved also leads us to the same conclusion. The motion and mobility of the sorbed water molecules will increase with increasing temperature, and so tend to leave the sorbent. Low temperatures favour more sorption, because with weaker thermal vibrations of the sorbed molecules the van der Waals forces will act more strongly. The influence of increasing pressure or concentration of the sorbate acts in the reverse sense, since the number of molecular collisions with the sorbent increases with increasing pressure. For proteins, if the amount of sorbed water is compared at constant relative vapour pressure values, then there is of the order 1 per cent reduction in hydration content per degree temperature rise.[122] It is of value to note that, contrary to the case of van der Waals' adsorption, chemisorption processes where the substance adsorbed reacts chemically with the adsorbent, by electron transfer for example, will lead to increased sorption with increasing temperature.

The amount of water sorbed by various proteins at a relative vapour pressure of 0.92 is given in Table 4.7. In the table, an estimate is given of the number of water molecules per amino-acid residue in the protein molecule, based on assigning a molecular weight value of 100 for a typical amino-acid residue. It is seen that at this high vapour pressure value, the water content corresponds to more than one water

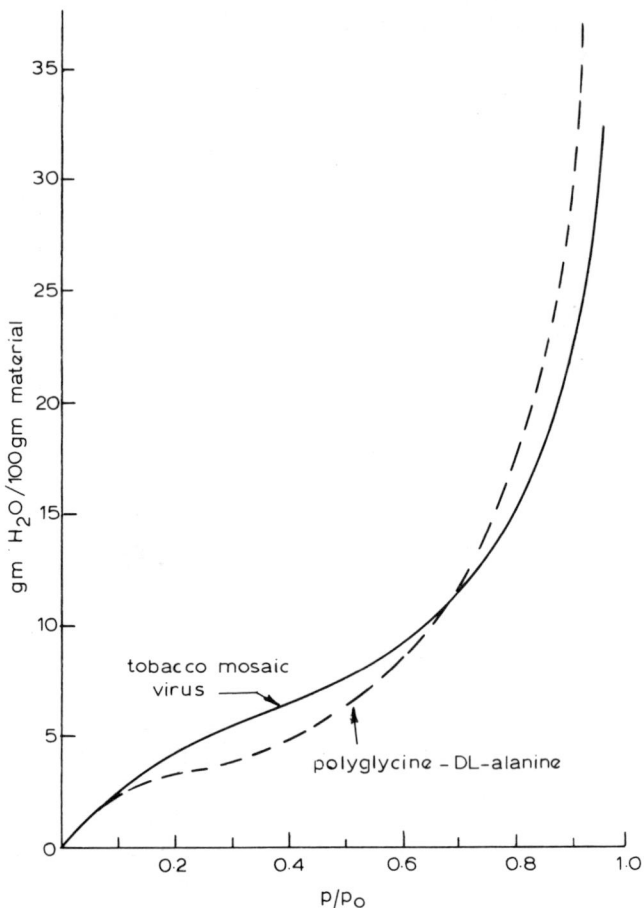

Figure 4.13 The water sorption isotherms for tobacco
mosaic virus and polyglycine-D L-alanine

molecule per amino-acid residue. From the concept of the 'oil-drop' model for
protein molecules, described in Chapter 3, where the non-polar, hydrophobic,
amino-acid residues are concentrated in the interior of the molecular structure, then
we will expect most of the water bound to hydrated proteins to be attached to the
polar, hydrophilic, amino-acid residues on the outside molecular surface. Early
X-ray studies confirmed this, where the diameter of tobacco mosaic virus particles
was found to be the same in both dry and in wet crystals.[124] This indicated that
water molecules did not penetrate appreciably into the interior of the biomolecules
to cause them to swell. The same was found to be true for insulin[125] and for
haemoglobin[126] crystals. The protein molecules themselves remain rigid while the
crystals change their volume with changing hydration. About one-third of the water
in horse methaemoglobin crystals, namely 0.3 g per g protein, is bound by the
protein and seems to form a single shell of water molecules around the haemoglobin

Table 4.7   Water uptake at 0.92 relative vapour pressure for proteins at 25 °C

Protein	$\dfrac{\text{g } H_2O}{\text{g protein}}$	$\dfrac{\text{Moles } H_2O}{\text{Moles monomer}}$	Reference
Bovine serum albumin	0.29	1.6	123
Casein	0.27	1.5	68[a]
Chymotrypsinogen	0.26	1.4	123
Collagen	0.44	2.4	122[b]
Cytochrome-c	0.35	1.9	123
Gelatin	0.44	2.4	122[b]
Haemoglobin	0.33	1.8	123
Insulin	0.21	1.2	123
Lysozyme	0.22	1.2	123
Myoglobin	0.38	2.1	123
Ovalbumin	0.30	1.7	122[b]
Ribonuclease	0.32	1.8	123
Salmine	0.74	4.1	122[b]

[a] Corrected from 24.2 °C by subtracting 0.8 per cent
[b] Corrected from 0.90 relative vapour pressure by adding 10 per cent

molecule. The remaining water in the crystals is 'free' in the sense that it can act as a solvent for diffusing electrolytes.[127] Furthermore, the swelling and shrinking of methaemoglobin crystals occurs in steps, the crystal changing from one lattice type to another without any intermediate state being observable, and Perutz[127] interpreted these volume changes as changes in the thickness of the water layer between the protein molecules. The thickness changes were in steps from 2.1 to 1.67 nm, and from 1.21 to 0.83 nm, indicating that the water itself had a layer structure so that not less than one layer of about 0.4 nm thickness could peel off at a time. Significantly, Perutz remarked that this is of the same order as that of a layer of water molecules in ice, namely 0.37 nm. The measurements by Bragg et al.,[128,129] show that the haemoglobin molecule in the dry state has dimensions 4.5 x 4.5 x 6.5 nm, and in its fully hydrated state is 5.3 x 5.3 x 7.1 nm, so that with a water molecule having a diameter of 0.28 nm, this difference corresponds to about two layers of water molecules bound to the surface of the haemoglobin molecule at water saturation.

    The greater water binding ability of the polar amino-acid residues is demonstrated in Table 4.8, where it can be seen that at 0.92 relative humidity the non-polar glycine and alanine peptide units bind only a fraction of a water molecule each, whereas the lysine residue, for example, binds nearly four water molecules. The molecular weight values used to determine the hydration content per amino-acid residue were obtained from Table 2.8. From Table 2.1 and Figure 3.2, we see that of the 20 common amino-acid residues, 13 of them are polar and hence capable of providing relatively strong water binding sites in protein molecules. Several workers have considered their protein hydration results in terms of comparing the moles of water sorbed with the polar residue content of the

Table 4.8   Water uptake at 0.92 relative vapour pressure for polypeptides

Polypeptide	Temp. (°C)	$\dfrac{\text{g } H_2O}{\text{g material}}$	$\dfrac{\text{Moles } H_2O}{\text{Moles monomer}}$	Reference
Tetraglycine	30	0.07	0.22	130
Pentaglycine	30	0.06	0.19	130
Hexaglycine	30	0.085	0.27	130
Polyglycine I	30	0.18	0.57	130
Polyglycine II	30	0.18	0.57	130
Polyalanine	31.5	0.14	0.55	131[a]
Polyglutamic acid	31.5	0.18	1.29	131[a]
Polylysine	31.5	0.53	3.77	131[a]

[a]Corrected from 0.90 relative vapour pressure by adding 10 per cent

proteins.[132-136] Pauling[136] suggested that each polar amino-acid residue could bind one water molecule, and indeed such an estimate gives the correct order of magnitude in predicting the monolayer (h_m) water content. Green[133] analysed this proposal in some detail and concluded that assigning one water molecule to each polar residue tends to overestimate the hydration content corresponding to h_m. This can be seen in Table 4.9 where the experimental values for h_m, derived from the BET plots (equation (4.8)) of the hydration isotherms for three proteins (see Table 4.4) are expressed as the number of bound water molecules per protein molecule to enable comparison with the total number of polar amino-acid residues per protein molecule. From this table it can be seen that the number of bound water molecules corresponding to the BET monolayer represents from 65 to 86 per cent of the total number of available polar sites in the protein molecule. Cardew and Eley[137] found that their h_m value for haemoglobin corresponded to 217 bound water molecules per protein molecule, representing a fraction of about 73 per cent of the total number that could theoretically bind, one each, to the total number of 296 polar residues. We have seen from Table 4.4 that the BET theory tends to underestimate the value for h_m for the proteins listed in Table 4.9. Values for h_m derived from

Table 4.9   A comparison of the number of water molecules bound in the monolayer hydration shell and the number of polar residues in the protein molecule*

Protein molecule	$h_m \left(\dfrac{\text{g } H_2O}{\text{100 g protein}} \right)$		Bound H_2O molecules / Protein molecule		$\dfrac{\text{Polar residues}}{\text{Protein molecule}}$
	BET	Eqn. (4.11)	BET	Eqn. (4.11)	
Bovine serum albumin	6.5	7.87	241	291	373
Cytochrome-c	7.5	8.27	56	62	65
Lysozyme	7.4	8.15	60	66	80

*The data is derived from reference 113

equation (4.11) are also included in Table 4.9, and although this does bring closer agreement it can be seen that the polar residues still exceed the number of bound H_2O molecules. This possibly reflects the fact that some of the polar amino-acid residues are buried in the protein structure in such a way as to be inaccessible to water molecules. From Figure 2.1(b) we can see that some of the polar amino-acid residues can take part in stabilizing interactions within the protein molecule. For example, cysteine residue pairs commonly form disulphide bridges, and aspartic acid, asparagine, serine, threonine, tryptophan and tyrosine can form intra-molecular hydrogen bonds between each other and the amino and carbonyl groups in the main polypeptide chain. A close inspection of the 0.28 nm resolution structure of horse cytochrome-c indicates that two cysteines are bound to the haem group and both a threonine and asparagine residue are buried within hydrophobic regions of the macromolecule. In the bovine serum albumin molecule, 34 of the cysteine residues form 17 disulphide bridges, and 8 cysteines form 4 disulphide bridges in lysozyme. Also, as depicted in Figure 3.3, some of the polar residues on the protein molecule surfaces may be closely spaced so that some of the bound water molecules may be able both to donate and accept a proton to form hydrogen bonds with two polar residues. High resolution X-ray studies indicate that molecules of globular proteins in crystals are generally in contact with each other at six or so points[103] and it is possible that a few water molecules could be bonded to two adjacent protein molecules. With all of these considerations in mind, it might be expected that the experimental monolayer hydration content will be slightly lower than indicated by the number of available polar binding sites, as is the situation shown in Table 4.9. Bull[122] remarked on the fact that for the fourteen protein samples he investigated, the surface area occupied by the BET monolayer water coverage represented only a fraction of the total protein molecule surface area, and Cardew and Eley[137] found that the h_m monolayer coverage for haemo-globin represented only a quarter of the total surface area. We therefore have strong support for the idea that most of the initial sorbed water on protein molecules is localized at specific sorption sites, namely accessible polar residue hydrogen-bonding sites, and that theories based on the Langmuir sorption mechanism have some validity at low hydration values.

We have seen that assigning one water molecule to each accessible polar residue in a protein molecule provides a good estimate for the primary, strongly bound, hydration coverage. But what of the water content at high humidities or at saturation in aqueous solution? From their measurements, Bull and Breese[123] proposed that by assigning six water molecules per polar residue, and a negative quantity, -7, for side-chain amides, then the resulting water content is in good agreement with the 92 per cent relative humidity experimental values. Although this 'formula' might lead to successful predictions, it is difficult to see how it can lead to a clearer understanding of the molecular processes involved in protein hydration. Fisher[138,139] has developed a so-called 'limiting law of protein structure' to predict the upper limit to the amount of hydration of a protein molecule. Starting from the basic premise that polar residues have a very limited solubility in the interior of a soluble protein molecule, and that non-polar residues have a very limited solubility in the protein surface, then Fisher has shown that knowledge of the amino-acid composition of the protein leads to a value for the limiting hydration content. Defining p as the ratio of the total volume of the polar residues to the total volume of non-polar residues, then it was shown[138] that

$$V_e = \frac{p}{(p+1)} V_t$$

where V_e is the volume of the external polar layer and V_t is the total protein volume. Also, assuming that the average thickness of the outermost layer of amino-acid residues is 0.4 nm, then the surface area A of the protein molecule is given by

$$A = \frac{p}{(p+1)} \frac{V_t}{0.4}$$

$$= 2.5 \, V_e$$

In these equations, if V_e is calculated from the amino-acid composition in nm^3 per g of protein, then A is obtained in terms of nm^2 per g of protein. Assuming the surface of each protein molecule to be covered with a monomolecular layer of the same density as the water of the bulk solvent, and that in this layer the water molecules have the same random distribution of orientations as in the rest of the solvent, then each bound water molecule may be considered to occupy a volume of $3.31 \times 10^{-2} \, nm^3$, equivalent to $0.103 \, nm^2$ of hydratable protein surface per water molecule. Defining the constant g as the grams of water in a monomolecular layer per nm^3 of protein surface, then

$$g = \frac{18 \text{ g/mole}}{(0.103 \text{ nm}^2/\text{molecule})(6.02 \times 10^{23} \text{ molecules/mole})}$$

$$= 2.9 \times 10^{-22} \text{ g/nm}^2$$

and the grams of water H_t bound per g of protein is given by

$$H_t = gA$$

$$= 7.25 \times 10^{-22} \, V_e$$

For some 34 different proteins, Fisher found the limiting hydration content H_t to be almost constant at a value of 0.28 ± 0.02, and to be well within the range of hydration values obtained experimentally. Here again, we are not directed towards a physical understanding of hydration, and although Fisher has provided a useful working estimate, some of the underlying assumptions are almost certainly not valid. For example, the bound water on a protein surface most likely involves more than just a monolayer coverage, and the orientation of these bound water molecules will not be as random as those in the surrounding bulk solvent.

By far the most useful indication of the specificity of sites of hydration for polypeptide systems has come from the low temperature n.m.r. measurements of Kuntz and his colleagues.[140] For temperatures down to $-80\,°C$ it is found that in

Table 4.10   Comparison of predicted and measured hydration contents of bound water for various proteins[140]

Protein	Hydration (g H_2O per g protein)	
	Predicted	Measured
Lysozyme	0.36	0.34
Myoglobin	0.45	0.42
Chymotrypsinogen	0.39	0.34
Chymotrypsin	0.36	0.33
Ovalbumin	0.37	0.33
Bovine serum albumin	0.45	0.40
(at pH 3)	0.32	0.30
(in urea)	0.45	0.44
Haemoglobin (denatured)	0.42	0.42

biomolecular water solutions a certain fixed amount of the water exhibits proton relaxations consistent with mobile water molecules whose thermodynamic properties have been sufficiently altered so as to be unable to freeze into normal ice. This water content is taken to represent the water bound to the biomolecules in the solution, and the amount determined in this way is in good agreement with values obtained by calorimetric and infrared measurements.[103] From work on polypeptide solutions to see whether the binding sites for the unfrozen water could be identified, a proposed assignment of amino-acid residue hydrations was made[103] as shown in Table 4.11. In accordance with isopiestic and infrared spectroscopic measurements, the ionized polar residue side chains were found to exhibit the largest hydration capacities, as indicated by Table 4.11. Also, it appears that both backbone and side-chain amide groups can bind about one water molecule per group, and the hydration capacities of the Gln, Ile, Leu, Met, Ser, Thr, and Trp residues were calculated on this basis for Table 4.11. The ionized histidine residue was assumed to have the same hydration properties as ionized lysine. Using Table 4.11, then good approximations for the bound water content of globular proteins can be obtained from knowledge of their amino-acid compositions, as shown in Table 4.10. The calculated hydration contents for Table 4.10 assume that all the residues are fully hydrated, and such an assumption will lead to an overestimation since some of the residues will be buried deeply within the protein molecule even when in its 'unfolded' state in solution. Such a small positive error is evident when the calculated values are compared with the n.m.r. freezing experiment results shown in Table 4.10. The denatured BSA and haemoglobin results are in better agreement as might be expected since there will be fewer buried groups. The result for bovine serum albumin in acid shows the effect of the carboxyl groups becoming uncharged at a pH value below their pK_a values. A simple rule of thumb proposed by Kuntz and Kauzmann[103] is that each exposed polar oxygen or nitrogen atom can hold one water molecule, while ionized groups average about six water molecules. Most of the binding will almost certainly involve hydrogen bonding between water molecules and the oxygen and nitrogen atoms, but in order to hold

Table 4.11   The hydrations (number of bound water molecules) of the amino-acid residues as indicated by n.m.r. measurements[140]

Residue	Hydration no.	Residue	Hydration no.
Polar		Ionized	
Asn	2	Asp$^-$	6
Gln	2	Glu$^-$	7
Pro	3	Tyr$^-$	7
Ser	2	Arg$^+$	3
Thr	2	His$^+$	4
Trp	2	Lys$^+$	4
Asp	2		
Glu	2	Non-polar	
Tyr	3	Ala, Gly	1
Arg	3	Ile, Leu	1
Lys	4	Phe	0
		Met, Val	1

six water molecules, the ionized groups must also exert dipole—dipole and electrostatic binding interactions.

DNA AND WATER

X-ray diffraction results indicate that DNA can exist in several distinct structural forms.[141-143] At 92 per cent relative humidity the helical B configuration is observed, in which the purine and pyrimidine base pair are stacked perpendicularly to the axis of the helix. Between 80 and 75 per cent relative humidity, the A helix form exists, which has a 70° angle between the helical axis and the planes of the bases, and as the relative humidity decreases from 75 to 55 per cent, a disordered molecular state is obtained. In the B helix conformation at 92 per cent relative humidity, the molar ratio of sorbed water to DNA has been estimated to be of the order of 20 water molecules per nucleotide.[144] Although it is evident that water plays a significant rôle in the various ordered structures of DNA, its precise rôle is not that well understood. It appears that in the absence of a sufficient number of water molecules the base—base interactions are not by themselves capable of maintaining an ordered overall configuration, but these effects do not reflect the expected relative stabilities of the various types of hydrogen bond. For example, base—base bonds would be expected to be weaker than bonds formed between water and the bases, and weaker than water—water bonds.[144]

Under favourable conditions infrared spectra are capable of showing specific interactions between particular chemical groups, and to this end several workers have studied the infrared spectra of DNA films as a function of relative humidity. In particular, Falk *et al.*[144] have studied the behaviour of water at different sorption sites in maintaining the overall conformation of NaDNA and LiDNA, with the studies essentially concerning the 20 water molecules per nucleotide that are intimately associated with the DNA macromolecule. The effects associated with the bulk water surrounding the macromolecule in an aqueous solution appear to constitute a separate problem, although they must be related to those associated with the strongly sorbed water molecules. It appears that no complete hydration shell is needed to stabilize the A or B helical conformations, so the 20 H_2O molecules per nucleotide are sufficient. Falk *et al.* find that the characteristic frequencies of water molecules sorbed by NaDNA are quite close to those for pure liquid water, but the small differences observed indicate stronger hydrogen bonding in the sorbed state than in the pure liquid, with the average strength of the hydrogen bonds decreasing with increasing water being sorbed. Also, no evidence is found for the occurrence of 'ice-like' water, since the observed spectra frequencies are relatively close to those observed for films of pure liquid water but quite different from those of ice. From studying the well-assigned antisymmetric PO_2^- stretching mode, it is concluded that at 65 per cent relative humidity the five to six water molecules absorbed per nucleotide form hydration shells about these outer phosphate ions.[144] Changes in the spectral bands associated with ring-stretching vibrations of purine and pyrimidine rings indicate that the hydration of the bases does not begin until near 65 per cent relative humidity, and at 80 per cent all the hydration sites are filled. The change from the A helix to the B helix as the relative humidity is decreased from 75 to 55 per cent is associated with the loss of four to five water molecules from the bases, with the water attached to the phosphate groups remaining in position. Finally, it can be recalled from Table 4.4 that the hydration activity parameter *b* for DNA has a value greater than unity. This implies that at relative humidities approaching 100 per cent the outermost layers of bound water have physical characteristics indistinguishable from that of normal bulk liquid water.

ELECTRIC AND DIELECTRIC PROPERTIES

The analysis and investigation of the dielectric properties of water and ice has provided unique problems for a long time. Towards the end of the nineteenth century, Kohlrausch and Heydweiler[145] spent many years purifying water to establish that its limiting upper resistivity at 18 °C was 2.6×10^5 ohm m. Today the same result can be established within a few hours using modern electrophoretic ion exclusion purification techniques which give the limiting resistivity value for water as 1.75×10^5 ohm m at 25 °C,[146] with a conductivity activation energy of 9.7 kcal/mole (0.42 eV).[147] The variation of the conductivity of water with pressure (up to 100 kbar) and temperature (up to 1000 °C) has been summarized by Holzapfel.[148] Establishing the limiting resistivity of ice has been more difficult. Resistivity values for ice as high as 10^9 ohm m have been measured by many investigators, but surface conduction effects[149,150] are considered to lead to an underestimation of the true bulk resistivity of ice, which may in fact be extremely high.

Considering the fact that the structures of water and ice have many common features, and that the mobility of protons in ice has been calculated[151] to be some hundred times greater than that for protons in water, then the very high resistivity of ice as compared with water may seem surprising. The conductivity activation energy of 9.7 kcal/mole for pure water can be taken to represent one-half the dissociation energy required to produce mobile H_3O^+ and OH^- ions, which in turn can be seen to be of the order of the energy required to break four hydrogen bonds maintaining the tetrahedral structure shown in Figure 4.1(b). We have already noted that the pH value of 7.0 for water at 24 °C is equivalent to there being 10^{-7} mole of H_3O^+ and OH^- ions in a litre. This corresponds to a concentration of 6.03×10^{19} m^{-3} of H_3O^+ and OH^- ions. The conductivity σ of water can then be calculated using the formula

$$\sigma = e(N_+\mu_+ + N_-\mu_-)$$

where N_+, N_- and μ_+, μ_- are the concentrations and mobilities of the cations and anions respectively. Using the concentration value of 6×10^{19} m^{-3} and the mobility values from Table 4.1 for H_3O^+ and OH^-, then the conductivity for pure water can be calculated to be of the order 5.4×10^{-6} mho/m, in good agreement with the experimentally derived values.[146,148] For the case of pure ice, the electrical and dielectric properties have often been interpreted in terms of Bjerrum defects.[152] Bjerrum envisaged that the rotation of a water molecule in its crystal lattice site could leave a neighbouring oxygen—oxygen link without a proton (L defect) and produce an oxygen—oxygen link having two protons (D defect) with another neighbouring water molecule. The energetics involved in producing such charged defects were considered to be appropriate for the molecular basis describing conduction and polarization in ice (see, for example, reference 153). However, such a model has been criticized[154] and an alternative mechanism is envisaged in which the transfer of a proton occurs by phonon excitation to an oxygen lone-electron-pair site within its own water molecule. For perfect ice, the formation of a pair of ionic defects requires more energy than assumed previously since a strong chemical bond has to be broken, and antipolarization effects hinder the transport of ions through the ice lattice. The high proton mobility values obtained experimentally for ice,[151] and which led to proton quantum mechanical tunnelling models for conduction, are now considered to have been influenced by surface leakage effects. This basic difference in the electrical conductivity of ice and water may have biological implications. We have seen that water associated with biomacromolecular systems has an 'ice-like' structure when compared with normal liquid water, and it may well be that the electrically insulating properties of such structured water helps to maintain the high electric fields of the order 10^7 V m^{-1} existing across cell membranes, by reducing the possibility of electrical breakdown occurring.

An understanding of the dielectric properties of water and ice has also presented some fundamental problems. The dielectric dispersion and loss characteristics for water and ice are shown in Figure 4.14. Water exhibits a simple relaxation spectrum that at first appears to fit in well with the Debye relaxation model described in Chapter 1. The dipole moment value for the water molecule was known from vapour phase measurements to be 1.84 debye units, and the challenge facing the theoreticians was to explain the high static permittivity value ϵ_s of the order 80 at

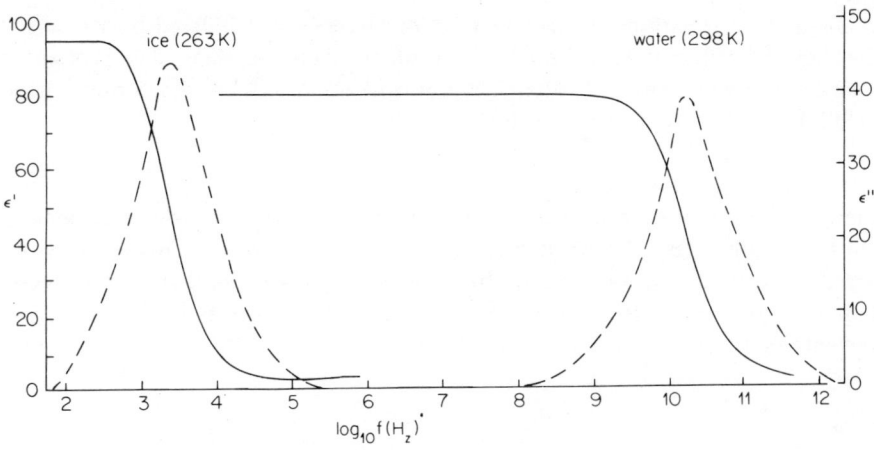

Figure 4.14 The dielectric dispersion and loss spectrum for liquid water and ice

room temperature. The original theory of Debye predicted the onset of a dielectric 'catastrophe' at around 1100 °C, and the improved Onsager theory gave a low value for the permittivity of only 27. When the hydrogen-bonded tetrahedral structure for water was recognized,[155] Kirkwood was able to introduce the empirical correlation factor g to take account of the dipole correlations involving the five water molecules in the tetrahedral structure. Although later refinements were made[156] to take into account bond bending and dipole—dipole interactions involving the next three shells of water molecules, it can be said that no real understanding yet exists of the molecular basis for the observed dielectric properties of water. Whereas the temperature variation of the static permittivity for water follows a simple $1/T$ law,[157] the temperature variation of the dielectric relaxation time[158] indicates that the associated dielectric relaxation activation energy is temperature dependent as shown by the following temperature and activation energy values; 75 °C (3.7 kcal/mole), 25 °C (4.5 kcal/mole) and −30 °C (11 kcal/mole). This may reflect a stiffening of the water structure as the temperature falls. By contrast, the dielectric relaxation activation energy for ice remains constant at 13.8 kcal/mole from −80 °C to 0 °C,[159] and possibly corresponds to a relaxation mechanism involving the breaking of three hydrogen bonds. Measurements on ice single crystals indicates that there is dielectric anisotropy.[160] More recent measurements on ice single crystals[161] indicate the existence of six relaxation components that can be assigned to one intrinsic polarization and to polarizations involving crystal defects and electronic space charges.

So far we have considered the properties of 'pure' water, which is not the case relevant to biological systems. The presence of dissolved salts and other molecules will modify the conductive and dielectric properties observed for pure water. The conductivity of water increases markedly when salts are added to it. For example, water containing 0.1 mole/litre of NaCl has a conductivity of the order 1 mho/m, a value some five decades of order of magnitude greater than for pure water. The conductivities of various electrolyte solutions have been tabulated by Parsons.[162] The dielectric properties of solutions of amino acids, peptides, and proteins has been described in Chapter 3, where according to equation (3.4) the permittivities of the dilute solution (ϵ) and pure solvent (ϵ_1) are given by

$$\epsilon = \epsilon_1 + \delta c$$

where c is the concentration of the solute and δ is the dielectric increment. For electrolyte solutions it has commonly been found (reference 9, Chapter 6) that the permittivity of the solution is *less* than that of pure water, so that δ being negative is referred to as the dielectric decrement. For electrolytic solutions equation (3.4) is conventionally written as

$$\epsilon = \epsilon_1 + 2\bar{\delta} c$$

where

$$\bar{\delta} = \frac{\delta_+ + \delta_-}{2}$$

Values for δ_+ and δ_- for various ions are given in Table 4.12, and apply to concentrations c less than 1 mole/litre. In deriving an estimate of the extent to which, for example, the addition of NaCl will reduce the static permittivity of water, then the decrement values for Na^+ and Cl^- are simply added together to give $\delta = -11$. In this way we can estimate that the static relative permittivity of a 0.5 mole/litre concentration of NaCl will have a value of the order 5.5 lower than that for pure water. This reduction in the permittivity results from the replacing of polar water molecules with non-polar atoms, together with the orienting effect of the local high electric fields around the solvated ions. The values for δ_+ and δ_- in Table 4.12 for various ions reflect differences in the ionic radii and in the number of water molecules that are oriented around the ions. This shell of oriented water molecules will be unable to respond so readily to the influence of externally applied electric fields, and so the effective polarizability of the solution will be reduced. Several theoretical analyses of this effect have been attempted.[163-165]

The original theory of Debye and Hückel[166] predicted that for dilute solutions the relative permittivity of an electrolyte solution would rise above that of the pure water solvent according to the square root of the electrolyte concentration. This was predicted from the polarization properties that would be associated with the

Table 4.12   Dielectric decrement values for some ions in water

Cation	δ_+ (± 1)	Anion	δ_- (± 1)
Na^+	-8	Cl^-	-3
K^+	-8	F^-	-5
Li^+	-11	I^-	-7
H^+	-17	SO^{2-}	-7
Mg^{2+}	-24	OH^-	-13

solvated ions and their surrounding atmosphere of counter-charged ions. Only recently has the observation of such an effect been reported[167] for various chlorides dissolved in water and methanol. For concentrations up to around 2×10^{-2} mole/litre the relative permittivity, as determined at frequencies between 5 and 20 MHz, was observed to be greater than that of the pure solvent. At higher concentrations the permittivity was observed to fall in accordance with all the previous measurements reported in the literature. A new theory was presented to account for the fall in permittivity in terms of the 'kinetic polarization deficiency'. According to this theory the dielectric decrement results from two effects associated with the migration of ions under the influence of the applied electric field. One effect is associated with the induced dipolar rotation of the surrounding solvent molecules as the ion migrates, and the other arises due to the finite time required for the ions to reach their steady terminal velocities. The total reduction $\Delta\epsilon$ in the permittivity due to these effects is calculated to be

$$\Delta\epsilon = \frac{(\epsilon_s - \epsilon_\infty)\tau\sigma_s}{\epsilon_0\epsilon_s}$$

where ϵ_s, ϵ_∞ are the limiting low and high frequency permittivity values of the pure solvent, ϵ_0 is the permittivity of free space, τ is the dielectric relaxation time of the pure solvent and σ_s is the low frequency conductivity of the electrolyte solution. Reasonable agreement with theory was found for the methanol solutions, but not for the aqueous solutions, although this may have resulted from the various simplifying approximations that were made in the theory.

Apart from a change in the permittivity of electrolyte solutions as compared with the pure solvent, a change is also observed for the value of the dielectric relaxation time. For low electrolyte concentrations the relaxation time τ varies linearly with concentration, and usually decreases in value. In this way a relaxation time decrement can be defined in the same manner as the permittivity decrement. It is usual to express the relaxation time decrement in terms of the wavelength λ corresponding to the frequency $(2\pi\tau)^{-1}$, so that

$$\lambda = \lambda_1 + 2\delta\lambda c$$

where

$$\delta\lambda = \frac{\delta\lambda^+ + \delta\lambda^-}{2}$$

Some wavelength decrement values for various ions are given in Table 4.13 and are derived from reference 9, Chapter 6. To a first approximation it can be considered that the relaxation time decrement reflects the degree to which the solvated ion disrupts the structure of normal water. This and related effects have been considered extensively by Hasted (see reference 9, Chapter 6) and in the review by Lestrade et al.[168]

For the case of the 'structured' water associated with biomacromolecules, we might expect that the dielectric dispersion for this structured water will occur in

142

Table 4.13   Relaxation wavelength decrement values for various ions in water

Cation	$\delta\lambda^+ (\pm2) \times 10^{-4}$ m	Anion	$\delta\lambda^- (\pm2) \times 10^{-4}$ m
H^+	+4	OH^-	-2
Li^+	-3	Cl^-	-4
Na^+	-4	F^-	-4
K^+	-4	SO_4^{2-}	-11
Mg^{2+}	-4	I^-	-15

the frequency region between the dispersions for ice and water shown in Figure 4.14. This is found to be the case. Buchanan *et al.*[169] examined the dielectric properties of six protein solutions at microwave frequencies between 3 and 24 GHz. On extrapolating their results to frequencies below that where the relaxation for normal bulk water occurs, it was found that the relative permittivity value was lower than that for normal bulk water. This difference was assumed to be primarily related to the existence of water 'irrotationally bound' to the protein molecules. Depending on the molecular shape assumed for the protein molecules, the amount of such bound water was calculated to be of the order 0.2–0.4 g water per g protein. These studies by themselves would suggest that a dielectric dispersion associated with the protein bound water occurs at a frequency below that for normal bulk water. This dispersion, now commonly referred to as the δ-dispersion,

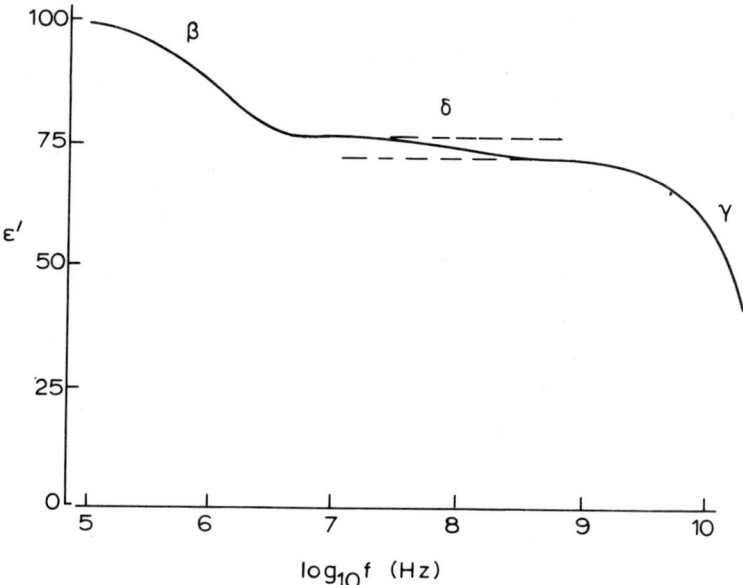

Figure 4.15 The dielectric dispersion spectrum commonly observed for aqueous protein solutions, showing the protein molecule orientational relaxation (β) and the bound (δ) and free (γ) water relaxations

was first observed by Schwan[170] for haemoglobin solutions and by Grant[171,172] for albumin solutions. The δ-dispersion occurs in the frequency region between the protein orientational relaxation (β-dispersion) and the bulk water relaxation (γ-dispersion), and the dielectric dispersion spectrum commonly observed for an aqueous protein solution is shown in Figure 4.15. Although it was assumed by these authors that the δ-dispersion resulted from the relaxation of the water bound to the protein molecules, Schwan[170] also considered the possibility that relaxations of the protein polar side chains might also contribute to the δ-dispersion. Later work[173] indicated that the relaxation of the polar side chains occurred in the frequency region between 10 and 100 MHz, whereas the bound water relaxations occurred between 100 and 1000 MHz, with a dielectric relaxation activation energy of 7.3 kcal/mole at 25 °C. This activation energy corresponds to that found for normal bulk water at −17 °C.[158] The magnitude of the δ-dispersion for haemoglobin solutions is considered to be consistent with relaxations occurring for a bound hydration layer of up to 0.45 g water per g protein.[174]

By investigating protein powders rather than protein solutions, complications associated with orientational polarizations of the protein molecules are avoided and the behaviour of the sorbed water molecules can be investigated more directly. Time domain reflectometry and microwave standing-wave measurements[175] over the frequency range 10 MHz to 25 GHz for lysozyme powders revealed two distinct dispersions centred around 250 MHz and 9.95 GHz. The dispersion at 250 MHz was

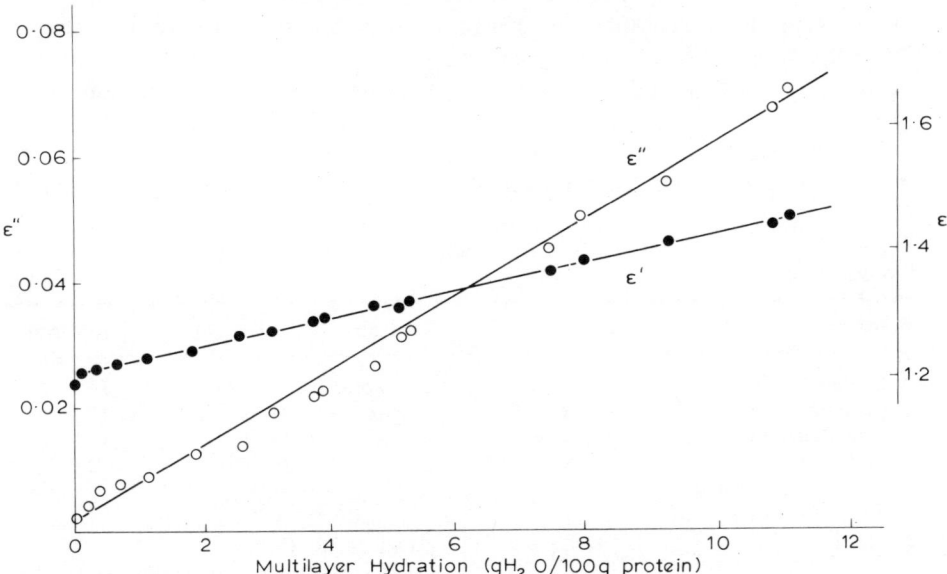

Figure 4.16 Variation of the permittivity ϵ' and dielectric loss ϵ'' at 10 GHz for lysozyme as a function of the water content in the secondary sorption sites. (Reproduced from S. Bone, P. R. C. Gascoyne and R. Pethig, *J.C.S. Faraday I*, **73**, 1605 (1977), by permission of The Chemical Society)

144

considered to be associated with a tightly bound primary monolayer of adsorbed water, and that at 9.95 GHz with a more loosely bound secondary hydration layer. Microwave resonant cavity measurements at 9.95 GHz for bovine serum albumin, cytochrome-c, and lysozyme powders[176] confirmed the earlier conclusions[175,177] that for hydrated proteins the dielectric loss at frequencies around 10 GHz is principally associated with water bound in the secondary sorption sites. From an analysis[113] of the hydration isotherms for the proteins studied at 9.95 GHz, the observed dielectric loss was found to be directly proportional to the water content of the secondary hydration sites, as shown in Figure 4.16 for lysozyme.

Such microwave measurements as these are relevant to the study of possible health hazards from microwave radiation, and perhaps also to cancer research. We have already seen that n.m.r. studies show that cancerous tissues exhibit larger water proton spin-lattice relaxation values than the corresponding normal tissues. Also, normal and virus-transformed tumour cells exhibit mm. microwave absorption differences, a result postulated to relate to differences in the way water molecules adsorb to biomolecules such as DNA in healthy and cancerous cells.[178] Dielectric measurements of cancerous and normal cellular material as a function of hydration should usefully complement such studies.

Apart from these studies of the dielectric relaxations associated with water bound to protein molecules, other dielectric studies have been made of water bound to other molecules and polymers. These studies are summarized in Table 4.14 which shows the relaxation time values derived for water bound to various proteins and other molecules as compared to that for normal bulk water and ice. From Table 4.14 it is seen that the dielectric relaxation times for water bound to biomolecules have values roughly midway between that for normal bulk water and ice at 25 °C, with a tendency for the values to be closer to that for normal liquid water.

Finally, Mascarenhas[182,183] has investigated the electret behaviour of water bound to various proteins, polynucleotides, and polysaccharides, by analysing the

Table 4.14 Dielectric relaxation time values for water bound to various biomolecules

Molecule	τ (s)	Remarks	Reference
Horse haemoglobin	2×10^{-10}	22% solution, 25 °C	173
Lysozyme	5×10^{-10}	powder, 33% H_2O, 25 °C	175
Bovine serum albumin	1.6×10^{-9}	1% solution, 25 °C	171
Egg albumin	5.3×10^{-10}	1% solution, 25 °C	171
Proline, hydroxyproline, ϵ-Amino-n-caproic acid	$\sim 10^{-10}$	2 M solutions, 20 °C	179
Starch	(a) 1×10^{-9} (b) 2×10^{-10}	powder, 25 °C (a) 15% H_2O, (b) 20% H_2O	180
D-Glucose	2×10^{-10}	2 M solution, 5 °C	181
Liquid water	8.2×10^{-12}	25 °C	158
Ice	2.8×10^{-5} 2.4×10^{-6}	-3.8 °C extrapolated to 25 °C	159

thermally stimulated current (t.s.c.) spectrum for such samples. Peaks in the t.s.c. spectrum could definitely be assigned to dipolar relaxation processes associated with bound water, and for gelatin at $30\,^{\circ}C$ the relaxation activation energy value was 9.2 kcal/mole (0.4 eV) and the relaxation time was of the order 10^{-5} s. This relatively simple technique could usefully supplement the dielectric and n.m.r. studies of the physical state of water associated with biomacromolecular systems.

References

1. A. Szent-Györgyi, *The Living State*, Academic Press, New York (1972).
2. T. H. Huxley, *Science Primers: Introductory*, p. 19. Macmillan & Co., London (1880).
3. Conference report: 'Forms of Water in Biologic Systems,' *Ann. N.Y. Acad. Sci.*, **125** (2), (1965).
4. *Water in Biological Systems*, L. P. Kayushin (Ed.). Consultants Bureau, New York (1969).
5. *Water − A Comprehensive Treatise*, F. Franks (Ed.), Volumes I−V, Plenum Press, London (1972−5).
6. B. V. Derjaguin, *Disc. Faraday Soc.*, **42**, 109 (1966).
7. F. J. Donahoe, *Nature*, **224**, 198 (1969).
8. B. V. Derjaguin and N. V. Churaev, *Nature*, **244**, 430 (1973).
9. J. B. Hasted, *Aqueous Dielectrics*, p. 5. Chapman & Hall, London (1973).
10. C. N. R. Rao, in reference 5, Chapter 3, pp. 93−114.
11. J. D. Bernal and R. H. Fowler, *J. Chem. Phys.*, **1**, 515 (1933).
12. J. Morgan and B. E. Warren, *J. Chem. Phys.*, **6**, 666 (1938).
13. D. Eisenberg and W. Kauzmann, *The Structure and Properties of Water*, Oxford Univ. Press, New York (1969).
14. J. L. Kavanau, *Water and Solute−Water Interactions*, Holden-Day, San Francisco (1964).
15. G. Nemethy and H. A. Scherage, *J. Chem. Phys.*, **36**, 3382 (1962).
16. A. Ben-Naim, in reference 5, Chapter 11, pp. 585−624.
17. R. P. Marchi and H. Eyring, *J. Phys. Chem.*, **68**, 221 (1964).
18. T. A. Litovitz and E. H. Carnevale, *J. Appl. Phys.*, **26**, 816 (1955).
19. A. Rahman and F. H. Stillinger, *J. Chem. Phys.*, **55**, 3336 (1971).
20. F. H. Stillinger and A. Rahman, *J. Chem. Phys.*, **57**, 1281 (1972).
21. J. Del Bene and J. A. Pople, *J. Chem. Phys.*, **52**, 4858 (1970).
22. D. Hankins, J. W. Moskowitz, and F. H. Stillinger, *J. Chem. Phys.*, **53**, 4544 (1970).
23. A. Karpfen, J. Ladik, P. Russegger, P. Schuster, and S. Suhai, *Theoret. Chim. Acta (Berl.)*, **34**, 115 (1974).
24. G. M. Bell and D. W. Salt, *J. C. S. Faraday II*, **72**, 76 (1976).
25. C. L. Prosser, in *Comparative Animal Physiology*, p. 79. C. L. Prosser (Ed.), W. B. Saunders Co., Philadelphia (1973).
26. R. M. Spanswick and E. J. Williams, *J. Exp. Botany*, **15**, 193 (1964).
27. R. A. Robinson and R. H. Stokes, *Electrolytic Solutions*, p. 15. Butterworth, London (1959).
28. H. J. C. Berendsen, *Fed. Proc. Am. Soc. Exp. Biol.*, **25**, 971 (1966).
29. J. M. Thorne and H. Slaughter, *Thermochim. Acta*, **3**, 181 (1972).
30. H. T. Edzes and H. J. C. Berendsen, *Ann. Rev. Biophys. Bioeng*, **4**, 265−85 (1975).
31. H. S. Harned and B. B. Owen, *The Physical Chemistry of Electrolytic Solutions*, Reinhold, New York (1958).

146

32. O. Ya. Samoilov, *Structure of Aqueous Electrolyte Solutions* (translator D. J. G. Ives) Consultants Bureau, New York (1965).
33. G. W. Brady and J. T. Krause, *J. Chem. Phys.*, **27**, 304 (1957).
34. G. W. Brady, *J. Chem. Phys.*, **33**, 1079 (1960).
35. R. W. Gurney, *Ionic Processes in Solution*, McGraw-Hill, New York (1953).
36. H. S. Frank and M. W. Evans, *J. Chem. Phys.*, **13**, 507 (1945).
37. H. S. Frank and W. Y. Wen, *Disc. Faraday Soc.*, **24**, 133 (1957).
38. H. L. Friedman and C. V. Krishnan, in reference 5, Vol. 1, p. 50.
39. P. H. von Hippel and T. Schleich, in *Structure and Stability of Biological Macromolecules*, pp. 417–574. S. N. Timasheff and G. D. Fasman (Eds.), Marcel Dekker, New York (1969).
40. R. A. Kohl, J. W. Cary, and S. A. Taylor, *J. Colloid Sci.*, **19**, 699 (1964).
41. E. W. Rushe and W. B. Good, *J. Chem. Phys.*, **45**, 4667 (1966).
42. R. Cini, G. Loglio, and A. Ficalbi, *Nature*, **223**, 1148 (1969).
43. W. Senghaphan, G. O. Zimmerman, and C. E. Chase, *J. Chem. Phys.*, **51**, 2543 (1969).
44. J. Timmerman and H. Bodson, *C. R. Acad. Sci. Paris* **204**, 1804 (1937).
45. B. A. Pethica, W. K. Thompson, and W. T. Pile, *Nature*, **229**, 22 (1971).
46. G. Peschel and K. H. Aldfinger, *Naturwiss*, **54**, 614 (1967); **56**, 558 (1969).
47. C. Salama and D. A. I. Goring, *J. Phys. Chem.*, **70**, 3838 (1966).
48. G. A. Johnson, S. M. A. Lecchini, E. G. Smith, J. Clifford, and B. A. Pethica, *Disc. Faraday Soc.*, **42**, 120 (1966).
49. W. Drost-Hansen, in *Chemistry of the Cell Interface* Part B, H. D. Brown (Ed.) Academic Press, New York (1971).
50. M. Dixon and E. C. Webb, *Enzymes*, Academic Press, London (1960).
51. T. Dalton and R. S. Snart, *Biochim. Biophys. Acta*, **135**, 1059 (1967).
52. W. Good, *Nature*, **214**, 1250 (1967).
53. S. M. Johnson and D. H. Bangham, *Biochim. Biophys. Acta*, **193**, 92 (1969).
54. O. C. Lippold, J. G. Nicholls, and J. W. T. Redfearn, *J. Physiol.*, **153**, 218 (1960).
55. J. Clifford, J. Oakes, and G. J. Tiddy, *Special Disc. Faraday Soc.*, **1**, 175 (1970).
56. F. W. Cope, *Biophys. J.*, **9**, 303 (1969).
57. J. A. Walter and A. B. Hope, *Prog. Biophys. Molec. Biol.*, **23**, 1 (1971).
58. J. A. Walter and A. B. Hope, *Aust. J. Biol. Sci.*, **24**, 497 (1971).
59. J. A. Glasel, *Nature*, **220**, 1124 (1968).
60. L. A. Abetsedarskaya, F. G. Miftakhutdinova, and V. D. Fedotov., *Biophysics*, **13**, 750 (1968).
61. M. M. Civan and M. Shporer, *Biophys. J.*, **12**, 404 (1972).
62. I. M. Klotz, *Arch. Biochem. Biophys.*, **116**, 92 (1966).
63. E. J. Cohn and J. T. Edsall, *Proteins, Amino Acids and Peptides*, Reinhold, New York (1943).
64. M. Dixon and E. C. Webb, *Advan. Protein Chem.*, **16**, 197 (1961).
65. F. Hofmeister, *Arch. Expt. Pathol. Pharmokol.*, **24**, 247 (1888).
66. S. P. Lovell, P. Firth, and J. B. Hasted, *Brit. J. Appl. Phys.*, **15**, 1439 (1964).
67. P. Batho and R. Pethig, Unpublished Work, cited in *Dielectric and Related Molecular Processes*, Vol. 3, The Chemical Society, London (1977).
68 E. Berlin, P. G. Kliman, and M. J. Pallansch, *J. Colloid Interf. Sci.*, **34**, 488 (1970).
69. R. E. Dehl, *J. Chem. Phys.*, **48**, 831 (1968).
70. C. Migchelsen and H. J. C. Berendsen, *J. Chem. Phys.*, **59**, 296 (1973).
71. M. Falk, A. G. Poole, and C. G. Goymour, *Can. J. Chem.*, **48**, 1536 (1970).
72. W. N. Lipscomb, J. A. Hartsuck, F. A. Quiocho, and G. N. Recke, *Proc. Nat. Acad. Sci. USA*, **64**, 28 (1969).

73. D. Hodgkin, in *De la Physique théorique à la Biologie*, *C.N.R.S.*, **1971**, 58–68.
74. R. E. Dickerson and I. Geis, in *The Structure and Action of Proteins*, p. 94. Harper & Row, New York (1969).
75. W. N. Lipscomb, J. A. Hartsuck, G. Recke, F. A. Quiocho, R. N. Bethge, M. L. Ludwig, T. A. Steitz, H. Muirhead, and J. C. Coppola, *Brookhaven Symp. Biol.*, **21**, 24 (1968).
76. A. Quiocho and W. N. Lipscomb, *Advan. Protein Chem.*, **25**, 1 (1971).
77. G. Navon, R. I. Shulman, B. J. Wylvda, and T. Yamane, *J. Molec. Biol.*, **51**, 15 (1970).
78. A. Liljas, K. K. Kannan, P. C. Bergsten, I. Waara, K. Fridborg, B. Strandberg, U. Carlbom, L. Järup, S. Lövgren, and M. Petef, *Nature, New Biol.*, **235**, 131 (1972).
79. H. W. Wyckoff, D. Tsernoglou, A. W. Hanson, J. R. Knox, B. Lee, and F. M. Richards, *J. Biol. Chem.*, **245**, 305 (1970).
80. G. N. Ramachandran and R. Chandrasekharan, *Biopolymers*, **6**, 1649 (1968).
81. I. A. Neiduszynski and R. H. Marchessault, *Biopolymers*, **11**, 1335 (1972).
82. J. J. Birktoft and D. M. Blow, *J. Molec. Biol.*, **68**, 187 (1972).
83. F. Friedberg and S. Bose, *Biochemistry*, **8**, 2564 (1969).
84. H. J. C. Berendsen in *Biology of the Mouth*, p. 145. P. Person (Ed.), American Association for the Advancement of Science, Washington (1968).
85. K. D. Watenpaugh, L. C. Sieker, J. R. Herriott, and L. H. Jensen, *Cold Spring Harbour Symp. Quantum Biology*, **36**, 359 (1972).
86. J. J. Birktoft, Private Communication in reference 88.
87. B. P. Schoenborn, *Cold Spring Harbour Symp. Quantum Biology*, **36**, 569 (1972).
88. R. Cooke and I. D. Kuntz, *Ann. Rev. Biophys. Bioeng.*, **3**, 95–126 (1974).
89. D. N. Glew, H. D. Mak, and N. S. Rath, in *Hydrogen-Bonded Solvent Systems*, pp. 195–210. Eds. A. K. Covington and P. Jones, Taylor & Francis, London (1968).
90. H. G. Hertz and C. Rädle, *Ber. Bunsenges. Phys. Chem.*, **77**, 521 (1973).
91. D. T. Warner, *Ann. N.Y. Acad. Sci.*, **125**, 605 (1965).
92. R. Damadian, *Science*, **171**, 1151 (1971).
93. C. F. Hazlewood, D. C. Chang, D. Medina, G. Cleveland, and B. L. Nichols, *Proc. Nat. Acad. Sci. USA*, **69**, 1478 (1972).
94. I. D. Weisman, L. H. Bennett, and L. R. Maxwell, *Science*, **178**, 1288 (1972).
95. D. C. Chang, C. F. Hazlewood, and D. E. Woessner, *Biochim. Biophys. Acta*, **437**, 253 (1976).
96. R. A. Floyd, T. Yoshida, and J. S. Leigh, *Proc. Nat. Acad. Sci. USA*, **72**, 56 (1975).
97. R. Damadian and F. W. Cope, *Physiol. Chem. Phys.*, **6**, 309 (1974).
98. S. Tokuoka and H. Morioka, *J. Cancer Res.*, **48**, 353 (1957).
99. C. D. Cone, *Trans. N.Y. Acad. Sci.*, **31**, 404 (1969).
100. G. N. Ling, *A Physical Theory of the Living State*, Blaisdell, New York (1962).
101. J. J. Katz, 'Chemical and Biological Studies with Deuterium', 39th Annual Priestly Lecture, Pennsylvania State University (1965).
102. A. D. McLaren and J. W. Rowen, *J. Polym. Sci.*, **7**, 289 (1951).
103. I. D. Kuntz and W. Kauzmann, *Advan. Protein Chem.*, **28**, 239 (1974).
104. W. Gibbs, *Collected Works*, Vol. 1, p. 230, Yale Univ. Press (1948).
105. J. J. Thomson, *Application of Dynamics to Physics and Chemistry*, p. 190. Macmillan Co., New York (1888).
106. S. R. Milner, *Phil. Mag.*, **13**, 96 (1907).

107. N. B. Weiser, *Colloid Chemistry*, pp. 18−19. Wiley, New York (1949).
108. E. S. Amis, *Kinetics of Chemical Change in Solution*, pp. 266−272. Macmillan, New York (1949).
109. I. Langmuir, *J. Am. Chem. Soc.*, **38**, 2221 (1916); **39**, 1848 (1917); **40**, 1361 (1918).
110. J. J. Hermans in *Colloid Science*, p. 515. Ed. H. R. Kruyt, Elsevier, New York (1949).
111. A. W. Adamson, *Physical Chemistry of Surfaces*, Wiley, New York (1967).
112. S. Brunauer, P. H. Emmett, and E. Teller, *J. Am. Chem. Soc.*, **60**, 309 (1938).
113. P. R. C. Gascoyne and R. Pethig, *J. C. S. Faraday I*, **73**, 171 (1977).
114. S. J. Gregg, *J. Chem. Soc.*, **1946**, 561.
115. M. G. Kennerley, *Polymer*, **10**, 833 (1969).
116. A. Clark, *The Theory of Adsorption and Catalysis*, Academic Press, New York (1970).
117. G. N. Ling, in *Water and Aqueous Solutions*, Ed. R. A. Horne, Wiley, New York (1972).
118. S. Brunauer, L. S. Deming, W. E. Deming, and E. Teller, *J. Am. Chem. Soc.*, **62**, 1723 (1940).
119. S. Brunauer, *The Adsorption of Gases and Vapors*, p. 5. Princeton University Press (1943).
120. S. J. Gregg, *Surface Chemistry of Solids*, pp. 94−119. Reinhold, New York (1951).
121. B. Katchman, J. Cutler, and A. D. McLaren, *Nature*, **166**, 266 (1950).
122. H. B. Bull, J. Am. Chem. Soc., **66**, 1499 (1944).
123. H. B. Bull and K. Breese, *Arch. Biochem. Biophys.*, **28**, 488 (1968).
124. J. D. Bernel and I. Fankuchen, *J. Gen. Physiol.*, **25**, 111 (1941).
125. D. Crowfoot and D. P. Riley, *Nature*, **144**, 1011 (1939).
126. J. Boyes-Watson, E. Davidson, and M. F. Perutz, *Proc. Roy. Soc. (London) A*, **191**, 83 (1947).
127. M. F. Perutz, *Research*, **2**, 52 (1949).
128. L. Bragg, E. R. Howells, and M. F. Perutz, *Proc. Roy. Soc. (London) A*, **222**, 33 (1954).
129. L. Bragg and M. F. Perutz, *Proc. Roy. Soc. (London) A*, **225**, 4315 (1954).
130. E. F. Mellon, A. H. Korn, and S. R. Hoover, *J. Am. Chem. Soc.*, **70**, 3040 (1948).
131. M. M. Breuer and M. G. Kennerley, *J. Colloid Interf. Sci.*, **37**, 124 (1971).
132. W. O. Baker and C. S. Fuller, *Ann. N.Y. Acad. Sci.*, **44**, 329 (1943).
133. R. W. Green, *Trans. Roy. Soc. N.Z.*, **77**, 313 (1948).
134. R. A. Robinson and D. A. Sinclair, *J. Am. Chem. Soc.*, **56**, 1830 (1934).
135. R. A. Robinson, *J. Chem. Soc.*, **1948**, 1083.
136. L. Pauling, *J. Am. Chem. Soc.*, **67**, 555 (1945).
137. M. H. Cardew and D. D. Eley, in '*Fundamental Aspects of the Dehydration of Foodstuffs*', Soc. Chem. Industry, 24−29 (1958).
138. H. F. Fisher, *Proc. Nat. Acad. Sci. USA*, **51**, 1285 (1964).
139. H. F. Fisher, *Biochim. Biophys. Acta*, **109**, 544 (1965).
140. I. D. Kuntz, *J. Am. Chem. Soc.*, **93**, 514 (1971).
141. J. D. Watson and F. H. C. Crick, *Nature*, **171**, 737 (1953).
142. R. E. Franklin and R. G. Gosling, *Nature*, **172**, 156 (1953).
143. R. Langridge, H. R. Wilson, C. W. Hooper, M. H. F. Wilkins, and L. D. Hamilton, *J. Molec. Biol.*, **2**, 19 (1960).
144. M. Falk, K. A. Hartman, and R. C. Lord, *J. Am. Chem. Soc.*, **84**, 3843 (1962); **85**, 387, 391 (1963).
145. F. Kohlrausch and A. Heydweiler, *Wied. Ann.*, **53**, 209 (1894).
146. W. Haller and H. C. Duecker, *J. Res. Nat. Bur. Standards*, **64A**, 527 (1970).
147. H. C. Duecker and W. Haller, *J. Res. Nat. Bur. Standards*, **66**, 225 (1962).

148. W. B. Holzapfel, *J. Chem. Phys.*, **50**, 4424 (1969).
149. *The Physics of Ice*, Eds. N. Riehl, B. Bullemer, and H. Engelhardt, Plenum Press, New York (1969).
150. M. A. Maidique, A. R. von Hippel, and W. B. Westphal, *J. Chem. Phys.*, **54**, 150 (1970).
151. M. Eigen, L. De Maeyer, and H. C. Spatz, *Ber. Bunsenges. Phys. Chem.*, **68**, 19 (1964).
152. N. Bjerrum, *Dansk. Mat. Fyz. Medd.*, **27**, 32, 41 (1951).
153. C. Jaccard, *Helv. Phys. Acta*, **32**, 89 (1959); *Ann. Nat. Acad. Sci.*, **125**, 391 (1965).
154. A. R. von Hippel, *J. Chem. Phys.*, **54**, 145 (1971).
155. J. D. Bernal and R. H. Fowler, *J. Chem. Phys.*, **1**, 515 (1933).
156. J. Lennard-Jones and J. A. Pople, *Proc. Roy. Soc. (London)*, **A205**, 163 (1951).
157. C. G. Malmberg and A. A. Maryott, *J. Res. Nat. Bur. Standards*, **56**, 1 (1956).
158. C. H. Collie, J. B. Hasted, and D. M. Ritson, *Proc. Roy. Soc. (London)*, **60**, 145 (1948).
159. R. P. Auty and R. H. Cole, *J. Chem. Phys.*, **20**, 1303 (1952).
160. F. Humble, F. Jona, and P. Scherrer, *Helv. Phys. Acta*, **26**, 17 (1953).
161. A. R. von Hippel, R. Mykolajewycz, A. H. Runck, and W. B. Westphal, *J. Chem. Phys.*, **57**, 2560 (1972).
162. R. Parsons, *Handbook of Electrochemical Constants*, Butterworths, London (1954).
163. H. Sack, *Phys. Z.*, **28**, 199 (1927).
164. D. M. Ritson and J. B. Hasted, *J. Chem. Phys.*, **16**, 11 (1948).
165. E. Glueckauf, *Trans. Faraday Soc.*, **60**, 1637 (1964).
166. P. Debye and E. Hückel, *Phys. Z.*, **24**, 305 (1923).
167. J. B. Hubbard, L. Onsager, W. M. van Beek, and M. Mandel, *Proc. Nat. Acad. Sci. USA*, **74**, 401 (1977).
168. J. C. Lestrade, J. P. Badiali, and H. Cachet, in *Dielectric and Related Molecular Processes*, Vol. 2, pp. 106–50, The Chemical Society, London (1975).
169. T. J. Buchanan, G. H. Haggis, J. B. Hasted, and B. G. Robinson, *Proc. Roy. Soc. (London)* **A213**, 379 (1952).
170. H. P. Schwan, *Ann. N.Y. Acad. Sci.*, **125**, Art. 2, 344 (1965).
171. E. H. Grant, *Ann. N.Y. Acad. Sci.*, **125**, Art. 2, 418 (1965).
172. E. H. Grant, *J. Molec. Biol.*, **19**, 133 (1966).
173. B. E. Pennock and H. P. Schwan, *J. Phys. Chem.*, **73**, 2600 (1969).
174. E. H. Grant, G. P. South, S. Takashima, and H. Ichimura, *Biochem. J.*, **122**, 691 (1971).
175. S. C. Harvey and P. Hoekstra, *J. Phys. Chem.*, **76**, 2987 (1972).
176. S. Bone, P. R. C. Gascoyne, and R. Pethig, *J. C. S. Faraday I*, **73**, 1605 (1977).
177. M. Kent, *J. Phys. (D)*, **5**, 394 (1972).
178. S. J. Webb and A. D. Booth, *Science*, **174**, 72 (1971).
179. J. C. W. Shepherd and E. H. Grant, *Proc. Roy. Soc. (London)*, **A307**, 345 (1968).
180. P. Abadie, R. Charbonniere, A. Gidel, P. Girard, and A. Guilbot, *J. Chim. Phys.*, **50**, C46 (1953).
181. M. J. Tait, A. Suggett, F. Franks, P. A. Quickenden, and S. Ablett, *J. Solution Chem.*, **1**, 131 (1972).
182. S. Mascarenhas, *J. Electrostatics*, **1**, 141 (1975).
183. S. Mascarenhas, 'Electrical Polarization Storage and Water Structuring in Biological Molecules', Paper presented at the Int. Symp. on Quantum Biology and Quantum Pharmacology, Sanibel Island, January 1976.

Chapter 5

Interfacial Dielectric Phenomena in Biological Systems

By their very nature biological materials are not homogeneous. At its simplest and most naive level, a typical biological material can be represented as a water-filled proteinous matrix. The overall dielectric and electronic properties of such a mixture will be governed by the physical nature and interactions of the separate components.

A simple example of a heterogeneous biological system is that of blood corpuscles suspended in blood serum. It has long been appreciated that the electrical conductivity of blood serum is much greater than that of the suspended blood cells, and that measurement of the conductivity of their mixture should be capable of providing a basis for the determination of the volume concentration of the blood cells. The first approach of this nature appears to be that of Bugarzky and Tangl[1], in 1897, who derived an empirical equation, for finding the percentage volume 'p' of serum in dog's blood, of the form

$$p = 92\sigma_b/\sigma_s + 13$$

where σ_b and σ_s are the conductivities of the red blood cells and of the blood serum, respectively. A little later, Stewart[2] reported more extensive measurements for dog's blood, and derived the empirical equation

$$p = \frac{\sigma_b}{\sigma_s} (180 - \sigma_b - \sqrt{\sigma_b})$$

Also, by this time Maxwell[3] and Lord Rayleigh[4] had considered the theoretical aspects of heterogeneous dielectrics, their work being extended later by Wagner.[5,6] It transpires that, in general, a heterogeneous medium exhibits frequency-dependent dielectric and conductive properties that differ from those of the constituent components. Such dielectric dispersions have become to be known as Maxwell–Wagner or interfacial polarizations.

The simplest of heterogeneous systems, and the earliest to have been analysed, is that of the two-layer capacitive system shown in Figure 5.1. The system consists of

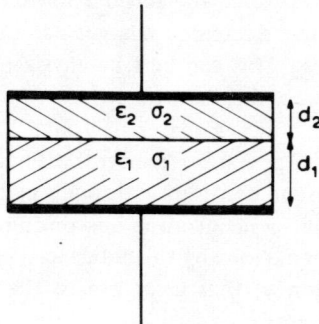

Figure 5.1 The two-layer
heterogeneous system

two parallel slabs of differing dielectric material of thicknesses d_1 and d_2 placed
together between two electrodes. What will be the dielectric behaviour of such a
two-layer assembly? We can simplify matters by allowing σ_2 to be negligibly small,
and assuming that ϵ_1, ϵ_2, and σ_1 are frequency independent. The two-layer model
of Figure 5.1 effectively represents two capacitors connected in series, with the two
complex capacitances given by

$$C_1 = A\epsilon_0(\epsilon_1 - j\sigma_1/\omega\epsilon_0)/d_1 \tag{5.1}$$

and

$$C_2 = A\epsilon_0\epsilon_2/d_2 \tag{5.2}$$

where A is the cross-sectional area of the two slabs. The total complex capacitance
is given by

$$\frac{1}{C} = \frac{1}{C_1} + \frac{1}{C_2}$$

which from equations (5.1) and (5.2) gives

$$C = \frac{A\epsilon_0\epsilon_2(\epsilon_1 - j\sigma_1/\omega\epsilon_0)}{d_2(\epsilon_1 - j\sigma_1/\omega\epsilon_0) + d_1\epsilon_2} \tag{5.3}$$

This shows that the effective permittivity of the two-layer system is a function of
frequency. At low frequencies, as $\omega \to 0$, then the limiting low frequency permittivity is given by

$$\epsilon_s = \epsilon_2 d/d_2$$

where d is the total thickness $d_1 + d_2$ as shown in Figure 5.1. Also, at high
frequencies, as $\omega \to \infty$, we have

$$\epsilon_\infty = \frac{\epsilon_1\epsilon_2 d}{d_2\epsilon_1 + d_1\epsilon_2}$$

From this we see that $\epsilon_s > \epsilon_\infty$, showing that the two-layer system exhibits a
dielectric dispersion.

The underlying physical mechanism for such dielectric behaviour is associated with the non-uniform distribution of free electronic charges across the interface between the dissimilar dielectric materials. This can best be envisaged by considering the situation under d.c. equilibrium conditions where there must be electrical current continuity across the boundary between materials 1 and 2 of Figure 5.1. Each material will have its own free charge carrier concentration and associated charge carrier mobility. To achieve current continuity through materials 1 and 2 there will have to be a charge carrier concentration discontinuity across the interface. It is essentially the dynamic behaviour of this interfacial charge build-up or polarization, as a function of frequency, that gives rise to the dielectric dispersion exhibited by inhomogeneous systems.

Full analyses of the two-layer model have been given by several workers.[3, 5-9] In particular, van Beek[10] has shown that the dielectric dispersions of two-layer systems can be described in terms of the Debye equations, described in Chapter 1, as

$$\epsilon' = \epsilon_\infty + \frac{\epsilon_s - \epsilon_\infty}{1 + \omega^2 \tau^2}$$

and

$$\epsilon'' = \frac{(\epsilon_s - \epsilon_\infty)\omega\tau}{1 + \omega^2 \tau^2}$$

with

$$\epsilon_s = \frac{d(\epsilon_1 d_1 \sigma_2^2 + \epsilon_2 d_2 \sigma_1^2)}{(\sigma_1 d_2 + \sigma_2 d_1)^2}$$

$$\epsilon_\infty = \frac{d\epsilon_1 \epsilon_2}{(\epsilon_1 d_2 + \epsilon_2 d_1)}$$

and

$$\tau = \frac{\epsilon_0(\epsilon_1 d_1 + \epsilon_2 d_2)}{\sigma_1 d_2 + \sigma_2 d_1}$$

The conductivity of the two-layer system as a whole is given by

$$\sigma = \frac{d\sigma_1 \sigma_2}{(\sigma_1 d_2 + \sigma_2 d_1)}$$

so that the dielectric loss factor contains an extra term owing to this conductivity $(\sigma(\omega) = \sigma + \omega\epsilon_0\epsilon'')$, and we have

$$\epsilon'' = \frac{\sigma}{\epsilon_0 \omega} + \frac{(\epsilon_s - \epsilon_\infty)\omega\tau}{1 + \omega^2 \tau^2}$$

A system composed of a number of layers of the same material (ϵ_1, σ_1) alternating with layers of another material (ϵ_2, σ_2) will have the same dielectric characteristics as the two-layer system, whereas a many-layer system of a multitude of materials will have the limiting permittivities in the Debye equations given according to Volger[9] by

$$\epsilon_s = d(\Sigma_i d_i \epsilon_i / \sigma_i^2)/(\Sigma_i d_i / \sigma_i)^2$$

and

$$\epsilon_\infty = d/(\Sigma_i d_i / \epsilon_i)$$

For the two-layer system, if the unlikely condition

$$\epsilon_1 \sigma_2 = \epsilon_2 \sigma_1$$

is satisfied, then $\epsilon_s - \epsilon_\infty = 0$, and no Maxwell–Wagner dispersion effect will be observed.

Some numerical examples are given in Table 5.1 for two-layer systems with characteristics approximating to those which may be considered relevant to hydrated biological samples. In this table, the value for ϵ_m'' is derived from the equation

$$\epsilon_m'' = \tfrac{1}{2}(\epsilon_s - \epsilon_\infty)$$

and the frequency f_m is calculated from the expression

$$f_m = 1/2 \pi \tau$$

Medium 1 (σ_1, ϵ_1) can be taken to represent a typical protein matrix, and medium 2 represents the adsorbed water. The first three rows of Table 5.1 correspond to normal 'free' liquid water of hydration, and the next two rows can be taken to represent the situation when the adsorbed water is strongly bound to the protein. As the table shows, the presence of 'free' water does not produce large dielectric losses in a two-layer system. Losses associated with bound water, which we take to have electronic and dielectric properties similar to that of ice, are also seen to be low. The system corresponding to the last row of Table 5.1 can be considered to represent water strongly bound to a protein matrix, with an effective conductivity for this bound water being much less than that corresponding to defect and ionic conduction in ice. It is seen that a large Maxwell–Wagner dispersion can exist at

Table 5.1   Values of the maximum loss factor ϵ_m'' and the frequency of maximum loss f_m calculated for the two-layer heterogeneous system of Figure 5.1, with $d_1 = 1$ mm and $\epsilon_1 = 2.7$

d_2 (m.m)	ϵ_2	σ_2 (mho/m)	σ_1 (mho/m)	ϵ_m''	f_m (Hz)
10^{-3}	80	10^{-4}	10^{-10}	5×10^{-5}	6.5×10^5
10^{-1}	80	10^{-4}	10^{-10}	5×10^{-3}	1.7×10^5
10^{-1}	79	1	10^{-10}	5.1×10^{-3}	1.7×10^9
10^{-3}	95	10^{-9}	10^{-10}	4×10^{-5}	6.6
10^{-1}	95	10^{-9}	10^{-10}	2.6×10^{-2}	1.5
10^{-3}	95	10^{-12}	10^{-10}	390	7.1×10^{-3}

154

Figure 5.2 A porous
heterogeneous system

sub-hertz frequencies for such a system. In practice this large dielectric loss at 7×10^{-3} Hz would be masked by the larger conductivity loss factor $\sigma/\omega\epsilon_0$. In general, large Maxwell–Wagner or interfacial dispersions occur for layered systems when one of the layers is very thin and of high resistivity compared with the other component.

A heterogeneous model which can be considered relevant to many real biological situations is that of a porous dielectric, which in its simplest form would be as shown in Figure 5.2, with water-filled pores (ϵ_2, σ_2) extending right through the supporting medium (ϵ_1, σ_1). For such a heterogeneous system, with $\sigma_2 \gg \sigma_1$, then van Beek[10] has shown that the dielectric dispersion is characterized by the dielectric parameters

$$\epsilon' = (1 - v)\epsilon_1 + v\epsilon_2$$

and

$$\epsilon'' = \sigma_2/\epsilon_0\omega$$

where v is the volume fraction of the water-filled pores in the total system. The more natural system would correspond to a model involving long pores that do not extend completely through the surrounding medium. These pores would then approximate to long prolate spheroids, with their major axes oriented parallel to the electric field direction. Van Beek[11] has provided experimental results for such a

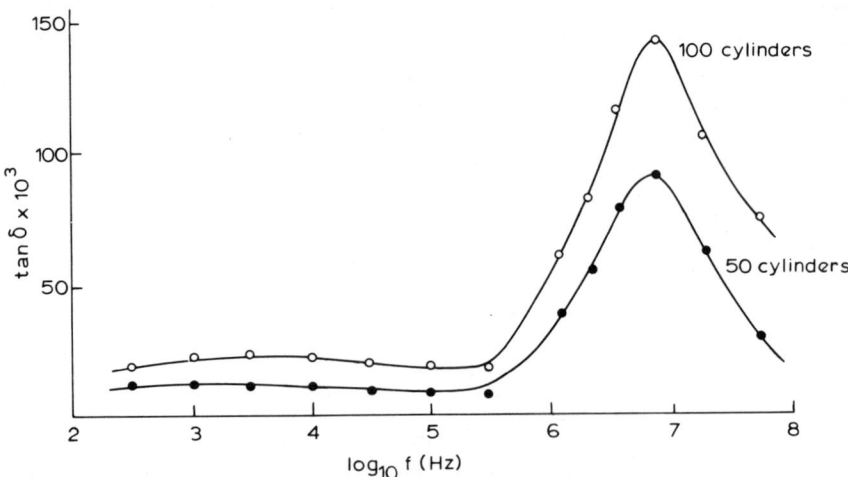

Figure 5.3 The dielectric dispersions obtained for water-filled cylinders embedded in paraffin wax. (Reproduced from L. K. H. van Beek, *Proc. Colloq. Ampere*, **11**, 229 (1962), by permission of the North-Holland Publishing Company)

model, where polyethylene cylinders, filled with water, were embedded in paraffin wax. The ratio of the major to minor axis for these water cylinders was 7:1, and the cylinders were orientated parallel to the electric field. The dielectric dispersions obtained are shown in Figure 5.3, where it can be seen that a significant dielectric loss peak occurs at around 10 MHz, whose magnitude is dependent on the volume concentration of water-filled cylinders. The broad loss peak occurring at lower frequencies was considered by van Beek to result from interactions between adjacent water cylinders.

The feature that should be stressed is that the dielectric loss peak shown in Figure 5.3 occurs at a frequency some four decades below the loss peak for bulk liquid water. The dielectric dispersion observed by van Beek was associated with the heterogenous nature of the water–paraffin system, and cannot be ascribed to either the water or paraffin components on their own.

In reality, the heterogeneous systems of biology will correspond to models lying somewhere between the two extreme cases depicted by Figures 5.1 and 5.2, where as a first approximation we can consider that one of the major components is dispersed randomly in the form of particles within a continuous supporting medium. The dielectric and electronic properties of such a mixture will be related to the average electric field E_i existing within the dispersed particles, and this field will differ from the field E in the surrounding medium.

It will be of value at this stage to review briefly the concepts and laws of electrostatics applicable to this situation. If we have two large parallel metal plates of area A held a small distance apart in vacuum, and if the plates carry total charges of $+q_a A$ and $-q_a A$, respectively, then the homogeneous electric field strength between them is given by

$$E_0 = q_a/\epsilon_0$$

where ϵ_0 is the permittivity of free space and q_a is the charge per unit area (surface charge density). If the space between the plates is now filled with a homogeneous dielectric, and the charge on the plates remains the same, then experimental experience shows us that the potential difference between the plates decreases. As a result, the field intensity must reduce, with the new value being given by

$$E = q_a/\epsilon_0 \epsilon$$

where ϵ is the relative permittivity of the dielectric medium. We can visualize this effect in terms of the appearance of induced charges (bound charges) on the dielectric surfaces, that partially neutralizes those charges (free charges) on the metal plates. Since the neutralized surface charges originally contributed to the potential difference opposing the voltage originally applied to the plates, then more charge can flow into this parallel plate capacitor arrangement to increase the effective charge stored. We can say that the decrease in electric field strength is due to the appearance of an induced bound charge polarization P of magnitude

$$P = q_a \left[1 - \left(\frac{1}{\epsilon} \right) \right]$$

In electrostatics account of both free and bound charge densities is taken by defining the total electric displacement D by

$$D = \epsilon_0 E_0 + P$$

For the dielectric medium we have

$$D = \epsilon\epsilon_0 E_0$$

so that

$$P = \epsilon_0(\epsilon - 1)E_0 = \chi\epsilon_0 E_0$$

where $\chi = (\epsilon - 1)$ is called the dielectric susceptibility of the medium and gives the effective ratio of bound to free charge density.

For the case of dispersed particles in the medium of the dielectric, we will not have the situation described above where the induced polarization in the dielectric medium is neutralized by charges bound at the electrode surfaces. There will be no electrodes associated with the dispersed particles and closure fields will exist at the particle ends. The macroscopic field inside the medium will be identical with the electric field determined by measurement of the potential difference between the electrodes, but the field E_i inside the particles will differ from this. To calculate the magnitude of E_i we require to know two basic laws of electrostatics concerned with the boundary between two dissimilar dielectrics. These two laws are:

(1) The normal component of the electric displacement D is continuous across the boundary; and

(2) The component of the electric field tangential to the boundary is continuous across the boundary.

We can apply these two laws to the simple situations shown in Figures 5.4(a) and (b). Figure 5.4(a) shows the dispersed particle in the form of a large flat plate having its major plane perpendicular to the field in the surrounding medium. Applying the first of the two laws above, we have for the normal components of the electric displacements

$$D_{\text{medium}} = D_{\text{particle}}$$

i.e.

$$\epsilon_0\epsilon_1 E = \epsilon_0\epsilon_2 E_i$$

giving

$$E_i = \frac{\epsilon_1}{\epsilon_2}E$$

That is, the field inside the particle is reduced by the factor ϵ_1/ϵ_2 relative to the original field E outside. We have neglected the small effect of field distortions occurring at the plate edges. The situation of Figure 5.4(b) corresponds to the particle being a long thin dielectric rod with its major axis parallel to the external field. Applying the second law given above, then in the central region of the rod

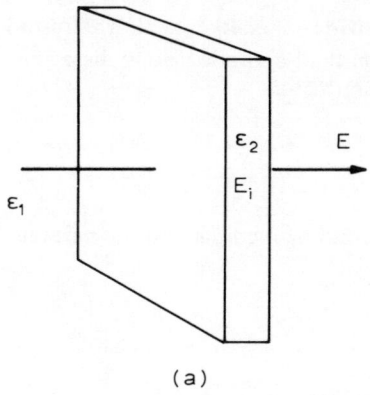

(a)

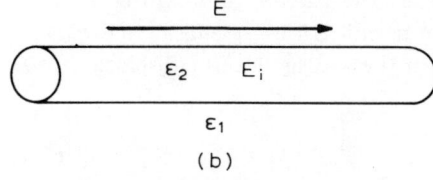

(b)

Figure 5.4 (a) Large flat plate with its major faces perpendicular to the external electric field. (b) Long thin rod with its major axis parallel to external electric field

away from the perturbing influence of the polarized rod ends, we have for the tangential components of the fields

$$E_i = E$$

In other words, the field inside the rod particle is the same as that in the surrounding medium.

In general, the field inside any particle will be of the form

$$E_i = E - E_{dep}$$

where E_{dep} is the depolarizing field resulting from the polarization charges on the particle surface, and is given by

$$E_{dep} = AP/\epsilon_0$$

where A is known as the depolarization factor. We can see for our flat plate particle that $A = 1$, whilst for the long rod parallel to the field, $A = 0$. These two particle geometries and orientations give the extreme values for A, since the values of A for all other particle geometries will lie between 0 and 1. Apart from the cases already given, a few other simple examples can easily be remembered. For a sphere $A = \frac{1}{3}$, for a very short rod with E parallel to its length $A = 1$, and for the case of a long rod with E perpendicular to its length $A = \frac{1}{2}$. Derivations of the induced field inside spheroids positioned in a homogeneous field are given in the many standard texts on electrical fields, as for example in J. A. Stratton's *Electromagnetic Theory*.[12] The concept of a depolarization factor is also used in magnetization studies, and Osborn[13] has provided charts and tables of the depolarization factor A for prolate and oblate spheroidal particles. Sillars[14] also provides similar data. For ellipsoid particles of axes a, b, and c, with a uniform external field E applied along the a-axis, then the uniform field E_a within the ellipsoid is given[15] by

$$E_a = \frac{\epsilon_1 E}{[\epsilon_1 + A(\epsilon_2 - \epsilon_1)]}$$

where ϵ_1 and ϵ_2 are the permittivities of the surrounding medium and the particle, respectively.

THE PERMITTIVITY OF MIXTURES

The derivation of a general formula for the permittivity of mixtures is quite straightforward. If the two components have permittivities ϵ_1 and ϵ_2 and occupy volume fractions v_1 and v_2 $(v_1 + v_2 = 1)$, then the average electric displacement $\bar{D}$ is given in terms of the components as[16]

$$\bar{D} = v_1 \bar{D}_1 + v_2 \bar{D}_2 \tag{5.4}$$

and the average electric field is given by

$$\bar{E} = v_1 \bar{E}_1 + v_2 \bar{E}_2 \tag{5.5}$$

Assuming that the permittivity of the mixture is given by

$$\epsilon_m = \bar{D}/\bar{E}$$

and that for each component we have

$$\bar{D}_1 = \epsilon_1 \bar{E}_1 \quad \text{and} \quad \bar{D}_2 = \epsilon_2 \bar{E}_2$$

then from equations (5.4) and (5.5) we have

$$\epsilon_m = \epsilon_1 v_1 f_1 + \epsilon_2 v_2 f_2$$

where $v_1 f_1 + v_2 f_2 = 1$, $f_1 = \bar{E}_1/\bar{E}$ and $f_2 = \bar{E}_2/\bar{E}$. This gives the two general formulae

$$\epsilon_m = \epsilon_1 + (\epsilon_2 - \epsilon_1) v_2 f_2 \tag{5.6}$$

or

$$(\epsilon_m - \epsilon_1) v_1 f_1 + (\epsilon_m - \epsilon_2) v_2 f_2 = 0 \tag{5.7}$$

The problem is basically one of finding appropriate values for f_1 and f_2, and to take account of possible interactions between particles. The two equations (5.6) and (5.7) are theoretically identical, but in practice this is not so owing to the approximations required for the factors f_1 and f_2. It would seem appropriate[16] to use equation (5.6) for the case of small particles dispersed in a continuous medium, and equation (5.7) for the case where the two component volume fractions v_1 and v_2 are nearly the same.

Following Reynolds and Hough,[16] then for the case of small spheroidal particles

$$f_2 = \sum_{i=a}^{c} \frac{\cos^2 \alpha_i}{(1 + A_i[(\epsilon_2/\epsilon_1) - 1])} \tag{5.8}$$

where α_i are the angles made by the ellipsoid axes a, b, c and the applied field E. The depolarization factors A_i depend on the axial ratios of the ellipsoid, and are related through the relationship

$$A_a + A_b + A_c = 1$$

For an ellipsoid of revolution about the a-axis, then

$$A_b = A_c = B, \quad \text{and} \quad A_a = 1 - 2B$$

A plot of A_a as a function of the axial ratio a/b is given in Figure 5.5, and is based on the values given for the factor B above by Reynolds and Hough.[16] For a random orientation of the spheroids

$$\cos^2 \alpha_a = \cos^2 \alpha_b = \cos^2 \alpha_c = \tfrac{1}{3}$$

and for the special case of long particles aligned with their long axes parallel but otherwise random, then

$$\cos^2 \alpha_a = \cos^2 \alpha_b = \tfrac{1}{2}; \quad \cos^2 \alpha_c = 0$$

Two limiting cases of the ellipsoids are infinitely long thin rods and infinitely thin circular discs, which provide good approximations for the rod- and plate-like particles often met with in practice. From Figure 5.5 we can see that if the field is directed along the a-axis, then the depolarization factor A_a for a rod and thin disc is of value zero and 1.0 respectively, in agreement with the values derived using Figures 5.4(a) and (b).

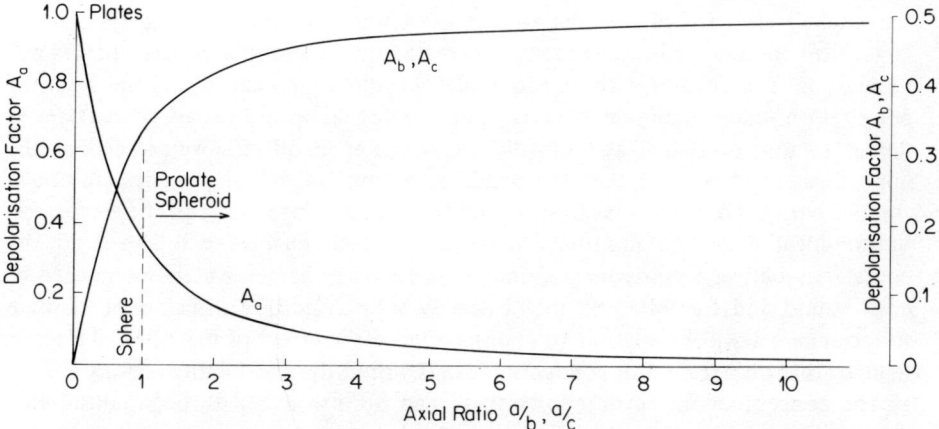

Figure 5.5 The depolarization factors for an ellipsoid of revolution about the a-axis

The development of a general theory for the dielectric properties of hetero-geneous systems has been attempted by many workers, and excellent reviews of the theory have been given by van Beek,[10] Reynolds and Hough,[16] Dukhin,[17] and Hasted.[18] The most consistent theories are those which consider the special case of spherical particles dispersed in a dielectric medium. The general consensus of opinion[17,18] indicates that for low volume fractions of dispersed spheres $(v_2 < 0.1)$, either in random or ordered arrangements, then the relationship first derived by Maxwell that

$$\epsilon_m = \epsilon_1 + \frac{3\epsilon_1(\epsilon_2 - \epsilon_1)}{2\epsilon_1 + \epsilon_2} v_2 \tag{5.9}$$

gives excellent agreement between experiment and theory for the permittivity for all values of ϵ_2/ϵ_1. We note, on passing, that equation (5.9) follows directly from equations (5.6) and (5.8) for the special case of random spheroids. As the volume concentration of the dispersed particles is increased (i.e. $v_2 \gtrsim 0.1$) then the formula, first obtained by Bruggeman[19]

$$\frac{\epsilon_m - \epsilon_2}{\epsilon_1 - \epsilon_2}\left(\frac{\epsilon_1}{\epsilon_m}\right)^{1/3} = 1 - v_2^2 \tag{5.10}$$

gives a more accurate expression for the mixture permittivity ϵ_m.

COMPLEX CONDUCTIVITY

So far in this discussion of the dielectric properties of mixtures of dielectric components, we have treated the components as being ideal dielectrics in the sense that we have ignored their conductive properties. In non-ideal dielectric mixtures there will be the usual polarization processes exhibited by ideal dielectric materials (dipolar relaxations, atomic and electronic polarizations) together with polariza-tions resulting from the displacement of free charge carriers under the influence of the applied electric field. If the size of the conducting dielectric components is large, then the induced dipole moments created by these displaced free charges will be enormous compared with normal molecular dipole moments, and the resulting polarization may dominate over all other polarization processes. The effect is similar to that described as 'nomadic' polarization in ref. 56. We are making the implicit assumption here that the conducting components or particles are not in contact with each other so as to form conducting pathways between the electrodes. In non-ideal dielectric mixtures, therefore, surface charges will appear at the boundaries between components, and these surface charges will be composed of both bound and free charges. In the steady state condition the amount of these surface charges will be such as to counterbalance the effect of the external electric field, so that no net field or conduction exists within the conducting regions.

The concept of dielectric conductivity can be introduced through considering the simple case of a dielectric-filled capacitor having a capacitance C and resistance R. When an a.c. voltage $V = V_0 e^{j\omega t}$ is applied to the plates of this capacitor, a resistive current I_c in phase with this a.c. voltage will be developed, together with a charging or displacement current I_D that leads the voltage in phase by $90°$. The total current is given by

$$I = I_c + I_D$$

$$= \frac{V}{R} + C\frac{dV}{dt}$$

$$= V\frac{A}{d}(\sigma + j\omega\epsilon_0\epsilon) \tag{5.11}$$

where σ and ϵ are the macroscopic values of the conductivity and permittivity of the medium, and A and d are the area and separation of the capacitor plates. Defining the current density J as I/A, and the field strength E as V/d, then

$$J = (\sigma + j\omega\epsilon_0\epsilon)E$$

so that (from Ohm's law) we can define a complex conductivity σ^* as

$$\sigma^* = \sigma + j\omega\epsilon_0\epsilon \tag{5.12}$$

Returning to equation (5.11), then we can write

$$I = \frac{j\omega\epsilon_0 A}{d}\left(\epsilon - \frac{j\sigma}{\omega\epsilon_0}\right)V$$

For a capacitor, the relationship between the experimentally observed current I and applied voltage V determines the permittivity of the dielectric medium, so considering only those intrinsic quantities controlled by the medium, we can define the complex permittivity ϵ^* as

$$\epsilon^* = \epsilon' - j\epsilon'' = \epsilon - \frac{j\sigma}{\omega\epsilon_0} \tag{5.13}$$

From equations (5.12) and (5.13) we can see that at high frequencies as $\omega \to \infty$ then $\epsilon^* \to \epsilon$, and at low frequencies as $\omega \to 0$, then $\sigma^* \to \sigma$. This stresses the fact that for high frequencies the dielectric properties of the material are important, whereas at low frequencies conduction phenomena are more important. This is the main justification for introducing the identities of equations (5.12) and (5.13), and for using the complex permittivity and conductivity parameters ϵ^* and σ^*, which can be expressed in terms of each other through the relationship

$$\sigma^* = j\omega\epsilon_0\epsilon^* \tag{5.14}$$

The complex analogue of Maxwell's equation (5.9) has been determined by Hanai,[20,21] who obtains the two expressions

$$\epsilon_m^* = \epsilon_{mh} + \frac{\epsilon_{ml} - \epsilon_{mh}}{1 + j\omega\tau} + \frac{\sigma_{ml}}{j\omega\epsilon_0} \tag{5.15}$$

and

$$\sigma_m^* = \sigma_{ml} + \frac{j\omega\tau(\sigma_{mh} - \sigma_{ml})}{1 + j\omega\tau} + j\omega\epsilon_0\epsilon_{mh} \tag{5.16}$$

where the suffixes 'ml' and 'mh' designate the low and high frequency mixture values, respectively. The four characteristic parameters of equations (5.15) and (5.16) are related according to the expression

$$(\sigma_{mh} - \sigma_{ml})\tau = (\epsilon_{ml} - \epsilon_{mh})\epsilon_0 \tag{5.17}$$

with the relaxation time being given by

$$\tau = \frac{2\epsilon_1 + \epsilon_2 + v_2(\epsilon_1 - \epsilon_2)}{2\sigma_1 + \sigma_2 + v_2(\sigma_1 - \sigma_2)} \epsilon_0 \tag{5.18}$$

Hanai also gives the following expressions for the limiting low and high frequency values of the permittivity and conductivity:

$$\epsilon_{mh} = \epsilon_1 \frac{2\epsilon_1 + \epsilon_2 - 2v_2(\epsilon_1 - \epsilon_2)}{2\epsilon_1 + \epsilon_2 + v_2(\epsilon_1 - \epsilon_2)} \tag{5.19}$$

$$\epsilon_{ml} - \epsilon_{mh} = \frac{9(\epsilon_1\sigma_2 - \epsilon_2\sigma_1)^2 v_2(1 - v_2)}{[2\epsilon_1 + \epsilon_2 + v_2(\epsilon_1 - \epsilon_2)][2\sigma_1 + \sigma_2 + v_2(\sigma_1 - \sigma_2)]^2} \tag{5.20}$$

$$\sigma_{ml} = \sigma_1 \frac{2\sigma_1 + \sigma_2 - 2v_2(\sigma_1 - \sigma_2)}{2\sigma_1 + \sigma_2 + v_2(\sigma_1 - \sigma_2)} \tag{5.21}$$

$$\sigma_{mh} - \sigma_{ml} = \frac{9(\sigma_1\epsilon_2 - \sigma_2\epsilon_1)^2 v_2(1 - v_2)}{[2\sigma_1 + \sigma_2 + v_2(\sigma_1 - \sigma_2)][2\epsilon_1 + \epsilon_2 + v_2(\epsilon_1 - \epsilon_2)]^2} \tag{5.22}$$

There are several points of interest arising from these formulae. From equations (5.19) and (5.21) we see that the high frequency permittivity of the mixture is not complicated by any conductivity effects, and also that the low frequency conductivity of the mixture does not involve any permittivity parameters. This is consistent with our earlier remarks concerning the implications of equations (5.12) and (5.13). From equation (5.20) we see that dielectric dispersions will always result unless we have the unlikely identity $\epsilon_1\sigma_2 = \epsilon_2\sigma_1$, and from equation (5.18) we see that the frequency of maximum dielectric loss ($f_m = 1/2\pi\tau$) is relatively insensitive to the volume fraction v_2 of the dispersed spherical component of the dielectric mixture.

Equations (5.18)–(5.22) apply only to dilute spherical dispersions with $v_2 < 0.1$. Hanai[20,21] has also developed the complex analogue of Bruggeman's formula (equation (5.10)) for more concentrated dispersions with $v_2 \gtrsim 0.1$, and obtains the following formulae relevant to the limiting low and high frequency parameters:

$$\frac{\epsilon_{mh} - \epsilon_2}{\epsilon_1 - \epsilon_2}\left(\frac{\epsilon_1}{\epsilon_{mh}}\right)^{1/3} = 1 - v_2 \tag{5.23}$$

$$\epsilon_{ml}\left(\frac{3}{\sigma_{ml} - \sigma_2} - \frac{1}{\sigma_{ml}}\right) = 3\left(\frac{\epsilon_1 - \epsilon_2}{\sigma_1 - \sigma_2} + \frac{\epsilon_2}{\sigma_{ml} - \sigma_2}\right) - \frac{\epsilon_1}{\sigma_1} \qquad (5.24)$$

$$\frac{\sigma_{ml} - \sigma_2}{\sigma_1 - \sigma_2}\left(\frac{\sigma_1}{\sigma_{ml}}\right)^{1/3} = 1 - v_2 \qquad (5.25)$$

$$\sigma_{mh}\left(\frac{3}{\epsilon_{mh} - \epsilon_2} - \frac{1}{\epsilon_{mh}}\right) = 3\left(\frac{\sigma_1 - \sigma_2}{\epsilon_1 - \epsilon_2} + \frac{\sigma_2}{\epsilon_{mh} - \epsilon_2}\right) - \frac{\sigma_1}{\epsilon_1} \qquad (5.26)$$

The relaxation time τ of the interfacial polarization will also be given by equation (5.17), and although a very complicated expression results in substituting into equation (5.17) the parameters ϵ_{ml}, ϵ_{mh}, σ_{ml}, and σ_{mh} using equations (5.23)–(5.26), it can be seen that the relaxation time τ increases slightly with increasing concentration of the dispersed component.

This feature of the frequency of maximum dielectric loss not varying significantly with increasing concentrations of the dispersed components appears to be an important characteristic of the interfacial polarization of such heterogeneous systems. It should also be remembered that if either of the components of the heterogeneous system exhibits dielectric dispersion phenomena of its own, then these 'intrinsic' dispersions will also appear, together with the interfacial dispersion, in the complete frequency spectrum of the dielectric behaviour of the system.

An essential feature of the Maxwell–Wagner polarization effect is the appearance of accumulated charge at the boundaries between regions of differing dielectric media as a result of electronic or ionic migration polarization processes. For the case of dispersed spherical particles, for example, this accumulated charge is taken to be in the form of a uniformly distributed surface charge layer. Thermal motions and concentration gradient diffusion effects result in this surface charge layer having a thickness of the order of the Debye dielectric screening length λ_D. The Debye screening length is given by the well-known equation (see Appendix to Chapter 5)

$$\lambda_D = \left(\frac{\epsilon_r \epsilon_0 kT}{2nq^2}\right)^{1/2}$$

where n is the free charge carrier concentration in a material of relative permittivity ϵ_r. We can envisage λ_D to represent the distance from the centre of a localized charge concentration at which the electrical potential has fallen to $1/e$ (i.e. $1/2.718 = 36.8$ per cent) of the value in the vicinity of the centre of the charge concentration. The interfacial polarization theories described so far assume that the dispersed particles have dimensions far exceeding λ_D. For dispersed metallic conducting particles in an insulating matrix, the free charge carrier concentration within highly conducting particles can easily exceed 10^{26} m^{-3}, so that the corresponding screening length λ_D for such a concentration of built-up charge can be less than 10^{-9} m — smaller than any possible particle dimension. For particles of much lower conductivity, the free charge carrier concentration can be less than 10^{18} m^{-3}, giving λ_D a value in excess of 10^{-6} m.

The effect on the dielectric properties of heterogeneous systems as a result of the appearance of surface charge layers having thicknesses of the order, or in excess, of the average dimension of one of the dielectric components, has been considered by Trukhan.[22] Trukhan considered two model systems, namely a lamellar system with alternating insulating and conducting planar regions, and a system of conducting spherical particles in an insulator medium. The major conclusion is that the dispersion frequency is shifted to higher frequency values than is predicted by the previously outlined Maxwell–Wagner theories. For example, for spheres of radius 'a' and permittivity ϵ_2, embedded in a dielectric medium of permittivity ϵ_1, then as the ratio a/λ_D decreases the dispersion frequency increases. Considerable deviations from the results of the more conventional theories begin for $\epsilon_2/\epsilon_1 = 0.1$ at $a/\lambda_D \lesssim 1$, for $\epsilon_2/\epsilon_1 = 1$ at $a/\lambda_D \lesssim 10$, for $\epsilon_2/\epsilon_1 = 10$ at $a/\lambda_D \lesssim 10^2$, and so on.

Trukhan shows that the relevant dispersion frequency is associated directly with the dispersed conducting particle size, and more precisely with the transit time for

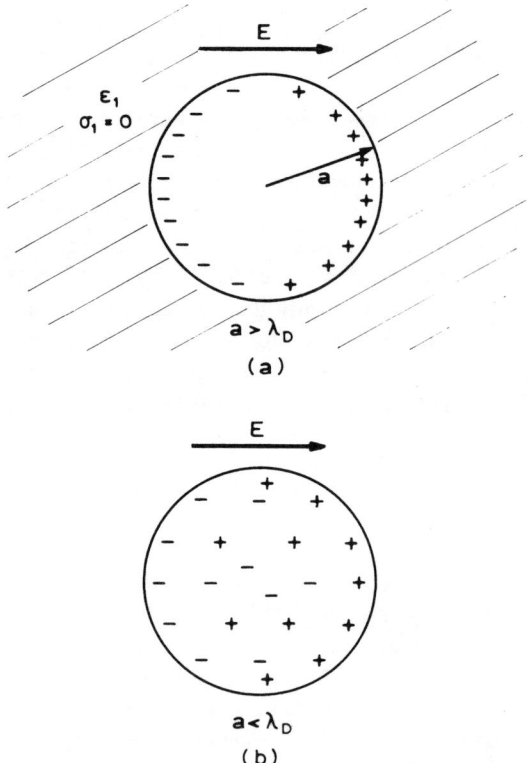

Figure 5.6 The polarization of free charges within a conducting spherical particle immersed in an insulating medium for the case: (a) where the particle radius greatly exceeds the Debye screening length, and (b) where the Debye screening length exceeds the particle radius

free carriers to diffuse back and forth within and between the boundaries of the conducting particles. This particle size and free charge carrier mobility effect does not appear in the conventional heterogeneous dielectric theories, and is of particular relevance to the interpretation of a.c. conduction and a.c. Hall effect measurements for organic and biological materials.[23,24] The situation can be demonstrated using Figure 5.6, where a conducting spherical particle is shown immersed in a non-conducting dielectric medium.

In Figure 5.6(a) the particle is shown to have different charge carrier concentrations, n and p, for negative and positive mobile charges, respectively. Also, the particle size is greater than the free charge Debye screening length λ_D. In Figure 5.6(b) the particle size is now less than λ_D. The more conventional interfacial or Maxwell—Wagner polarization theories consider the situation represented by Figure 5.6(a), whereas Trukhan[22] considers the case of Figure 5.6(b) and takes into account space—charge effects within the conducting particle.

The problem associated with the situation shown by Figure 5.6(b) is the calculation of the spatial localization of the polarization charges within the particle. In an alternating electric field applied along the x-axis, the field can be written as

$$E(x, t) = E(x)e^{j\omega t}$$

and the corresponding electric potential is given by

$$V(x, t) = V(x)e^{j\omega t}$$

The mobile charge carriers will redistribute themselves within the conducting particle according to the so-called continuity equations

$$\frac{\partial p}{\partial t} = \frac{\partial}{\partial x}\left[D_p \frac{\partial p}{\partial x} + \mu_p p \frac{\partial V}{\partial x}\right]$$

$$\frac{\partial n}{\partial t} = \frac{\partial}{\partial x}\left[D_n \frac{\partial n}{\partial x} - \mu_n n \frac{\partial V}{\partial x}\right]$$

and according to the Poisson equation

$$\frac{\partial^2 V}{\partial x^2} = -\frac{(p - n)q}{\epsilon_2 \epsilon_0}$$

where 'q' is the electronic charge, and D_p and D_n are the charge carrier diffusion coefficients. The diffusion coefficients are given according to the Einstein relationship

$$D_p = \mu_p kT/q \quad \text{and} \quad D_n = \mu_n kT/q$$

where μ_p and μ_n are the positive and negative free charge carrier mobilities, respectively. Essentially, these equations describe the electric field driven motions, and the diffusion motions due to carrier concentration gradients, of the mobile charges in the conducting spherical particle. On solving these equations for a system of conducting spherical particles of radius 'a', dispersed in an insulating medium,

Trukhan obtains for the permittivity of the overall mixture the expression

$$\epsilon_m = \epsilon_1 \left(1 + 3v_2 \frac{\epsilon_2 - \epsilon_1 + \beta}{2\epsilon_1 + \epsilon_2 - 2\beta}\right) \qquad (5.27)$$

where

$$\beta = \frac{\lambda_D^{-2}\,[3 + (\gamma a)^2]\tanh\gamma a - 3\gamma a}{\gamma^2\,[2 + (\gamma a)^2]\tanh\gamma a - 2\gamma a}$$

with, for the case of negative charge carrier concentrations,

$$\gamma^2 = \lambda_D^{-2} + \frac{j\omega}{D_n}$$

Likewise, the frequency f_m, corresponding to the maximum dielectric loss associated with this interfacial polarization, is given by

$$f_m = \frac{1}{2\pi} \cdot \frac{2\sigma_1 + \sigma_2}{(2\epsilon_1 + \epsilon_2)\epsilon_0} \cdot \frac{(a/\lambda_D)^2\tanh(a/2\lambda_D) - 4[(\epsilon_1/\epsilon_2) - 1]\left[2\tanh(a/2\lambda_D) - \dfrac{a}{\lambda_D}\right]}{(a/\lambda_D)^2\tanh(a/2\lambda_D) + 6\left[2\tanh(a/2\lambda_D) - \dfrac{a}{\lambda_D}\right]}$$

where, as before, the suffices 1 and 2 refer to the surrounding medium and to the particle, respectively. The way in which the dispersion frequency f_m varies as a function of particle size and charge carrier mobility is indicated in Figure 5.7. The

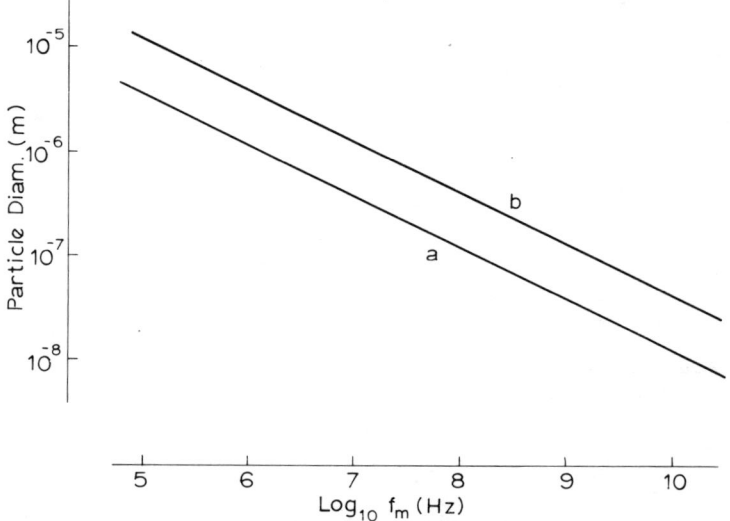

Figure 5.7 The variation of the dispersion frequency for conducting spheres in an insulating medium with free charge carrier mobility and particle size (a) $\mu = 1 \text{ cm}^2/\text{V s}$; (b) $\mu = 10 \text{ cm}^2/\text{V s}$. (Reproduced from R. Pethig, *Disc. Faraday Soc.*, **51**, 214 (1971), by permission of The Chemical Society)

system corresponding to the values given in Figure 5.7 is that of conducting spherical particles of permittivity $\epsilon_2 = 3$ and free charge carrier density of 10^{22} m^{-3}, suspended in an insulating medium of relative permittivity $\epsilon_1 = 1$. From Figure 5.7 it can be seen that with decreasing size of the spherical particles, the dispersion frequency rapidly increases. For constant particle size, the increase in dispersion frequency with increasing mobility reflects the fact that the dispersion frequency is directly related to the transit time for the free charge carriers to diffuse back and forth across the conducting particles.

Another important result of Trukhan's theoretical work, is that according to equation (5.27), measurements of the low frequency permittivity of a system consisting of spherical conducting particles dispersed within an insulating medium should lead to a value of the free charge carrier concentration within the particles, especially if this carrier concentration is low.

If our spherical particle had in fact been a biological object, a protein molecule or bacteria *in vivo*, for example, then the surrounding dielectric medium would have been better described as a conducting aqueous electrolyte rather than as an insulating dielectric material. Our simple models of Figure 5.6(a) and (b) would then be further complicated by having the possibility of there being adsorbed ions on the outer surface of the particle. In order to achieve overall electrical neutrality, there will also be an associated atmosphere of counter-charge ions, forming an electrical double layer around the particle.

SURFACE IONIC CONDUCTIVITY AND COUNTER-ION EFFECTS

Let us consider a suspension of spherical protein molecules in an aqueous electrolytic solution. At a pH value away from its isoionic point, then as a result of ionized surface chemical groups or adsorbed ions, such protein particles will be electrically charged. To retain overall electrical neutrality these charged particles will attract counter-charges. As a result of the formation of a surrounding 'atmosphere' of such counter-charges (counter-ions), an electrical double layer will be formed as shown schematically in Figure 5.8. The particle in Figure 5.8(a) is assumed to have an excess of negatively charged adsorbed ions or ionized chemical groups, and the counter-charge is provided by a surrounding atmosphere of cations. Surrounding the particle with its stabilized electrical double layer is the bulk volume of the electrolytic solution. The concept of the electrical double layer is given in more detail in the Appendix to this chapter.

It has commonly been found, and reported in the scientific literature, that systems of non-polar insulating particles dispersed in electrolytes exhibit very large dielectric dispersions at frequencies of the order 10 kHz and lower. Effectively, similar results have been observed for kaolin,[25] polystyrene,[26] and biological particles[27,28] in dilute KCl solutions, for example. Typically, for spherical particles of diameter of the order 10^{-6} m, suspended in dilute electrolytes, the relative permittivity of the system can exceed values greater than 10^4 at frequencies below 1 kHz. We shall see that such large dielectric dispersions can be associated with the electrical double layer surrounding such particles.

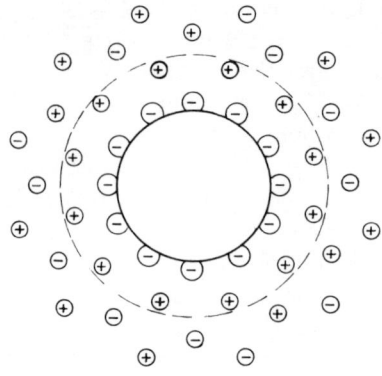

Figure 5.8 The formation of an electrical double layer around a particle, with fixed electronic charge, immersed in an electrolyte

   A possibility that comes readily to mind, perhaps, is that if there is a high concentration of ions and counter-ions on the particle surface, then a large surface conductivity could exist associated with the displacement of these ions under the influence of applied electric fields. We saw essentially this type of effect in Chapter 3 with the model of Kirkwood and Shumaker[29] where protons were envisaged to fluctuate between the various ionizable chemical groups on the surface of a protein molecule. In Figure 5.9(a) we see bound anions distributed discreetly over the surface of a protein molecule, and the associated counter-cations. These cations are held to the vicinity of the protein surface by the electrostatic attraction of the bound anions. In other words they tend to be located at the bottom of potential energy wells, as shown in Figure 5.9(b). We can derive an estimate of the potential energy barrier E_s restraining a counter-ion from leaving the vicinity of a charged particle, by assuming the particle has just one single bound ion and counter-ion each of unity electronic charge q (1.6×10^{-19} C). The coulombic electrostatic force $F(x)$ of attraction between the bound ion and counter-ion when

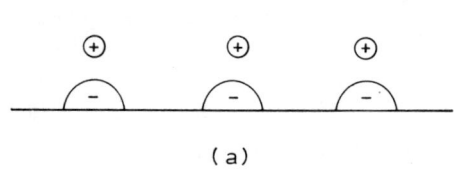

(a)

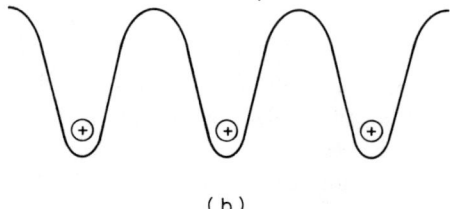

(b)

Figure 5.9 (a) Counter-ions neutralizing the fixed surface charge. (b) Counter-ions are electrostatically bound to the charged particle surface, and are located within coulombic potential wells

distance x apart is given by

$$F(x) = \frac{-q^2}{4\pi\epsilon_0\epsilon x^2}$$

where ϵ is the permittivity of the electrolytic medium between the ion pair. The potential energy barrier E_s, representing the energy required to remove the counter-ion from the surface of the charged particle into the bulk of the surrounding electrolyte, is obtained by mathematically integrating the product force x distance from the point x to infinity, i.e.

$$E_s = \int_x^\infty F(x)\,\mathrm{d}x$$

$$= -\frac{q^2}{4\pi\epsilon_0\epsilon x} \tag{5.28}$$

The problem is now in assigning a suitable value for ϵ, the permittivity of the electrolyte medium between the bound ion and the counter-ion. This is by no means straightforward. Apart from requiring to know what effect solvated ions have on the dielectric properties of water, we also need to ask about the significance of ϵ in equation (5.28) for small values of x.

As described in Chapter 4, it has been found (see, e.g. reference 30) that the low frequency permittivity ϵ_i of aqueous electrolytes is lower than the value ϵ_w for pure water. For low electrolyte concentrations c of less than 1 mole/litre, we have seen that ϵ_i varies linearly with concentration c according to the relationship

$$\epsilon_i = \epsilon_w - \delta c \tag{5.29}$$

where δ is the dielectric decrement factor. Typical values for δ are $\delta(\mathrm{NaCl})$ and $\delta(\mathrm{KCl}) = 11$, $\delta(\mathrm{LiCl}) = 14$ and $\delta(\mathrm{MgSO_4}) = 31$. The differing dielectric decrement values are related to the differences in the radii of the various dissociated ions and in the number of water molecules that can be oriented by the large electrostatic field associated with the ions.

The permittivity value ϵ_i given by equation (5.29) corresponds to the bulk electrolyte value. Distances of less than about 0.6 nm from a hydrated ion correspond to the region of the orientated water molecules, and it is no longer valid to represent the aqueous surroundings in terms of a continuous aqueous dielectric medium. In this way, as the value of x in equation (5.28) falls below a value of 0.6 nm, we can expect the value of the permittivity to decrease rapidly. The lower limit of ϵ will correspond basically to just the atomic and electronic polarizations of the water molecules, since the dipolar orientational effects will be greatly reduced for the strongly orientated water molecules around the bound surface charge. This lower limit for ϵ should be of the order of the value of 4.7, corresponding to the high frequency value ϵ_∞ for bulk liquid water (see reference 18, p. 58). The problem regarding the polarizability of water molecules in electrolytes has been approached both experimentally and theoretically by Collie et al.[30] and Hasted (see reference 18, p. 141), and based on their work the variation of the permittivity with distance from a hydrated ion is shown in the inset of Figure 5.10. We have

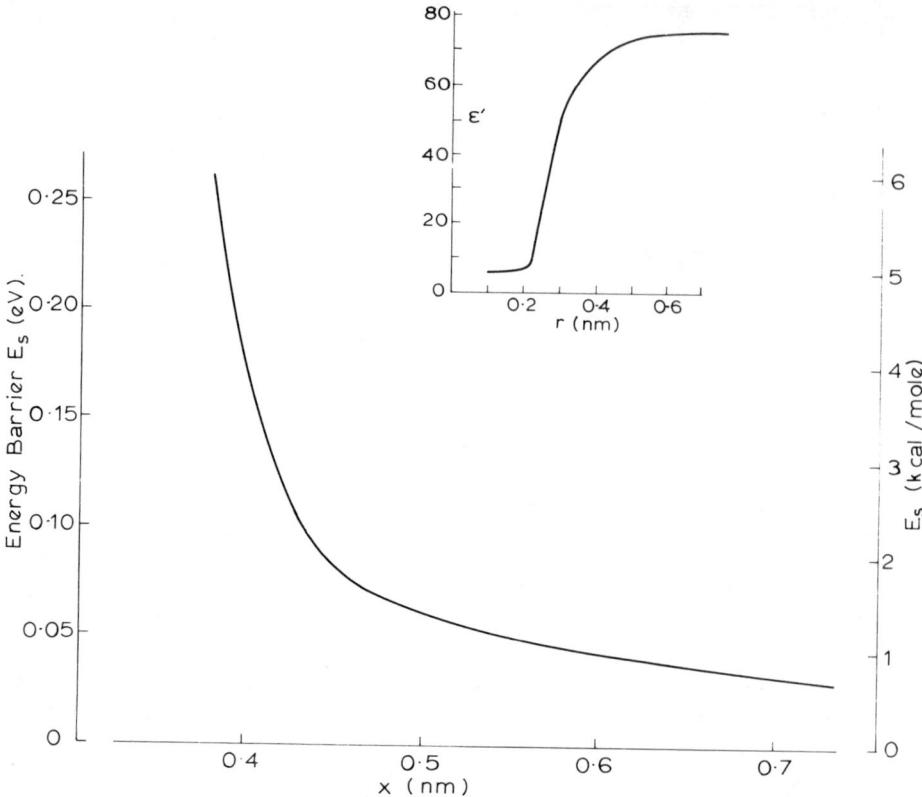

Figure 5.10 The variation of the potential energy barrier E_s, which retains a counter-ion to the vicinity of a charged particle, with distance of the counter-ion from the particle surface. The inset shows the variation of the electrolyte permittivity with distance from a free ion

assumed the ionic concentration to be 0.15 M, corresponding to the approximate concentration of K^+ salts in the cytoplasm of cells and of NaCl in blood plasma, for example. In calculating the values (using equation (5.28)) of the potential energy barrier height E_s shown in Figure 5.10, the model assumed was that of a K^+ ion in the vicinity of a negatively charged particle. The separation distance x took into account the ionic radius value of 0.13 nm for K^+ ions.[31] If we wish to retain the concept of the dielectric medium between the two charges being that of an aqueous electrolyte, then their closest approach will be of the order of 0.44 nm, corresponding to their being positioned centrally either side of a water molecule (H_2O diameter = 0.31 nm). From Figure 5.10 it can be seen that the corresponding potential barrier heights E_s are small, of the order $3.4 kT$ for $x = 0.44$ nm. When the counter-ion is at a distance of 0.86 nm from the bound surface charge, then $E_s = kT$ and the counter-ion can be considered to be beyond the influence of the charged particle and able to exchange freely with other like-charged ions in the electrolyte bulk.

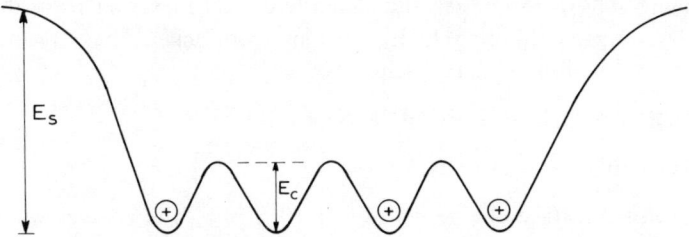

Figure 5.11 The overlapping of adjacent potential energy wells for a particle with a large surface charge density

If the density of charges on the surface of the particle is large enough, then the potential energy well associated with such a charge and its counter-ion will overlap with those of neighbouring charge pairs. The potential energy barrier E_c between two such overlapping coulombic wells will therefore be reduced compared with the barrier height E_s, as shown in Figure 5.11. The barrier height E_c can be calculated using the relationship[32]

$$E_c = E_s - \frac{q^2}{\pi\epsilon_0 \epsilon d} \tag{5.30}$$

where d, for our purpose, is the distance between neighbouring surface charges. Polystyrene spheres[33] have been found to have a surface charge density of 1.9×10^{17} m^{-2}, and at pH 7, the equine cytochrome-c molecule has an excess of 12 positively charged ionized surface groups, which corresponds to a surface charge density of 1.1×10^{17} m^{-2}. In fact, polystyrene latex particles can be taken to represent a good model for protein molecules such as casein and bovine serum albumin, because like these proteins, polystyrene particles have carboxyl groups as the dominant ionizable chemical group. Above about pH 5.0 the carboxyl groups at the ends of the polystyrene polymer chains will lose protons and become ionized, leaving the polystyrene particles with a net negative charge.[34] We can treat these ionized surface chemical groups as bound surface ions, around which will form a counter-ion layer of cations. In the Appendix to this chapter it is shown that this description of the build-up of surface charge is only one of several possibilities. For some particles, the form of the electrical double layer is controlled by the surface electrostatic potential arising from the difference between the electro-chemical potentials of the particle material and the surrounding aqueous medium. For protein macromolecules there is also the possibility that the net surface charge is controlled by the adsorption of chemicals. This effect can be observed if, for example, 1 ml of 0.1 N HCl is added to 100 ml of a 1 per cent protein solution. The resulting pH will be much higher than if the acid had been added to 100 ml of pure water. Similarly, if 0.1 N NaOH is added to a protein solution, the resulting pH will be much lower than if we had diluted the alkali with the same volume of pure water. In other words proteins can act as buffers, which is an effect of very great biological importance. This buffering action results from the carboxyl groups of the

172

protein being able to react with the added hydroxyl ions, and from the ability of the amino groups in binding hydrogen ions from acids. These reactions can be summarized by the following two schemes:

$$-COOH + NaOH \longrightarrow -COONa + H_2O$$

$$-NH_2 + HCl \longrightarrow -NH_3Cl$$

For a uniform surface charge density of 10^{17} m^{-2}, the charges will on average be 3.16 nm apart. Using this separation distance, then from equation (5.30) the corresponding coulombic potential barrier E_c between neighbouring charges has been plotted in Figure 5.10 as a function of the distance between the surface charge and its counter-ion. Although the local environment of a counter-ion near to a charged surface will differ from that of an ion in free solution, we can as a first approximation describe the counter-ion surface motion in terms of an effective ionic mobility μ_s, with

$$\mu_s = \mu_0 \exp - \frac{E_c}{kT} \tag{5.31}$$

where μ_0 is the ionic mobility in free solution. Typical μ_0 values at room temperature are: 36×10^{-8} for H$^+$; 5×10^{-8} for Na$^+$; and 8×10^{-8} m^2/V s for K$^+$

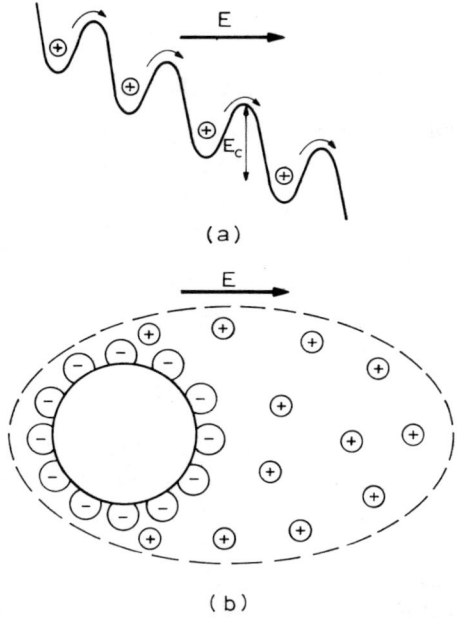

(a)

(b)

Figure 5.12 (a) The activated hopping motion of counter-ions along a charged surface. (b) Polarization resulting from net displacement of counter-ions as a result of the influence of an external electric field

ions. From Figure 5.10 we can see for surface charge concentrations of the order 10^{17} m^{-2}, that the value for E_c is of the order 0.5 kcal/mol for counter-ions positioned between 0.5 and 0.6 nm from the charged surface. The surface mobility of these counter-ions will therefore be comparable with the mobility in free solution, and we can expect these ions to respond freely to the influence of external electric fields. Such a surface ionic conductivity could very easily exceed the conductivity of the particle itself, an effect that is not taken into account by conventional Maxwell—Wagner theories. The mode of activated hopping motion of the counter-ions along the particle surface under the action of an applied electric field is shown in Figure 5.12(a). As a result we can envisage that there will be a net displacement of counter-ion charge relative to the centre of symmetry of the spherical particle, as shown in an exaggerated manner in Figure 5.12(b). The net effect will be the appearance of a large field-induced dipole, the existence of which could easily account for the large dielectric dispersions often found for colloidal suspensions in electrolyte solutions. The distribution of these surface counter-ions under equilibrium conditions will essentially involve a counterbalance of field and concentration gradient diffusion forces. The re-establishment of the original surface charge distribution will be diffusion controlled. Schwarz[35] has given the corresponding diffusion limited relaxation time τ for such a surface ion polarization effect as

$$\tau = \frac{a^2}{2D_s} \tag{5.32}$$

with

$$D = \mu_s kT/q$$

and where 'a' is the particle radius. For a surface potential barrier height of the order 0.5 kcal/mol and a particle radius of 10^{-6} m, then at 300 K the dispersion associated with a surface counter-ion distribution of K$^+$ ions will be centred at a frequency of the order 0.3 kHz. This is about a decade below the frequency region where the large dielectric dispersions of colloidal particle suspensions are commonly found. If the dispersed particles are not of uniform size, then there will be a distribution of associated relaxation times and the dielectric dispersion will be broad, which is a feature observed experimentally. Furthermore, both the double layer and free solution ionic mobilities will involve activated processes, so from equations (5.31) and (5.32) it will follow that the dispersion mid-frequency point f_m will obey an activated process law, with f_m increasing with increasing temperature.

Apart from Schwarz[35] others have considered the concept of the polarizability of colloid particles in electrolytes being associated with surface counter-ion conductivity effects. Some of the notable contributions have been made, in chronological order, by Miles and Robertson,[36] Eigen and Schwarz,[37] O'Konski,[38] Ingram and Jerrard,[40] Schwarz,[35,39] Schurr,[41] Stoylov,[42] and Dukhin.[17] The various theories proposed by these authors differ somewhat as to the nature and dynamic properties of the electrical double layer. One popular model considers a particle with a double layer to be a sphere with a concentric surface layer of elevated conductivity, with this surface layer being regarded as infinitesimally thin.[36-38] The polarization process is considered to result from a redistribution of the charges in this double layer, with some of the models considering only field driven ionic movements tangential to the particle surface.[35-41] In some of the theories the possibility of ion exchange between the double layer and the electroneutral volume of the electrolyte is taken into account[17,36,38,41] whereas sometimes no such exchange is considered.[35,37,39] We have seen from Figure 5.10 that the forces holding the counter-ions to the charged particle can be very small, so we will expect there to be a certain degree of ion exchange between the electrolyte and the double layer. We can imagine the borderline between the strongly bound and diffused counter-ions to occur in the neighbourhood of the so-called plane of shear associated with the zeta potential, as described in the Appendix to this chapter. A more sophisticated model was proposed by Schurr[41] who regarded the double layer as consisting of such regions of strongly bound and loosely bound counter-ions, with the loosely bound counter-ions forming a diffuse layer able to exchange with the bulk electrolyte. In this model by Schurr, concentration diffusion fluxes are considered for the bound counter-ions, but are considered negligible for the diffuse layer. In the most sophisticated model attempted to date, Dukhin[17] extends Schurr's model to include both tangential and normal components of ion fluxes at the boundary of the diffuse outer region of the double layer and the bulk electrolyte. Conduction fluxes due to the effect of induced electric fields arising from the polarizations are also considered. It is found in general, that when there are free counter-ions together with bound ones, then the double layer polarization process cannot be completely characterized by the relaxation time given by Schwarz in the form of equation (5.32). Relaxation of the polarization process will also be influenced by deformations of the diffuse outer regions of the double layer, where diffusion and exchange effects will be important. The net result of Dukhin's work can be summarized by citing the results obtained for the dispersion relaxation time τ and the maximum dispersion $\Delta\epsilon$ that will be observed when the particle surface charge density becomes sufficiently large. The relaxation time is given by

$$\tau = \frac{a^2}{2D_s} \frac{1}{M} \tag{5.33}$$

with $M = 1 + Z^+ Z^- (Z^+ + Z^-) n_b a / n_0 \lambda_D$ where Z^+ and Z^- are the electrovalencies of the cations and anions, and n_b and n_0 represent the equilibrium density of the bound counter-ions and the total counter-ion density respectively. The factor λ_D is the Debye screening length associated with the surface charge density, and 'a' is the particle radius. Equation 5.33 differs from the one derived by Schwarz,[35] through the factor M. This factor M can have values ranging from 1 up to 10^3, which indicates that equation (5.32) can yield an overestimate of the relaxation times. With this in mind, we see that the dispersion frequency of the order 0.3 kHz, calculated earlier for particles of radius 10^{-6} m using Schwarz's formula, will be an underestimate by a factor possibly of the order of two decades in frequency. The maximum dielectric dispersion $\Delta\epsilon$, given by Dukhin,[17] for sufficiently large surface charge densities, is found from the relationship

$$\Delta\epsilon_{max} = \frac{9av_2}{4\lambda_D [1 + (v_2/2)]^2}$$

For a suspended particle volume fraction $v_2 = 0.1$, and $a/\lambda_D = 10^3$, then $\Delta\epsilon_{max} \simeq 200$, which indicates that dispersions arising from polarizations of the electrical double layer can be very large.

At this point, it is pertinent to mention that under average experimental conditions, it is quite possible for such double layer dispersion effects to be masked by anomalous dispersions arising from electrode polarization processes. Schwan, in particular, has considered the problem regarding electrode polarization effects[43,44] and developed experimental techniques to overcome them.[45] Basically, electrode effects are minimized by having the sample impedance much larger than the electrode interface impedance, which is normally accomplished by having large electrode separation. Taking measurements for different electrode spacings is by far the most satisfactory method of separating electrode polarization effects from bulk phenomena.[45]

SOME EXPERIMENTAL RESULTS FOR BIOLOGICAL MATERIALS

We have seen that a protein—water mixture can be expected to exhibit interfacial polarization effects. In fact, it is most likely that such interfacial phenomena will dominate the dielectric properties of hydrated biological systems, especially for frequencies below about 100 kHz. Bayley[46] was the first to describe in detail the increase in dielectric permittivity of materials, such as amino acids and proteins, with increasing hydration. This work, together with that of later investigators, can be summarized as shown in Figures 5.13(a) and (b). The most remarkable feature of the various results reported in the literature is the wide range of values found for the total dielectric increment $\Delta\epsilon$, defined in Figure 5.13, even for similar protein systems at similar hydration levels and frequencies of measurement. This can be seen for some experimental results summarized in the table below:

$\Delta\epsilon$	Reference
$1 \sim 10$	Rosen[47]
$50 \sim 220$	Takashima and Schwan[48]
$70 \sim 600$	Brausse et al.[49]

In the experiments outlined in this table, Rosen, and Takashima and Schwan, used lightly compressed samples, whereas Brausse et al. used highly compressed and dense samples. Although this might be expected to lead to differing dielectric properties, it is difficult to imagine that the differences should be as large as those experimentally found. A most likely explanation is that the different results reflect differences in the magnitude of both interfacial and electrode polarization effects. The intrinsic dielectric properties of protein and water on their own would not give rise to permittivity values in the hundreds, for example. Also, Takashima and

176

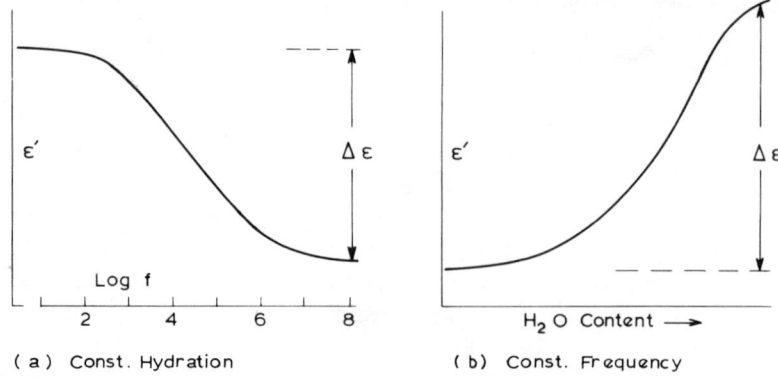

(a) Const. Hydration   (b) Const. Frequency

Figure 5.13 The characteristic low frequency dielectric behaviour of hydrated biological materials

Schwan[48] found that the relaxation time of the observed dielectric dispersions decreased as the size of the biological particles (i.e. powder grain size) decreased. Such a result is compatible with surface conductivity effects described in this chapter, where as shown by equations (5.32) and (5.33) the relaxation time is directly proportional to the particle radius. From Figure 5.13(b) it can be seen that in general the variation of permittivity with water content is characterized by two regions, one where the permittivity remains fairly insensitive to water content, followed at higher hydration values by a region where the permittivity increases rapidly with increasing hydration. The critical hydration point representing the borderline between these two experimental characteristics possibly corresponds to the hydration value at which a conducting layer of water is just formed to completely cover each particle. At lower hydrations we can imagine that the adsorbed water molecules are strongly bound to localized sites and do not form a conductive layer. The interpretation of such hydration effects on compressed powder samples is complicated by the fact that it is by no means certain as to the extent to which adsorbed water penetrates into the biological particles or forms multilayer films covering the particles.

A more specific example of dielectric losses associated with the heterogeneous nature of hydrated biomacromolecules is given by the work of Chang and Chien[50]

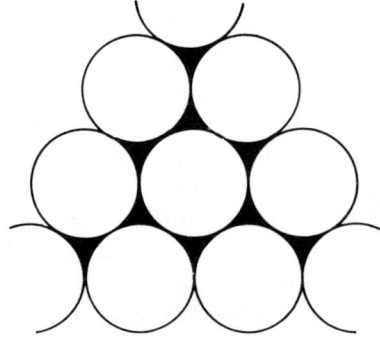

Figure 5.14 The packing of collagen fibres

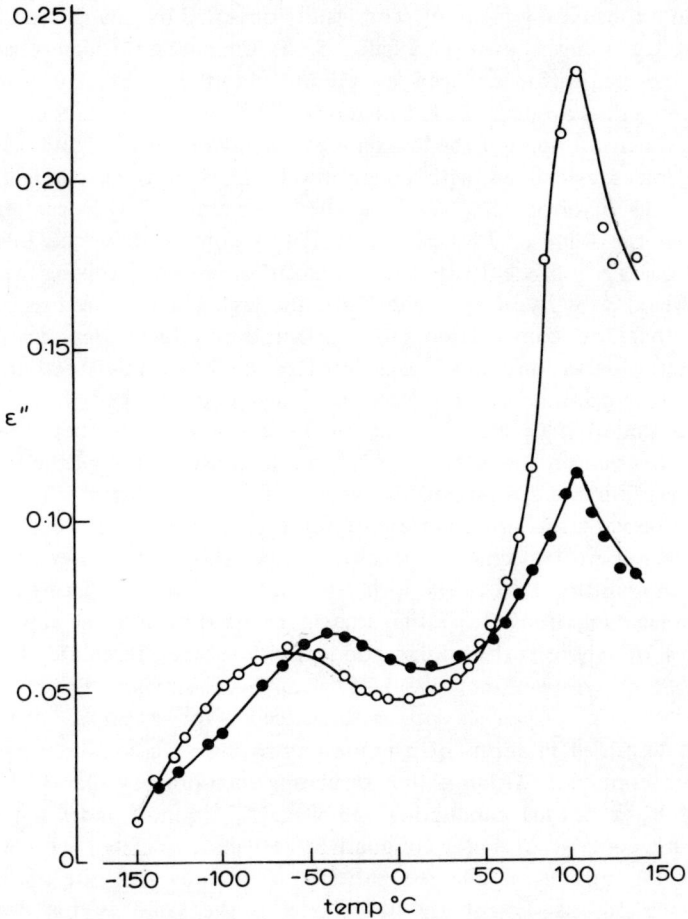

Figure 5.15 The variation of the dielectric loss factor ϵ'' with temperature for calf-skin collagen. ○, 1 kHz; ●, 10 kHz. (Reproduced from E. P. Chang and J. C. W. Chien, *J. Polym. Sci. Polym. Phys.*, **11**, 737 (1973), by permission of John Wiley & Sons, Inc.)

for measurements on collagen and gelatin. The collagen macromolecules have a fibre or rod-like superstructure, probably associated with linearly polymerized tropocollagen molecules, with average fibre lengths of the order 5 to 10 μm. In the nearly dehydrated state the collagen fibres can be envisaged to be packed together as shown schematically in Figure 5.14, with open channels running along the fibre direction into which chains of water molecules can fit. The hydrated collagen system, consisting essentially of low conductivity protein cores separated by conducting layers and channels of water molecules, certainly represents an ideal heterogeneous model as far as the likelihood of observing interfacial polarizations is concerned. A simple analogy would be that of prolate conducting spheroids dispersed in an insulating medium. The results obtained by Chang and Chien[50] for calf-skin collagen are shown in Figure 5.15. Chang and Chien observed two dielectric loss peaks for collagen containing about 12 per cent water, a low temperature peak at around 220 K and another at high temperatures around 370 K. Chang and Chien interpreted the low temperature loss peak of Figure 5.15 in terms of dipolar losses associated with collagen side-chain motions coupled to water molecules, and involving the breaking and reforming of hydrogen bonds. The activation energy value of 7.8 kcal/mole (0.34 eV) obtained for this low temperature loss process is consistent with such a relaxation motion involving the reorientation and breaking of hydrogen bonds. For the high temperature loss peak it was concluded that the contribution from orientation effects was small, and the experimental observations were considered to be best understood in terms of interfacial polarizations of the Maxwell–Wagner type. This basic difference between the underlying physical origins of the low and high temperature dielectric dispersions for collagen and gelatin was further demonstrated[50] by the fact that the low temperature losses correspond to dynamic mechanical dispersions observed for these materials, whereas no counterpart for the high temperature dielectric loss peak is observed with mechanical measurements. This gives support to the viewpoint that the contribution to the high temperature dielectric dispersion does not involve dipolar relaxations, but rather that it results from interfacial polarizations.

Examples of where surface ionic conductivity effects dominate the dielectric properties of microorganisms have been given by Einolf and Carstensen.[51] The observed dielectric properties of bacterial cells, for example, can be quite adequately described in terms of a model consisting of a conducting inner cytoplasmic core contained within a thin insulating membrane wall, which in turn is surrounded by a porous conducting cell wall.[52–54] Einolf and Carstensen have extended the concept of ionic conductivity effects associated with counter-ion motions at the surface of charged particles to include the case of the particle surfaces being porous. Essentially, the model is the same as that described by Schwarz,[35] with the essential difference being that layers of ions both inside and outside the particle are considered to move parallel to the particle surface in response to external electric field variations. Such a phenomenon leads to very high permittivity values at low frequencies, and has been used as the physical model to describe the dielectric properties of ion exchange resins.[55] The application of this model to the dielectric behaviour of bacterial cells follows from the fact that it has been shown[53] that at low frequencies the bacterial cell membrane effectively provides an electrically insulating layer around the cytoplasmic core. Therefore, a bacterial cell at low a.c. frequencies can be represented as an insulating core surrounded by a thick conducting layer containing fixed charges. If this conducting layer (which in fact corresponds to the cell wall) is porous, then the counter-ions will be distributed within and outside this layer.

For the case of non-porous charged particles suspended in an electrolyte, the number of counter-ions which are electrostatically bound to the particle surface and take part in surface conductivity effect, can to a first approximation be equated to the fixed surface charge density on the particle. For the case of porous particles where there will be counter-ions on both sides of the particle surface/electrolyte interface, Einolf and Carstensen[55] give the estimate for the counter-ion concentration as

$$N_i \simeq (N_s \epsilon_i k T)^{1/2} \ln(N_s k T / \epsilon_i)/2q \qquad (5.34)$$

where ϵ_i is the permittivity of the porous particle region. The particle surface charge density N_s is taken to be much greater than the equilibrium density of ions in the surrounding electrolyte solution. Following a mathematical treatment closely resembling that first derived by Schwarz,[35] Einolf and Carstensen show that the effective, homogeneous, complex conductivity of the suspended particle of radius 'a' is given by

$$\sigma_2^* = \sigma_0^* + j\omega\epsilon_0 \ \frac{\Delta\epsilon_2}{1 + j\omega\tau} \qquad (5.35)$$

where

$$\Delta\epsilon_2 = q^2 a N_i / \epsilon_0 k T \qquad (5.36)$$

and represents the low frequency permittivity of the particle (cf. equation (5.12)). The relaxation time τ is identical to that derived by Schwarz, as given by equation (5.32). In equation (5.35), σ_0^* is the effective, homogeneous, complex conductivity of the particle in the absence of surface ionic conductivity effects and includes the intrinsic particle wall conductivity.

Although no direct chemical method is known for accurately determining the fixed charge density in intact bacterial walls, acid—base titrations of isolated wall fragments of *Micrococcus lysodeikticus* have provided charge density values in good agreement with values estimated from the effective conductivities of the cell walls.[53] For isolated cell walls, the titration measurements yield surface charge densities in the range 24—82 mass equiv./litre, whilst the conductivity values correspond to densities in the range 70—95 mass equiv./litre. In Table 5.2, which compares the experimental results for $\Delta\epsilon_2$, as measured by Einolf and Carstensen,[51] with the theoretical values derived using equations (5.34) and (5.36), the fixed surface charge density N_s for *Micrococcus lysodeikticus* cells is taken as 80 mass equiv./litre.

Table 5.2   Comparison of the theoretical and experimental low frequency permittivity values for two bacterial species, based on equations (5.34) and (5.35), and on the work of Carstensen *et al.*[51-55]

Bacteria	N_s (mass equiv./litre)	N_i (m^{-2})	Cell radius (μm)	Permittivity	
				Theory	Experiment
M. lysodeikticus	80	1.3×10^{18}	0.4	3.2×10^5	$3.2 \pm 0.2 \times 10^5$
Micrococcus species	250	2.1×10^{18}	0.4	5.3×10^5	$6.0 \pm 0.4 \times 10^5$

Table 5.3   Theoretical and experimental relaxation frequencies for two species of *Micrococcus*

Bacteria	Relaxation frequency (kHz)		
	Theory	Expt. permittivity	Expt. conductivity
M. lysodeikticus	3	0.8	5
Micrococcus species	3	2	10

Also included in Table 5.2 are values for an unnamed species of *Micrococcus* bacteria which Carstensen[53] has reported as having the high value of 1.1 mho/m for its cell wall conductivity. Using this cell wall conductivity value, and assuming the mobility of the counter-ions to have the value of free sodium ions in solution, then the fixed surface charge density can be estimated to have a value of the order 250 mass equiv./litre. From Table 5.2 it can be seen that there is close agreement between the theoretical and experimentally derived values for the low frequency permittivity of the suspended bacteria.

In Table 5.3, the experimental values for the frequencies f_m of maximum dispersion loss, obtained from permittivity and conductivity measurements, are compared with the theoretical value $f_m = 1/2\pi\tau$ obtained from Schwarz's formula given by equation (5.32). In deriving τ, the counter-ion mobility is again assumed to have the value 5.4×10^{-8} m^2/V s, corresponding to that for free sodium ions in solution. From this table it can be seen that the theoretical relaxation frequency values are in good agreement with those observed experimentally. Einolf and Carstensen[51] comment on the fact that there is likely to be a distribution of counter-ion mobility values, and conclude that the permittivity measurements will tend to reflect the low relaxation frequency phenomena, whereas the conductivity data will tend to be dominated by the high relaxation frequencies. The fact that the theoretical relaxation frequencies of Table 5.3 have values between those derived from the permittivity and conductivity measurements would tend to confirm this viewpoint, although the precise physical explanation for this is not clearly evident.

Electron microscope examinations of *Micrococcus* bacterial cell walls indicate that they have comparatively simple structures.[51] In this way, the assumption used in deriving the theoretical values in Tables 5.2 and 5.3 that the cell walls form a homogeneous porous layer around the cytoplasmic membrane is probably adequate, as demonstrated by the close agreement between experiment and theory. If the cell walls had in fact been highly inhomogeneous, such close agreement would possibly not have resulted. Complications associated with inhomogeneous cell wall structures represent a useful area of work for those wishing to develop models describing the dielectric properties of bacterial suspensions and other microorganisms. The development by Einolf and Carstensen of a theoretical model involving ion porous membranes is of great value, since it is likely that the low frequency dielectric and electronic properties of many real biological systems will be dominated by counter-ion motions at the interfaces between such porous materials having differing electrochemical properties, where electrical double layers will be formed.

References

1. S. Bugarzky and F. Tangl, *Centralbl. f. Physiol. (Leipzig u. Wien)*, **1897**, 297–300; *Arch. Physiol. (Bonn)*, **1898**, 531–65.
2. G. N. Stewart, *J. Physiol.*, **24**, 356 (1899).
3. J. C. Maxwell, *A Treatise on Electricity and Magnetism*, Vol. 1, p. 435. Clarendon Press, Oxford (1881).
4. Lord Rayleigh, *Phil. Mag.*, **34**, 481 (1892).
5. K. W. Wagner, *Arch. Elektrotech.*, **2**, 378 (1914); **3**, 100 (1915).
6. K. W. Wagner, *Die Isolierstoffe der Elektrotechnik*, Ed. H. Schering. Springer, Berlin (1924).
7. J. B. Whitehead, *Lectures on Dielectric Theory and Insulation*, McGraw-Hill, New York (1927).
8. A. R. von Hippel, *Dielectrics and Waves*, J. Wiley, New York (1954).
9. J. Volger, *Progress in Semiconductors*, **4**, 205, Heywood, London (1960).
10. L. K. H. van Beek, *Progress in Dielectrics*, **7**, 69, Heywood, London (1960).
11. L. K. H. van Beek, *Proc. Colloq. Ampere*, **11**, 229 (1962).
12. J. A. Stratton, *Electromagnetic Theory*, Chapter 3. McGraw-Hill, New York (1941).
13. J. A. Osborn, *Phys. Rev.*, **67**, 351 (1945).
14. R. W. Sillars, *J. Instn. Elect. Engrs.*, **80**, 378 (1937).
15. D. Polder and J. H. van Santen, *Physica*, **12**, 257 (1946).
16. J. A. Reynolds and J. M. Hough, *Proc. Phys. Soc. B*, **70**, 769 (1957).
17. S. S. Dukhin, *Surface and Colloid Sci.*, **3**, 83 (1973).
18. J. B. Hasted, *Aqueous Dielectrics*, Chapter 5, Chapman & Hall, London (1973).
19. D. A. G. Bruggeman, *Ann. Physik. (Leipzig)*, **24**, 636 (1935).
20. T. Hanai, *Kolloid Z.*, **171**, 23 (1960).
21. T. Hanai, *Bull. Inst. Chem. Res. Kyoto Univ.*, **39**, 341 (1961).
22. E. M. Trukhan, *Fiz. tverd. tela*, **4**, 3496 (1962).
23. E. M. Trukhan, *Biofizika*, **11**, 412 (1966).
24. R. Pethig, *Disc. Faraday Soc.*, **51**, 214 (1971).
25. H. Fricke and H. J. Curtis, *J. Phys. Chem.*, **41**, 729 (1937).
26. H. P. Schwan, G. Schwarz, J. Maczuk, and H. Pauly, *J. Phys. Chem.*, **66**, 2626 (1962).
27. H. P. Schwan, *Advan. Biol. Med. Phys.*, **5**, 147 (1957).
28. H. P. Schwan and J. Maczuk, *Proc. 1st Nat. Biophys. Conf., Columbus, Ohio*, p. 348 (1959).
29. J. G. Kirkwood and J. B. Schumaker, *Proc. Nat. Acad. Sci. USA*, **38**, 855 (1952).
30. C. H. Collie, D. M. Ritson, and J. B. Hasted, *J. Chem. Phys.*, **16**, 1 (1948).
31. *CRC Handbook of Chemistry and Physics*, 53 Edition, F-177 (1972–73).
32. R. M. Hill, *Phil. Mag.*, **23**, 59 (1971).
33. C. L. Sieglaff and J. Mazur, *J. Colloid Sci.*, **15**, 437 (1960).
34. R. H. Ottewill and J. N. Shaw, *Kolloid Z.Z. Polym.*, **34**, 218 (1967).
35. G. Schwarz, *J. Phys. Chem.*, **66**, 2636 (1962).
36. J. B. Miles and H. R. Robertson, *Phys. Rev.*, **40**, 583 (1932).
37. M. Eigen and G. Schwarz, *Z. Physik. Chem.*, **4**, 516 (1955).
38. C. T. O'Konski, *J. Chem. Phys.*, **23**, 1559 (1955); *J. Phys. Chem.*, **64**, 605 (1960).
39. G. Schwarz, *Z. Physik.*, **145**, 563 (1956).
40. P. Ingram and H. Jerrard, *Sci. Progress*, **49**, 651 (1961).
41. J. M. Schurr, *J. Phys. Chem.*, **68**, 2407 (1964).
42. S. P. Stoylov, *Proc. 4th Int. Cong. Surface Active Substances*, Vol. 2, p. 171. Gordon & Breach, N.Y. (1967).
43. H. P. Schwan, *Z. Naturforsch.*, **6(b)**, 121 (1951).

44. H. P. Schwan, *Physical Techniques in Biological Research*, **6**, 336 (1963).
45. H. P. Schwan, *Ann. N.Y. Acad. Sci.*, **148**, 191 (1968).
46. S. T. Bayley, *Trans. Faraday Soc.*, **47**, 509 (1951).
47. D. Rosen, *Trans. Faraday Soc.*, **59**, 2178 (1963).
48. S. Takashima and H. P. Schwan, *J. Phys. Chem.*, **69**, 4176 (1965).
49. G. Brausse, A. Mayer, T. Nedetzka, P. Schlecht, and H. Vogel, *J. Phys. Chem.*, **72**, 3098 (1968).
50. E. P. Chang and J. C. W. Chien, *J. Polym. Sci: Polym. Phys.*, **11**, 737 (1973).
51. C. W. Einolf and E. L. Carstensen, *Biophys. J.*, **13**, 8 (1973).
52. E. L. Carstensen, H. A. Cox, W. B. Mercer, and L. A. Natale, *Biophys. J.*, **5**, 289 (1965).
53. E. L. Carstensen, *Biophys. J.*, **7**, 493 (1967); **8**, 536 (1968).
54. C. W. Einolf and E. L. Carstensen, *Biophys. J.*, **9**, 634 (1969).
55. C. W. Einolf and E. L. Carstensen, *J. Phys. Chem.*, **75**, 1091 (1971).
56. H. A. Pohl, *J. Biol. Phys.*, **2**, 113 (1974).

APPENDIX: THE ELECTRICAL DOUBLE LAYER

When an electrically neutral object is immersed into a liquid medium the surface of this object can acquire a net electrical charge. This surface charge can arise from several possible processes. A most basic cause of all will be that arising from a difference in thermodynamic or electrochemical potential (referred to as the Fermi energy in solid-state physics) between the solid and the solution. This effect has been studied for the case of silver iodide particles in aqueous solution for example.[1,2] Basically, we can say that two material phases in contact will have different affinities for electrons, and that the thermodynamic equilibrium condition is realized as a result of the flow of electrons or ions from one phase to the other. As early as 1898, Coehn[3] formulated the general rule, that in a disperse system composed of two non-conductors the material with the larger permittivity ϵ will acquire a positive charge, whereas the phase with the smaller ϵ value will become negative. Thus glass ($\epsilon \simeq 5-6$) against water ($\epsilon = 80$) or acetone ($\epsilon = 21$) becomes negatively charged, but against benzene ($\epsilon = 2.2$) becomes positively charged. The electrochemical potential difference between water and vacuum is about 0.5 V, with the water potential negative compared with vacuum (reference 1, p. 127). Another effect occurs for clay particles,[4] which attain a net charge as a result of the replacing of one of its constituent ions (e.g. Al^{3+}) with one of lower valency, such as Mg^{2+}. The situation most relevant for biological particles in weak electrolytes is where the surface charge is influenced by the adsorption of ionizable chemicals, or by the presence of constituent ionizable surface chemical groups such as COOH or NH_2.

As a result of the build-up of such surface charges, an electrostatic potential will be present in the locality of the particle, whose value decreases to that of the bulk liquid medium with increasing distance from the particle. As a result of this electrostatic potential, ions of opposite charge (counter-ions) to that of the surface charge will be attracted towards the particle, whilst ions of like charge (co-ions) will be repelled into the bulk liquid medium. The presence of a negatively charged particle, for example, surrounded by an 'atmosphere' of positively charged counter-ions, provides the concept of an electrical double layer.

The way in which the counter-ions are distributed in the double layer depends on the spatial variation of the electrostatic potential associated with the fixed charge on the particle. If the equilibrium volume density of the ions in the surrounding electrolyte is given as n_0, then the variation of the counter-ion volume density n^- with distance x from the particle surface is given by the Boltzmann formula

$$n^- = n_0 \exp[z^- q V(x)/kT]$$

where z^- is the valency of the counter-ion. The surface is assumed to be positively charged and the spatial variation of the electrostatic potential $V(x)$ is taken to be in a direction normal to the surface. For the sake of simplicity we shall take the surface as being planar. Likewise, the spatial variation of the co-ion volume density n^+ is given by

$$n^+ = n_0 \exp[-z^+ q V(x)/kT]$$

where z^+ is the valency of the co-ion. From these equations we see that everywhere $n^- n^+ = n_0^2$. The spatial variation of the volume charge density $\rho(x)$ in the ionic cloud surrounding the charged particle is given, according to Poisson's equation, as

$$\frac{\partial^2 V}{\partial x^2} = -\frac{\rho(x)}{\epsilon_0 \epsilon}$$

where ϵ_0 and ϵ are the permittivity of free space and the electrolyte medium, respectively, and $\rho(x)$ can be written as

$$\rho(x) = q(z^+ n^+ - z^- n^-)$$

If the surface charge density on the particle has an effective volume density greater than the equilibrium volume density n_0 of ions in the electrolyte, then the number of counter-ions near to the particle surface will greatly exceed the co-ion density. On the other hand, for electrolyte concentrations greater than around 0.1 M, then it is possible for n_0 to exceed the particle surface charge density, in which case the counter-ion and co-ion densities near to the particle will be of the same order of magnitude, with their density difference just counterbalancing the excess charge on the particle.

If we assume that $z^+ = z^- = z$, then the volume charge density is given by

$$\rho(x) = zq(n^+ - n^-)$$
$$= zqn_0 [\exp[-zqV(x)/kT] - \exp[zqV(x)/kT]]$$
$$= -2zqn_0 \sinh[zqV(x)/kT]$$

Poisson's equation then becomes

$$\frac{\partial^2 V}{\partial x^2} = \frac{2zqn_0}{\epsilon_0 \epsilon} \sinh[zqV(x)/kT]$$

Taking the boundary conditions that as $x \to 0$, $V(x) \to V_0$, and that as $x \to \infty$,

184

$V(x) \to 0$ and $(\partial V/\partial x) \to 0$, then the solution of Poisson's equation gives

$$V(x) = \frac{2kT}{zq} \ln\left[\frac{1 + \gamma \exp(-x/\lambda_D)}{1 - \gamma \exp(-x/\lambda_D)}\right]$$

where

$$\gamma = \frac{\exp(\alpha/2) - 1}{\exp(\alpha/2) + 1}$$

$$\alpha = zqV_0/kT$$

and

$$\lambda_D = \left(\frac{\epsilon_0 \epsilon kT}{2z^2 q^2 n_0}\right)^{1/2}$$

For the common case that $zV_0 \ll 5 \times 10^{-2}$ volts, then to a very good approximation

$$V(x) = V_0 \exp(-x/\lambda_D)$$

This indicates that the electrostatic potential decreases exponentially with distance from the particle surface, an effect which has commonly resulted in the electrical charge situation around a charged particle being described as the 'diffuse double layer'. Furthermore, when $x = \lambda_D$, then $V(x) = V_0/e = V_0/2.72$ and $V(x)$ has fallen by 63.2 per cent from its value V_0 at the particle surface. In this way, the factor λ_D can be used as a convenient indication of the spatial extent of the electrical double layer. For a 0.15 M KCl electrolyte, the equilibrium concentration of free K^+ and Cl^- ions is 6.5×10^{26} m^{-3}, so that at $T = 300$ K, assuming a permittivity value for the aqueous electrolyte of $\epsilon = 78.7$, the factor λ_D has the value 0.3 nm. A concentration of around 0.15 M corresponds to that of K^+ salts in the cytoplasm of cells and of NaCl in blood plasma. The factor λ_D, as mentioned in Chapter 5, is often referred to as the Debye, or Debye–Hückel screening length.

The description of the electrical double layer has been extended to include the possibility that some of the counter-ions form a monolayer adjacent to the charged surface, producing what is called a Stern layer.[5,6] Assuming that the Stern layer and the surface charge take the form of planar arrays of uniformly distributed charges at a distance δ apart, then as this represents the classical capacitor situation, the electrostatic potential will fall linearly between the surface and the Stern layer, as shown in the figure below. Beyond the linear fall of the potential to the value $V(\delta)$, the potential will then fall off exponentially with distance into the bulk electrolyte medium.

In electrophoretic measurements on colloidal systems, a parameter often cited is the so-called zeta potential. When there is relative motion between the charged particle and the surrounding liquid medium, it is assumed that part of the diffuse double layer remains firmly attached to the particles, whilst the remainder moves away from the particle. This separation of the diffuse double layer is taken to occur

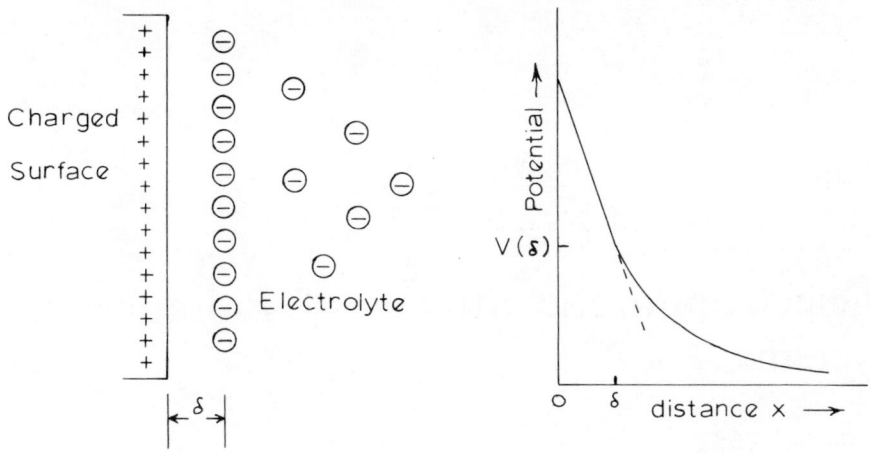

The electrical double layer formed according to the Stern layer model

at an infinitely thin plane termed the plane of shear, and the potential at this plane is what is referred to as the zeta potential.[1] As yet, no adequate theory appears to exist which definitely relates the zeta potential to the electrostatic potential associated with the charged particle.

References

1. H. R. Kruyt, *Colloid Science*, Vol. 1, Elsevier, Amsterdam (1952).
2. J. Lyklema, *Med. Electron. Biol. Eng.*, **2**, 265 (1964).
3. A. Coehn, *Ann. Phys. Chem.*, **64**, 217 (1898).
4. H. van Olphen, *An Introduction to Clay Colloid Chemistry*, Wiley, N.Y. (1963).
5. O. Stern, *Z. Elektrochem.*, **30**, 508 (1924).
6. E. J. W. Verwey and J. T. G. Overbeek, *Theory of the Stability of Lyophobic Colloids*, Elsevier, Amsterdam (1948).

Chapter 6

Dielectrophoretic Studies of Biomolecular Systems

In this chapter we shall consider the phenomenon of dielectrophoresis and its possible application for the study of small biological organisms.

'Dielectrophoresis' is defined[1] as the motion of matter caused by polarization effects in a non-uniform electric field, where the most polar matter moves towards the region of greatest field intensity. As such, dielectrophoresis should not be confused with the phenomenon of electrophoresis. Both terms imply the study of motion, as the Greek word *phoresis*, meaning motion, implies. The essential difference is that unlike electrophoresis, dielectrophoretic phenomena are concerned with the motion of electrically *uncharged* or neutral particles in *non-uniform* electric fields.

In Chapter 1 we were led to expect that for the simple arrangement of parallel planar electrodes separated by a dielectric medium, a uniform electric field would be produced between such electrodes on application of a voltage difference between them. That such an idealized uniform field is difficult or not readily achieved in practice, even between two very large flat and parallel electrodes, has been elegantly shown by the work of Felici[2] and Cassidy *et al.*[3] These workers employed the Kerr effect to study molecular orientations in high purity chlorobenzene and nitrobenzene used as the dielectric between two flat parallel metal electrodes. Within a time period of microseconds after the application of a potential difference between the electrodes, the electric field was observed to become highly distorted and warped, an effect which persisted for periods up to hours after the application of the steady voltage.

Field strengths up to values of 10^7 volts/meter (0.1 volt across 100 Å) are not uncommon in biological membranes. If supposedly ideal geometries for uniform fields do not in reality produce ideal results, then the large fields in biological systems are even less likely to provide the ideal uniform state. In this way, the consideration of non-uniform field effects, as manifested by dielectrophoresis, will be meaningful. Apart from providing a useful tool for the investigation of biological organisms, the understanding of dielectrophoretic phenomena may also give us an increased insight into the physical mechanism underlying certain biological functions.

By comparison with uniform field effects, the deliberate study of non-uniform field phenomena has been attempted by relatively few workers, and of this work most has been concerned with macromolecular sized particles rather than with molecules. Wrede[4] in the late 1920s studied the use of non-uniform electric fields to deflect molecular beams, and in 1938 Müller[5] presented theoretical evidence to show that non-uniform field effects would not be appreciably large for molecular sized particles. More recently, the subject has gained fresh momentum, chiefly as a result of the considerable and significant work of H. A. Pohl and his co-workers.[1,6–10] This work, together with that of Lösche and Hultschig,[11] has given both a theoretical and practical foundation for the study of dielectrophoresis. The effects are easily measurable for large particles of solid matter, or liquid, but as the size of the electrically neutral particles approaches that of small molecules, randomizing thermal processes begin to mask dielectrophoretic phenomena.

It will be our purpose here to outline the physical principles of dielectrophoresis, and to indicate its relevance to the study of biological materials.

THE PHYSICAL PRINCIPLES

In understanding the action of non-uniform electric fields on electrically neutral particles, it will be helpful first to consider the action of electric fields on electrically charged bodies. If a particle possessing a net positive charge Q is placed in a uniform electric field of intensity E volts/meter, it will experience a force $F = Q \cdot E$, and will be pulled along the field lines towards the electrode of opposite polarity, in this case the cathode, as shown in Figure 6.1(a). If placed in a non-uniform field, this charged particle will behave in essentially the same manner and will be attracted towards the electrode of opposite polarity to its own net charge, as shown in Figure 6.1(b). The behaviour of a neutral particle is different. In a uniform field the neutral particle will merely become polarized, with a positive charge being induced on the side nearest the cathode and negative charge on the opposite side nearest the anode, as shown in Figure 6.1(a). Since the particle is electrically neutral, then these two regions of induced charge will be of equal magnitude. The extent of polarization will depend upon its total polarizability, as discussed in Chapter 1. If the particle is composed of anisotropic material, or the particle is non-symmetric in shape (e.g. is elongated), then this polarization may produce a torque acting upon the particle to orientate or align it along the field direction. No such torque will result for a uniformly symmetric or isotropic particle. Whether or not an induced torque results, no net force will act upon this neutral particle in a uniform field, and it will remain motionless unless subjected to forces arising from other effects.

In a non-uniform electric field, however, the behaviour of a neutral particle is different. The neutral particle will again become polarized, but now a net force will act upon the particle so as to give it a translational motion towards the region of strongest electric field. This effect can be described using Figure 6.1(b), where we see that the field E_l to the left of the neutral particle is greater than the field E_r to the right (the field intensity is proportional to the number of electric field lines per

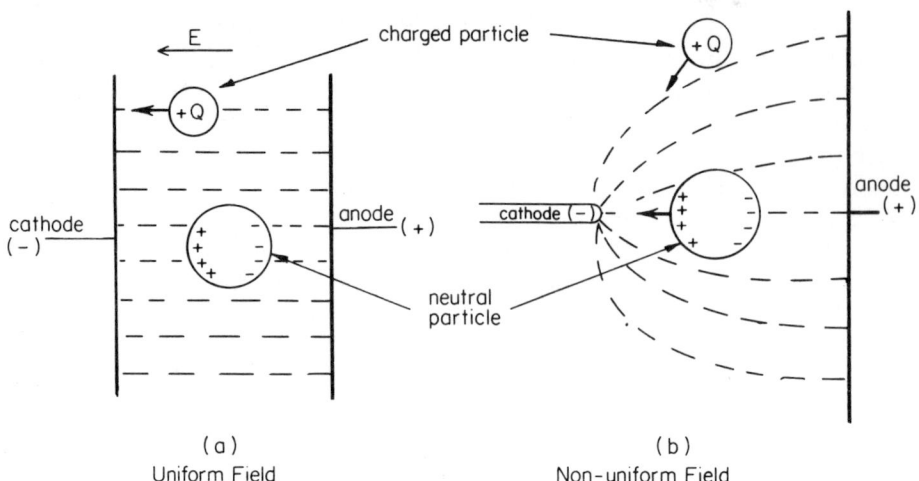

Figure 6.1 The behaviour of electrically neutral and charged bodies in (a) uniform electric field and (b) non-uniform field. Neutral bodies only experience a net translational force in non-uniform fields

unit area). The induced positive and negative charge δq on the sides of the particle will still have equal magnitudes, so that the net force, proportional to $E_1 \delta q$, pulling the particle towards the cathode will exceed that pulling it towards the low field direction of the anode. This net force producing the translational motion is termed the dielectrophoretic force.

We should now immediately forget that we have specified the anode and cathode locations regarding the behaviour of neutral particles in non-uniform fields. If the polarity of the electrode arrangement shown in Figure 6.1(b) were to be reversed, the neutral particle would still move towards the left, in other words towards the region of greatest field intensity. This now indicates another fundamental difference between the behaviour of charged and neutral particles in electric fields. In an alternating field, the charged particles will always tend to move towards the electrode having the opposite polarity to its own net charge (it appears, in fact, that a more detailed analysis shows that a gentle force acts upon a charged particle in an alternating field so as to force it away from the region of highest field intensity[1 2]). A neutral particle, on the other hand, will tend towards the field region of maximum intensity, no matter what is happening to the polarity of the electrode producing the region of maximum field intensity. We have seen in Chapter 1 that the degree of polarization of a dielectric material can vary with the frequency of the applied alternating voltage, with characteristic dielectric dispersion regions occurring as the frequency is steadily increased from near d.c. frequencies through to optical frequencies. We will expect, therefore, the translational motion of typical neutral dielectric particles in non-uniform fields to vary with the frequency of the field. We shall see later in this chapter how such effects can lead to dielectric experiments on living (and deliberately killed) cells, and to possible *in vivo* experiments on biomacromolecular systems.

THEORY

The theory for the force exerted by a non-uniform field on an electrically neutral body suspended in a fluid medium will now be outlined. Provided certain assumptions are made, this theory can be reduced to a relatively simple analytical expression. The treatment given here is based on that given by Pohl.[6,12]

In a static field, the net translational force on a neutral small body at equilibrium is given by

$$F = (\mu \cdot \nabla)E$$

where μ is the dipole moment vector (induced or permanent), ∇ is the del vector operator, and E is the external electric field. For the case where the neutral dielectric body is homogeneously, linearly, and isotropically polarizable, then

$$\mu = \alpha VE$$

where α is the polarizability, and V is the volume of the body. This gives us

$$F = \alpha V(E \cdot \nabla)E$$

$$= \frac{\alpha V}{2} \nabla |E|^2 \tag{6.1}$$

If we now consider the body to be a sphere of radius a, composed of an ideal (zero conductivity) dielectric of relative permittivity ϵ_2, suspended in an ideal dielectric fluid medium of infinite extent and relative permittivity ϵ_1, then the field interior to the small spherical body is given by

$$E_{in} = \left(\frac{3\epsilon_1}{\epsilon_2 + 2\epsilon_1} \right) E \tag{6.2}$$

The induced polarization per unit volume is[13]

$$P = \epsilon_0(\epsilon_2 - \epsilon_1)E_{in} \tag{6.3}$$

and the induced dipole moment is given by

$$\mu = VP = \alpha VE$$

The polarizability α per unit volume is therefore given from equations (6.2) and (6.3) as

$$\alpha = \frac{P}{E} = \epsilon_0(\epsilon_2 - \epsilon_1) \frac{E_{in}}{E}$$

$$= \frac{3\epsilon_0\epsilon_1(\epsilon_2 - \epsilon_1)}{\epsilon_2 + 2\epsilon_1}$$

and from equation (6.1), the total dielectrophoretic force F acting on the small sphere of volume $V = 4\pi a^3/3$, is

$$F = 2\pi a^3 \epsilon_0 \epsilon_1 \left(\frac{\epsilon_2 - \epsilon_1}{\epsilon_2 + 2\epsilon_1} \right) \nabla |E|^2 \tag{6.4}$$

We have used the term 'small sphere' here deliberately, so that we can make the approximation that although the field is divergent, it does not vary so strongly across the body as to alter appreciably the degree of polarization throughout the volume of the sphere.

For ellipsoidal bodies in general, having axes a, b, and c, the uniform field E_a within the ellipsoid, when the external field E is applied along the a-axis, is given by[14]

$$E_a = \frac{\epsilon_1 E}{\epsilon_1 + A(\epsilon_2 - \epsilon_1)} \tag{6.5}$$

where A is the depolarization factor as described in Chapter 5. For ellipsoidal bodies, the polarizability is then given by

$$\alpha = \frac{P}{E}$$

$$= \epsilon_0(\epsilon_2 - \epsilon_1)\frac{E_a}{E}$$

so that from equation (6.5)

$$\alpha = \frac{\epsilon_0 \epsilon_1(\epsilon_2 - \epsilon_1)}{[\epsilon_1 + A(\epsilon_2 - \epsilon_1)]}$$

and from equation (6.1) the dielectrophoretic force F is given as

$$F = \frac{2\pi abc}{3} \cdot \frac{\epsilon_0 \epsilon_1(\epsilon_2 - \epsilon_1)}{[\epsilon_1 + A(\epsilon_2 - \epsilon_1)]} \nabla|E|^2 \tag{6.6}$$

For spheres, $a = b = c$, and $A = \frac{1}{3}$, and equation (6.4) is seen to follow directly from the more general equation (6.6).

From equation (6.6) we see that the dielectrophoretic force depends directly upon the volume and polarizability of the body, and upon the square of the electric field intensity. This field square law dependence reminds us that dielectrophoresis is independent of the sign of the field, and that it can take place in both an alternating and static field. Furthermore, for isotropic dielectric materials we have the following vector relationship that

$$\nabla|E|^2 = 2E(\nabla \cdot E)$$

which is useful in that it indicates to us that to maximize the dielectrophoretic force given by equation (6.6), we should use both as strong and as rapidly a divergent electric field as possible. Equation (6.4) also indicates to us that the dielectrophoretic force depends upon the relevant permittivities in a way which does not, as might initially have been thought, increase without limit as the polarizability (hence permittivity) of the small body increases. Instead, the force is limited by the permittivity of the surrounding dielectric fluid medium, as is indicated by Figure 6.2, where the factor $\epsilon_1(\epsilon_2 - \epsilon_1)/(\epsilon_2 + 2\epsilon_1)$ of equation (6.4)

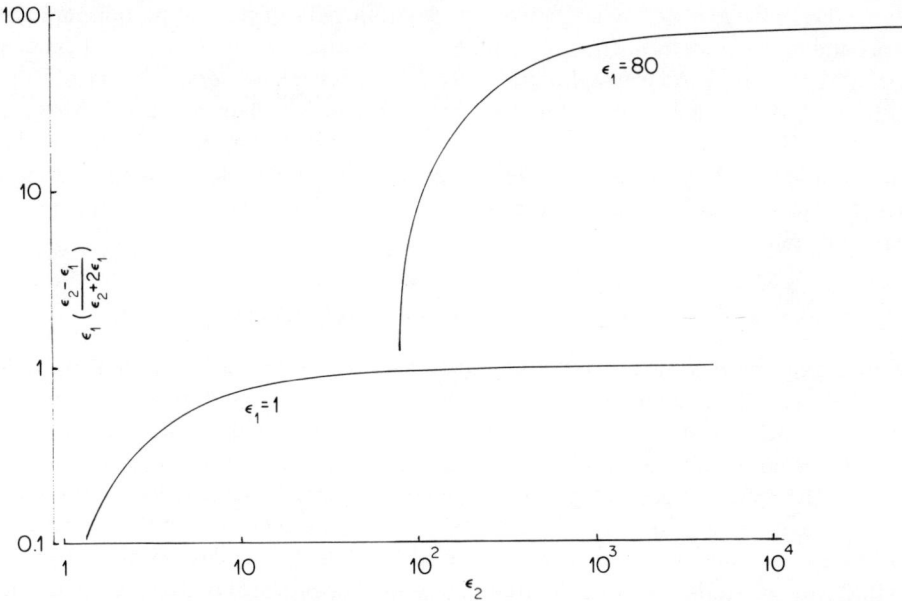

Figure 6.2 The variation of the dielectric factor $\epsilon_1(\epsilon_2 - \epsilon_1)/(\epsilon_2 + 2\epsilon_1)$ with the permittivity ϵ_2 of the immersed body, to show that this factor tends to the permittivity ϵ_1 of the surrounding fluid as ϵ_2 increases

is given against the permittivity values for small bodies immersed in air ($\epsilon_1 = 1$) and water ($\epsilon_1 = 80$). It can be seen that as the body tends to an infinitely high polarizability approaching that of the metallic state ($\epsilon_2 = \infty$), the dielectric term factor of equation (6.4) tends to the value ϵ_1 of the surrounding dielectric medium. From equation (6.4) we can also see that the net force on the body will be zero when $\epsilon_1 = \epsilon_2$, and that when the permittivity of the surrounding fluid exceeds that of the spherical body (i.e. $\epsilon_1 > \epsilon_2$), then the net force is reversed, and now the body is directed away from the greatest field intensity towards the *weakest* field intensity.

Before proceeding with an extension of dielectrophoretic theory to the study of biological materials, it will be of value once again to draw attention to how the phenomenon differs from that of electrophoresis. By contrast with dielectrophoresis, electrophoresis does not necessarily depend on particle volume or the total polarizability, but on the net excess charge on the particle. Electrophoresis can be observed equally well for large particles as well as molecular ions, and the effect can be observed for relatively low applied field strengths. In fact, high field effects such as dielectric breakdown, space charge, and electrode injection phenomena, can complicate the observation of electrophoretic phenomena. Dielectrophoretic phenomena, on the other hand, require strongly divergent fields, large particles, and relatively high field strengths. From equation (6.4) and Figure 6.2 it can also be seen that the dielectrophoretic force is maximized by having the surrounding dielectric medium with as high a permittivity value ϵ_1 as possible, and by having as large a difference between the permittivity of the test particle and the surrounding medium as possible. A simple, and essential, test to distinguish between dielectrophoretic and electrophoretic effects is the checking of whether or not the induced translational motion of the test particle is independent of the field polarity. If the translational motion direction depends on the polarity of the electrodes, then electrophoretic effects associated with electrically charged bodies are being observed. Dielectrophoretically induced motions are independent of electrode polarity.

APPLICATION OF DIELECTROPHORESIS

With a given particle material and size, then apart from the choice of the surrounding dielectric fluid medium, the only other parameter to be decided upon is the form of the electric field distribution. Two experimentally reliable electrode geometries are shown in Figure 6.3, together with the resulting expressions for $\nabla|E|^2$. The spherical geometry can experimentally be approximated by a rounded wire tip extending into a spherical hollow of an outer electrode, and the cylindrical geometry is easily achieved using a central wire held coaxially within an outer cylindrical electrode, or more approximately by a wire-plate combination with the two components parallel to each other. We have then, for these two field distributions, resulting dielectrophoretic forces given by

$$F = -4\pi a^3 \epsilon_0 \epsilon_1 \left(\frac{\epsilon_2 - \epsilon_1}{\epsilon_2 + 2\epsilon_1} \right) \frac{r_1^2 r_2^2 V^2}{r^5 (r_2 - r_1)^2} \qquad \text{(spherical field)}$$

and

$$F = -4\pi a^3 \epsilon_0 \epsilon_1 \left(\frac{\epsilon_2 - \epsilon_1}{\epsilon_2 + 2\epsilon_1} \right) \frac{V^2}{r^3 (\log_e r_1/r_2)^2} \qquad \text{(cylindrical field)}$$

The negative signs indicate that the force is directed towards the field axis, and the resultant particle motion will be strictly radial. It is worthwhile calculating what the magnitude of these forces will be for a typical experimental arrangement, where we choose the following parameters:

$\epsilon_1 = 3$ (fluid permittivity)
$\epsilon_2 = 10$ (particle permittivity)
$V = 5$ kV (outer electrode at earth potential)
$r_1 = 5 \times 10^{-4}$ m (central electrode radius)
$r_2 = 10^{-2}$ m (outer electrode radius)
$r = 10^{-3}$ m (particle distance from axis)

With these parameters, then the force exerted on a particle of radius 1 micron (10^{-6} m) is 1.01×10^{-12} newtons in a spherical field, and 4.07×10^{-13} newtons in a cylindrical field. The variation of dielectrophoretic force with particle radius for these two field distributions is shown in Figure 6.4.

If there is a concentration of particles suspended in the dielectric fluid medium,

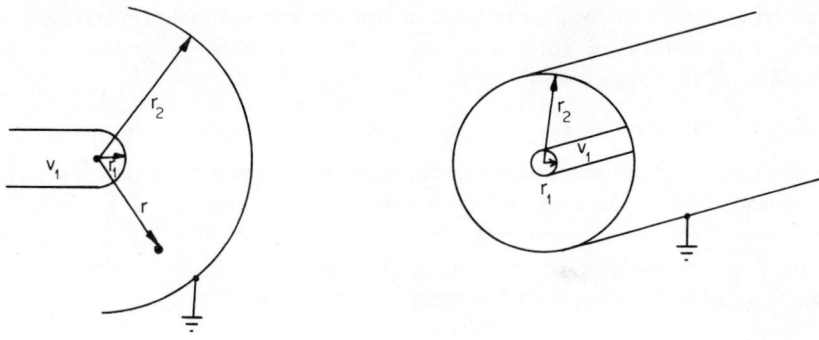

Spherical geometry

$$\nabla |E|^2 = -\frac{(2r_1^2 \, r_2^2 \, v_1^2) \, \bar{r}_0}{r^5 (r_2 - r_1)^2}$$

Cylindrical geometry

$$\nabla |E|^2 = \frac{-2v_1^2 \, \bar{r}_0}{r^3 (\log_e r_1/r_2)^2}$$

$\bar{r}_0 =$ unit radius vector

Figure 6.3 Spherical and cylindrical electrode geometries that can be used to generate non-uniform fields

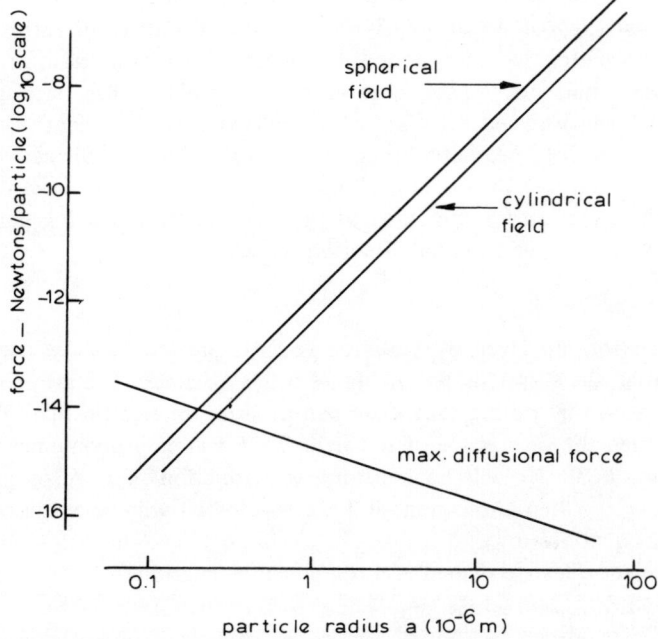

Figure 6.4 The variation of the dielectrophoretic and diffusional force with particle radius

then a diffusional force will be present in regions where the concentration is not uniform. This diffusional force $F_{\delta n}$ due to a concentration gradient, can be calculated using the relationship[8]

$$F_{\delta n} = kT(\Delta n/n)(1/\Delta r)$$

where $\Delta n/n$ is the fractional change in concentration along r for any volume under consideration. The maximum relative change possible for $\Delta n/n$ is unity (the presence versus absence of particles). The minimum change in radial direction r to have any physical meaning corresponds to the diameter, $2a$, of the particle. The maximum possible diffusional force is then given by

$$F_{\delta n}(\text{max}) = kT/2a$$

which for $a = 10^{-6}$ m has the value of the order 2×10^{-15} newtons at room temperature. The variation of this maximum diffusional force possible, as a function of particle radius, is also plotted in Figure 6.4. For particle sizes above about 0.5 micron, it can be seen from Figure 6.4 that the dielectrophoretic force, for our set of experimental parameters, far exceeds any forces that might arise from non-uniform particle concentrations.

In these two examples of possible field distributions, we see from Figure 6.3 and equation (6.1) that the dielectrophoretic force F can vary enormously with the position r of the test particle in the field. For example, using spherical electrode arrangements then $F \propto r^{-5}$, while for cylindrical electrodes $F \propto r^{-3}$. Such force variations would make the study of a collection of particles impractical. For example, two particles of equal size lying at radial distances of ratio 4:1 in a spherical field geometry, would experience dielectrophoretic forces different by a factor of greater than 10^3! The solution to this problem has been given by Pohl,[10,15,16] who introduced the concept of the isomotive field electrode system. In such an isomotive field geometry, there is practically the same dielectrophoretic force acting on a particle over a large range of possible particle locations. Pohl shows[16] that the equation for the constant potential surfaces which generate such an isomotive field along a given radial direction is given by

$$V \propto r^{3/2} \sin(3\theta/2) \tag{6.7}$$

When the relationship given by equation (6.7) is plotted to show the lines of constant potential, the result shown in Figure 6.5 is obtained. The rationale of the theory is understood by noting that if we can produce an electric field E proportional to $r^{1/2}$, then the divergence of the field $\nabla \cdot E$ will be proportional to $r^{-1/2}$, and the product $E(\nabla \cdot E)$ will be constant as a function of r. Also, since E is proportional to $r^{1/2}$, then the potential V ($E = -\nabla \cdot V$) will be proportional to $r^{3/2}$, as given by equation (6.7).

When metal electrodes are shaped according to the equation

$$r = r_{60} [\sin(3\theta/2)]^{-2/3}$$

as shown in Figure 6.5, then the desired isomotive field is produced anywhere in the θ = constant radial direction. The theory[16] indicates that in practice a solid

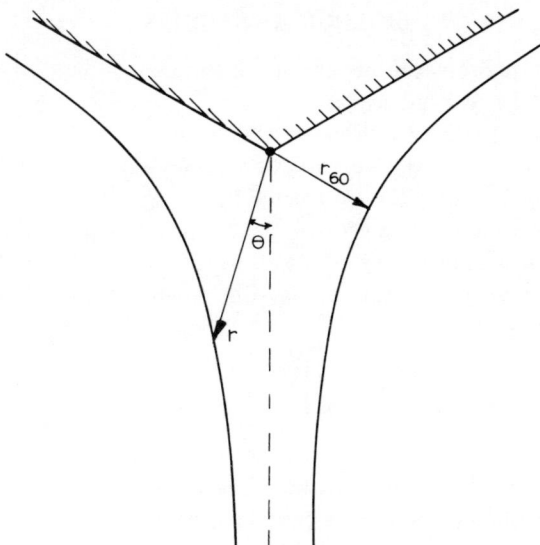

Figure 6.5 The isomotive field geometry

conductor should extend within the region bounded by $120° < \theta < 240°$. The simplest practical electrode system to provide the isomotive field distribution is shown in Figure 6.6, where only one curved electrode is employed and the test samples are located in the $\theta = 0$ direction.

In Figure 6.6, the curved electrode is maintained at a potential V_1 (d.c. or a.c.) and the V-shaped electrode is held at earth potential. With such an electrode system, particles in the xz plane will experience an essentially constant positive dielectrophoretic force in the positive x-direction over the electrode region OA. The curled ends to the electrodes are made to minimize field distortions at the electrode edges. In practice, it is often helpful to channel the test particles in their fluid medium into a groove cut into an insulating plate covering the electrode,[10] as shown in Figure 6.6.

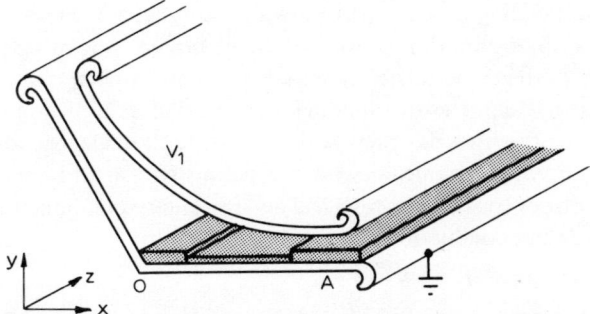

Figure 6.6 A simple electrode system providing isomotive conditions

BIOLOGICAL STUDIES

Although dielectrophoresis can be, and has been, used to study both animate and inanimate matter, we will restrict ourselves here to biological organisms. The most direct observation of dielectrophoresis is that of the yield of the test material collected at one or other of the electrodes producing the non-uniform field. When a suspension of cells or organelles is subjected to a non-uniform field, formed by two pin electrodes, for example, the cells or organelles are found to form long whiskers or chains attached radially to one of the electrodes. For a spherical field configuration, the effective yield y, expressed as the length of the chains of organelles formed, is given by Pohl[1,6] as

$$y = \frac{8\pi a^4 CVr_2}{9r_1(r_2 - r_1)} \left[\frac{2t\epsilon_1(\epsilon_2 - \epsilon_1)}{\eta(\epsilon_2 + 2\epsilon_1)} \right]^{1/2} \tag{6.8}$$

where C is the particle suspension concentration, η is the viscosity of the surrounding medium, and t is the elapsed time. The application of equation (6.4) to biological systems shows good agreement in general with experimental findings. However, no account is given implicitly in equation (6.8) for the frequency dependence of the dielectric behaviour of the suspended body or the surrounding medium.

It is found that different cellular species show their own distinct yield versus field frequency spectra, with up to three characteristic peaks or relaxation regions. Representative curves obtained by Pohl and his colleagues are shown for various organisms and organelles in Figures 6.7 and 6.8. The yield characteristic has also been shown to depend on the physiological state of the organism, as exemplified by the behaviour of live and dead yeast cells shown in Figure 6.9.

It is possible to interpret these yield spectra in terms of changes of the relative magnitude of the permittivity of the suspended body and the surrounding medium. This can be demonstrated using the results obtained for chloroplasts,[12] as shown in Figure 6.10(a), together with the frequency variation of the permittivity of the chloroplasts and surrounding aqueous medium that could produce such a yield spectrum, as outlined in Figure 6.10(b). It will be recalled from equation (6.4) that a positive collection rate only occurs when $\epsilon_2 > \epsilon_1$ (particle permittivity greater than that of the suspending medium), as shown by the hatched areas of Figure 6.10(b).

But such an interpretation is not sufficient, because the assumption that we are dealing with perfect dielectric materials remains implicit. In reality both the suspended particle and the surrounding medium will exhibit non-perfect dielectric properties. In other words, they will exhibit both dielectric displacement and conduction current phenomena, with the permittivity ϵ and conductivity σ being complex quantities (possessing both real and imaginary components). The dielectric displacement D and conduction current J_c are given by

$$D = \epsilon E \quad \text{and} \quad J_c = \sigma E,$$

with the total current given, according to Maxwell's equations, as

$$J = J_c + j\omega D$$

$$= (\sigma + j\omega\epsilon)E \tag{6.9}$$

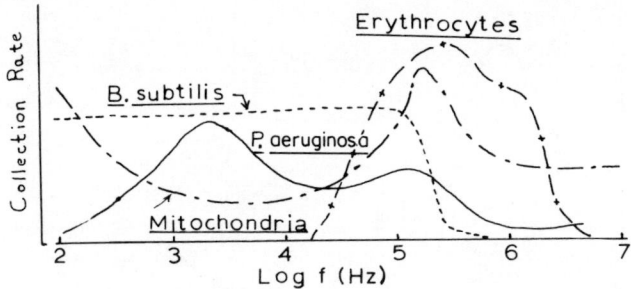

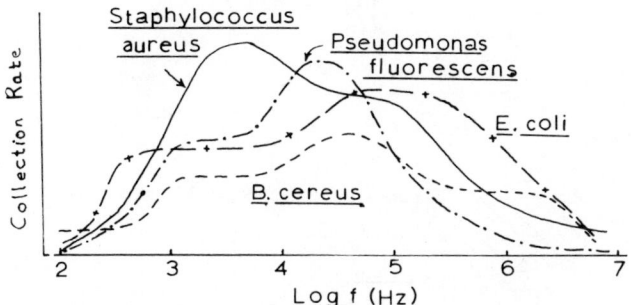

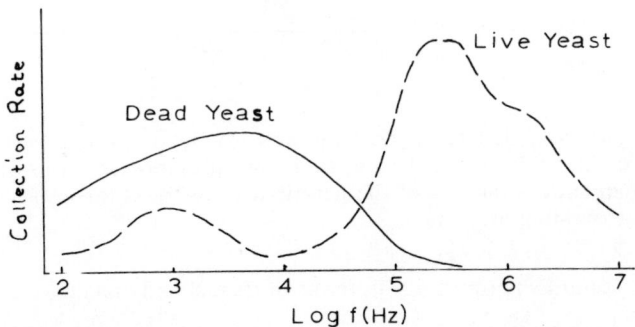

Figures 6.7–6.9 The dielectrophoretic collection rate
frequency spectra for various organisms, cells, and bacteria.
(Reproduced from H. A. Pohl, *J. Biol. Phys.*, **1**, 1 (1973),
by permission of Physical Biological Sciences, Inc.)

To a first approximation the conductivity and permittivity components of
equation (6.9) can be considered to be independent of each other. For example, the
low frequency conductivity of a liquid will depend upon the type and concentration of dissolved ionic impurities, whereas its relative permittivity will depend
essentially upon its molecular structure. As described in Chapter 1, we can

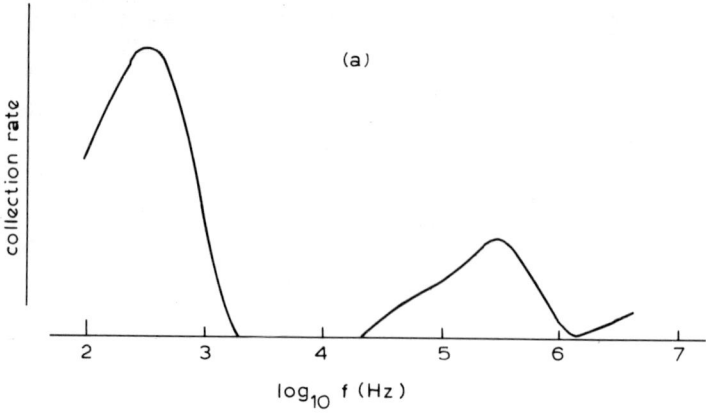

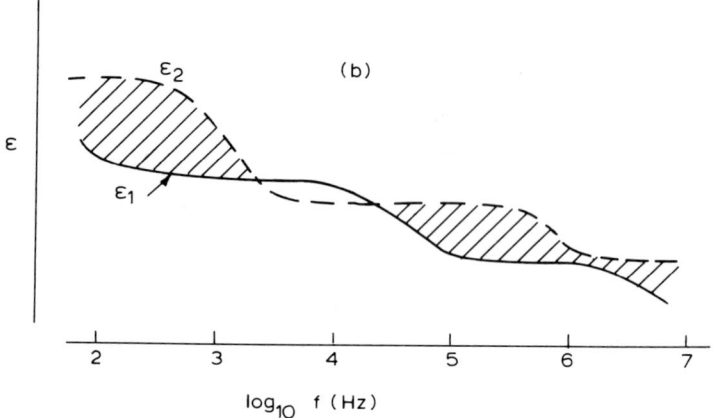

Figure 6.10 (a) The dielectrophoretic collection rate spectrum for chloroplasts and (b) a possible interpretation in terms of the frequency variations of the permittivity of the chloroplasts and surrounding medium

represent the complex permittivity in terms of its real and imaginary components as

$$\epsilon^* = \epsilon' - j\epsilon''$$

In the same way the complex conductivity can be written as

$$\sigma^* = \sigma' + j\sigma''$$

At this point Pohl and Crane[17] define a complex conduction factor K as

$$K \equiv \sigma^* + j\omega\epsilon^*$$

and a complex dielectric factor ξ as

$$\xi \equiv K/j\omega$$

From the determination of a general expression for the total electric energy of a suspension of particles in a dielectric medium, following similar lines to the procedure used by Schwarz[18] for non-conducting dielectrics, Pohl and Crane[17] have derived a generalized force equation, which gives the dielectrophoretic force as

$$F = -\frac{1}{4} \int_{body} \text{Real}\left\{ \nabla \left[\xi_m^* \left(1 - \frac{\xi}{\xi_m} \right) E_0^* \cdot E \right] \right\} dV \qquad (6.10)$$

The symbol ∇ is again the vector del operator, ξ_m and ξ are the complex dielectric factors for the surrounding medium and the body respectively, E_0 is the original impressed field, E is the resultant field throughout the suspended body, and the symbol $*$ indicates that the complex conjugate value is used for the particular complex parameter.

Crane and Pohl[19] have applied equation (6.9) for the case of living and dead yeast cells in aqueous media. The model used by these workers to describe a yeast cell is shown in Figure 6.11. The model consists of three concentric spheres, chosen as an approximation to the spherical shape of many yeast cells, surrounded by the fluid medium. The inner sphere represents the cell cytoplasm, with the next sphere defining the extent of the cell membrane. The outer sphere corresponds to an ion atmosphere surrounding the cell and anchored to the cell wall, and finally the fourth region represents the supporting aqueous dielectric medium.

As the measurements involved frequencies below 10^8 Hz, dielectric dispersions associated with the aqueous medium were considered to be absent. Also, dispersions associated with the lipid and protein structures of the membrane were neglected. The dominant dispersions were considered to be associated with frequency-dependent changes in the conductivity of the ionic atmosphere surrounding the cell, and in the conductivity of the membrane. As there are at present no known methods for obtaining the purely conductive properties of cell membranes, then no experimental values can be attached to the conductivity parameters of the cell membrane. Also, there are no accepted experimental values for the permittivity, conductivity, and thickness of the ionic atmosphere layer. As a first approximation, Crane and Pohl treated the ionic layer as essentially a more concentrated solution of the surrounding aqueous medium, where its permittivity would be somewhat less than that of pure water and the conductivity higher.

Using either accepted literature values for the parameters of Figure 6.11, or computer-derived best fits to the experimentally derived data, Crane and Pohl[19]

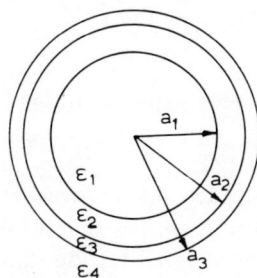

Figure 6.11 The theoretical model used for the yeast cell

Table 6.1   Parameters used by Crane and Pohl[19] for the calculations for living yeast cells

Parameter (see Figure 6.11)	Value
a_1	4×10^{-6} m
$a_2 - a_1$	10^{-8} m
$a_3 - a_2$	4×10^{-9} m
ϵ_1	60
ϵ_2	9
ϵ_3	40
ϵ_4	80
σ_1	1.0 mho/m
σ_2 (low frequency)	10^{-6} mho/m
σ_2 (high frequency)	9.9×10^{-6} mho/m
Ionic shell relaxation time	10^{-3} s
σ_3/σ_4	4.9×10^{3}

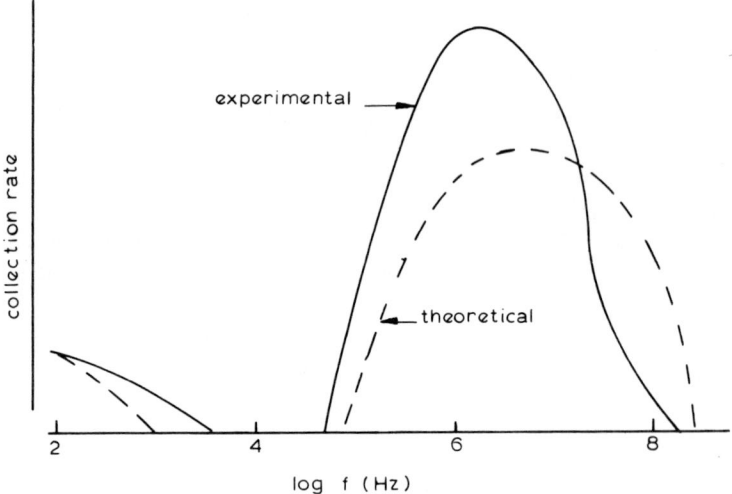

Figure 6.12 A comparison of the experimental and theoretical collection rate spectrum for live yeast cells in a suspension of conductivity 10^{-2} mho/m. (Reproduced with permission from J. S. Crane and H. A. Pohl, *J. Theoret. Biol.*, **37**, 15 (1972). Copyright by Academic Press Inc. (London) Ltd.)

were able to make a very good agreement between experiment and theory for the cell collection spectrum at one particular aqueous medium conductivity. It was found that using this same cell characteristics data, the theory also gave good predictions of the cell collection spectra for all the other experimental values used for the conductivity of the aqueous medium. This is an elegant test for checking the applicability of the theoretical data. The values adopted for the various characteristic cell parameters of Figure 6.11 are given in Table 6.1. A comparison of experimental and theoretical cell collection yields for live yeast cells is shown in Figure 6.12, for a suspension conductivity of 10^{-2} mho/m.

Considering the various approximations that were used, and the complicated nature of the problem, the agreement between experiment and theory shown by Figure 6.12 is remarkably good. Furthermore, the theory also had considerable success in describing the behaviour of dead yeast cells. For this purpose Crane and Pohl[19] reduced the value for the cell radius to $2 \mu m$, as microscopic evidence indicated, and the membrane was assumed to have been made slightly porous by the killing process (autoclaving), so allowing the conductivity of the central region to be equal to that of the suspending aqueous medium. All the other parameters were as those used for the living cells, as given in Table 6.1. The theory produced the two essential characteristics found for dead yeast cells, namely the absence of a low frequency minimum in the yield spectrum, and a cut-off point occurring at a lower frequency than for living yeast cells.

This work is extremely encouraging. Although as yet the theoretical model has not the refinements of including dispersions associated with bound water or the protein structure of the membrane, and has ignored the existence of the cell nucleus and the various intracellular bodies, the theory in its present form can teach us a great deal about the essential dielectric properties of biomacromolecular systems. The discovery that the dielectrophoretic properties of cellular bodies depends upon their physiological state also indicates that it should be possible through dielectrophoretic measurements to study and decipher the action of administered external chemical agents to such biological organisms. Recent developments[20] allow the investigation of single cell behaviour in non-uniform electric fields. Using a procedure roughly analogous to the famous Millikan oil-drop experiment, single cells can be suspended in mid-air with the dielectrophoretic force being adjusted to just cancel out the gravitational force. Using such a technique, the changes in dielectric properties of yeast cells were able to be correlated with the life cycle of the cells.[20]

The application of dielectrophoresis could also find great use in the separation and isolation of cells from cell mixtures. The general equation (6.6) shows us that the dielectrophoretic force depends on the shape of the test body, according to the depolarization factor A. Spherical cells or organelles will experience a different force than lamellae or rods having the same dielectric properties, for example. Also, by choosing the permittivity ϵ_1 of the suspending medium to have a value intermediate between that of the two cells constituting a two-cell mixture, we can arrange for the cells to be physically separated. One cell type (with $\epsilon_2 > \epsilon_1$) will head towards the electric field region of greatest intensity, whereas the other cells (with $\epsilon_2 < \epsilon_1$) will head towards the weakest field region. In cancer research great interest is directed towards isolating the so-called T-cells from bone marrow. These T-cells have concentrations some millionfold less than the other cellular matter of bone marrow, and great difficulty is experienced in isolating them to any great concentrations using conventional methods. The application of dielectrophoretic techniques could prove to be of great value in such work.

Recently, Szent-Györgyi has expressed the interesting viewpoint that a basic difference between cells in the normal and cancerous state may be associated with the electronic properties of the structural, membrane-bound proteins of the cell.[21,22] According to his scheme, the protein molecules are made electronically active as a result of charge-transfer interactions with electron-accepting molecules such as the aldehydes. As a result of such interactions the valence bands of the proteins become electronically desaturated and because of the mobility of the electron holes in the valence band energy states the protein molecules are converted from insulators into electronic conductors. Healthy cells correspond to the situation where the membrane-bound proteins are in such a conducting state. If the correct electron-donor/acceptor balance is disturbed either as a result of chemical impurity interference or enzyme (e.g. glyoxalase) action, the protein valence bands may become electronically saturated and hence non-conducting, with the result that the cell reverts back to a primitive, uncontrolled proliferative (cancerous), state. If this difference in the electronic conductivity of the membrane-bound proteins is reflected as an overall change in the conductivity of the membrane wall itself, then dielectrophoretic studies such as those described by Crane and Pohl[19] may be able to quantify such a basic difference. It may also be possible to be able to separate and isolate cancerous cells from a general cell population.

The dielectrophoretic behaviour of biological macromolecules could also be extended to include *in vivo* measurements. For example, it should be possible to study erythrocytes not only in aqueous media, but also in their own plasma environment. A factor that would complicate such *in vivo* studies would be the relatively high conductivity of the plasma, although the challenge is obviously a worthy one.

MUTUAL INTERACTIONS

So far we have considered just the effect of the non-uniform field on electrically neutral particles, and have ignored any possible interactions between these particles. Two inter-particle interactions are possible: a straightforward dipole–dipole interaction which will tend to make the particle arrangement of Figure 6.13(b) more stable than that of Figure 6.13(a); and a more subtle effect resulting in the effective 'bunching together' of the electric field lines in the neighbourhood of a highly polarizable particle, as shown in Figure 6.13(c). A second neutral particle, approaching the first particle, will now see an enhancement of the local field divergence and will tend to move towards the first particle, since the field strength will be greater nearer this first particle, as shown schematically by Figure 6.13(d).

As a result of these interactions, we might expect neutral particles in non-uniform fields to tend to form chain-like aggregates. Such neutral particle formations, commonly referred to as pearl-chain formations, have long been known and studied. Muth,[23] and much later Manegold,[24] observed such a phenomenon when fatty ester particle emulsions were subjected to high frequency fields, and Liebesny[25] appears to be the first to observe such an effect in biological particles, namely erythrocytes. More recently, Teixeira-Pinto *et al.*[26] studied pearl-chain

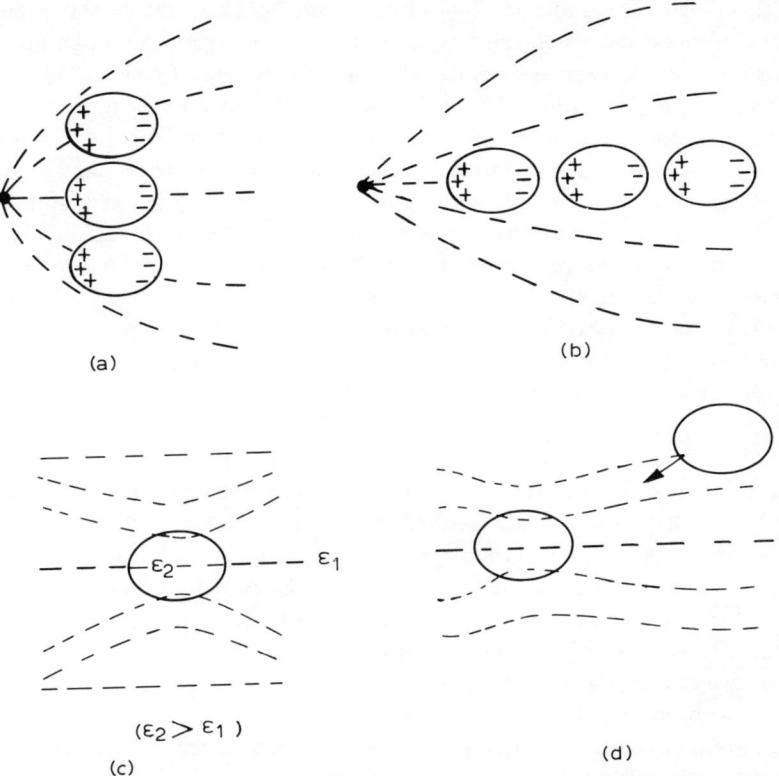

(a)

(b)

(c)

$(\varepsilon_2 > \varepsilon_1)$

(d)

Figure 6.13 Mutual interaction effects for neutral bodies in non-uniform electric fields

formation, and more complicated orientational and motional effects, of various organisms when subjected to high fields in the frequency range 100 kHz–100 MHz. Other studies of electric field induced particle aggregation for non-biological and biological particles includes the work of Füredi and Valentine[27] (powders, potato starch grains), Griffin and Stowell[28] (*Euglena* cells), and Griffin[29] (erythrocytes). Schwan and his colleagues[30,31] have presented a general theory, applicable to spherical and non-spherical particles in a.c. fields to account for the phenomenon of pearl-chain formation, and their theoretical results are in reasonable agreement with the experimental findings of Teixeira-Pinto *et al.*[26] Recently, Fomcenkov and Gabreliok[32] have investigated the dielectrophoretic behaviour of erythrocyte, bacterium, and water-microplant suspensions, paying particular attention to the relationship between cell collection rates and the appearance of pearl-chain formations and other field orientational effects. It was found that the orientational and dielectrophoretic spectra of the erythrocytes depended upon the pH and conductivity of the suspension medium, with a decrease in pH resulting in an increased collection rate at the electrodes and an increased tendency to orientate in a direction normal to the applied electric field, especially at the higher frequencies around 10 MHz. The addition of bovine serum albumin to the erythrocyte suspensions led to the disappearance or appearance of the transverse cell orientation in the field at the high frequencies depending upon the number of times the cells had been washed in sucrose solution. The well-washed cells exhibited a more pronounced orientation effect on addition of small amounts (~0.01 per cent) of albumin. The native erythrocyte and *Pseudomonas* cell suspensions exhibited similar dielectrophoretic spectra as those obtained by Pohl *et al.* and outlined in Figures 6.7 and 6.8. The erythrocytes that had been chemically fixed did not exhibit the native erythrocyte spectrum centred around 100 kHz, but gave a weaker frequency dependence with a small collection peak centred at a frequency below 100 Hz.

Pohl[33] has described how non-uniform field effects can be used to produce desired structures of cellular aggregates. Suspensions of the desired organisms are prepared in a non-toxic copolymer solution at $4-5\,^{\circ}C$ and then collected at electrodes used to produce the non-uniform field. The copolymer solution is then gently heated to $25-30\,^{\circ}C$ until it becomes a solid gel, so trapping the organisms. Using rod-shaped *Bacillus subtilis*, for example, gelled matrices of these cells can be produced, with the cells forming either monolayers or thick multilayered structures with a variety of cell orientations. Such a procedure is obviously most useful for preparing tissue models for dielectric and other investigations. The action of large non-uniform fields may also have relevance to the way in which membrane components arrange themselves and segregate with living cells. The ways in which electric fields may be involved in cell differentiation and membrane component segregation have recently been discussed by several authors.[34-36] Although these authors were basically considering the possible effects of uniform electric fields and electrophoretic effects, it is almost certain that non-uniform fields and dielectrophoretic phenomena must also be relevant.

The study and use of non-uniform electrical field effects for biological materials can therefore be considered a most important pursuit. Using non-uniform fields, living cells in aqueous media can be moved, collected, and even separated according to their species through the phenomenon of dielectrophoresis. Distinction can be made between old and young cellular organisms, between native and chemically fixed cells,[32] as well as between live and dead ones.[19,37-39] Dielectrophoresis provides perhaps the only known purely physical technique that is capable of such discriminations. The results of Crane and Pohl in the separation of living and dead yeast cells has been confirmed by Mason and Townsley.[39] Furthermore, these workers were able through dielectrophoresis to separate two nutritional types of yeast cells. In these ways the investigation of dielectrophoretic effects in biological organisms not only can lead us to increased knowledge regarding the dielectric properties of living cells, but also to the nature of their gross molecular structures and interactions. Both *in vitro* and *in vivo* experiments are possible, and the study of 'bunching' effects in non-uniform fields may also lead to an insight into the way organelles and cellular bodies aggregate to form the regular biological structures of living matter.

Other theories and applications that should be mentioned include those of Daba, and of Jones and Bliss. Daba has considered the action of dielectrophoretic forces on dielectric liquids, and has been able to make a quantitative assessment of the so-called Sumoto effect which describes the way dielectric liquids climb up electrodes producing stationary non-uniform electric fields.[40] Daba has also used classical electrodynamics to show that the excess pressure due to the dielectrophoretic forces exerted by an electric field on a dielectric liquid can be expressed in terms of the mass density of the liquid in the absence of the field.[41] Jones and Bliss[42] have investigated both the theoretical and experimental aspects of dielectrophoretic forces acting on small gas bubbles in insulating dielectric liquids, and have shown that a relatively simple model which neglects bubble elongation effects is able to describe adequately both the static and dynamic behaviour of the bubbles. They also have provided an experimental demonstration of a simple bubble trap which levitates and positions bubbles using dielectrophoretic forces, and which may have possible applications in laser target fabrication and zero-gravity space-processing procedures. The application of dielectrophoresis is also of use for dielectric measurements on irregular shaped bodies, such as mineral rocks or polymeric solids. Using an isomotive force cell, it is possible to determine the permittivity of a test sample by measuring the force exerted by a known field gradient, using a conventional beam balance weighing assembly.[43] The advantage of such a technique is that the use of electrodes to the test sample is not required.

Finally, it would not seem amiss to describe what (to the author at least) appears to represent a particularly interesting example of dielectrophoresis. In Woolaroc Lodge, the home of the late Mr & Mrs Frank Phillips, in the Woolaroc Ranch, Bartlesville, Oklahoma, there is a large room packed with the preserved heads and bodies of various animals that have died from causes other than that of the hunter's gun. Amongst the various horns, tusks, antlers, and snouts, can be seen some fine examples of the *Aquila chrysaetos* (golden eagle). On enquiring about the cause of death of these beautiful birds one is told that they were electrocuted by nearby electric power cables, and that according to the gamekeeper of the ranch, eagles have a peculiar substance in their wing structures that causes these birds to be drawn, as if by magnetism, into the overhead power cables! From a count of the number of golden eagles that appear to have been killed in this manner, it is possible to estimate that the (dielectrophoretic?) collection rate is of the order 2 eagles per decade per mile of cable. With the aid of equation (6.8), one could perhaps be able to estimate the dielectric properties of these large, winged, uncharged bodies!

References

1. H. A. Pohl, *J. Appl. Phys.*, **22**, 869 (1951).
2. N. J. Felici, *Révue Générale de L'Electricité*, **76**, 786 (1967).
3. E. C. Cassidy, H. N. Cones, and S. R. Booker, *IEEE Trans. Instrum. Meas.*, **IM–19**, 395 (1970).
4. E. Wrede, *Z. Physik.*, **44**, 261 (1927).
5. F. H. Müller, *Wiss. Veröffentl. Siemens-Werken*, **17**, 20 (1938).
6. H. A. Pohl, *J. Appl. Phys.*, **29**, 1182 (1958); **32**, 1784 (1961).
7. H. A. Pohl and J. P. Schwar, *ibid.*, **30**, 69 (1959).
8. H. A. Pohl, *J. Electrochem. Soc.*, **107**, 386 (1960).

206

9. H. A. Pohl and J. P. Schwar, *ibid.*, **107**, 383 (1960).
10. H. A. Pohl and C. E. Plymale, *ibid.*, **107**, 390 (1960).
11. A. Lösche and H. Hultschig, *Kolloid Z.*, **141**, 177 (1955).
12. H. A. Pohl, *J. Biol. Phys.*, **1**, 1 (1973).
13. A. R. Von Hippel, *Dielectrics and Waves*, p. 39. J. Wiley & Sons, Inc., New York (1962).
14. D. Polder and J. H. Van Santen, *Physica*, **12**, 257 (1946).
15. H. A. Pohl, *Sci. American*, **203**, 107 (1960).
16. H. A. Pohl, *J. Electrochem. Soc.*, **115**, 155C (1968).
17. H. A. Pohl and J. S. Crane, *J. Theoret. Biol.*, **37**, 1 (1972).
18. G. Schwarz, *J. Chem. Phys.*, **39**, 2387 (1963).
19. J. S. Crane and H. A. Pohl, *J. Theoret. Biol.*, **37**, 15 (1972).
20. C. S. Chen and H. A. Pohl, *Ann. N.Y. Acad. Sci.*, **238**, 176 (1974).
21. A. Szent-Györgyi, *Life Sciences*, **15**, 863 (1974).
22. A. Szent-Györgyi, *Electronic Biology and Cancer*, Marcel Dekker Inc., New York (1976).
23. E. Muth, *Kolloid Z.*, **41**, 97 (1927).
24. E. Manegold, *Kolloid Z.*, **111**, 11 (1950).
25. P. Liebesny, *Arch. Phys. Ter.*, **19**, 736 (1939).
26. A. A. Teixeira-Pinto, L. L. Nejelski, J. L. Cutler, and J. H. Heller, *Exp. Cell. Res.*, **20**, 548 (1960).
27. A. A. Füredi and R. C. Valentine, *Biochim. Biophys. Acta*, **56**, 33 (1962).
28. J. L. Griffin and R. E. Stowell, *Exp. Cell. Res.*, **44**, 684 (1966).
29. J. L. Griffin, *Exp. Cell. Res.*, **61**, 113 (1970); *Nature*, **226**, 152 (1970).
30. G. Schwarz, M. Saito, and H. P. Schwan, *J. Chem. Phys.*, **43**, 3562 (1965).
31. M. Saito, H. P. Schwan, and G. Schwarz, *Biophys. J.*, **6**, 313 (1966).
32. V. M. Fomcenkov and B. K. Gabreliok, *Studia Biophysica (Berlin)*, **65**, 35 (1977).
33. H. A. Pohl, *J. Colloid Interf. Sci.*, **39**, 437 (1972).
34. B. A. Horwitz and L. P. Horwitz, *J. Theoret. Biol.*, **42**, 169 (1973).
35. L. F. Jaffe, *Nature*, **265**, 600 (1977).
36. M. Poo and K. R. Robinson, *Nature*, **265**, 602 (1977).
37. J. S. Crane and H. A. Pohl, *J. Electrochem. Soc.*, **115**, 584 (1968).
38. H. A. Pohl and J. S. Crane, *Biophys. J.*, **11**, 711 (1971).
39. B. D. Mason and P. M. Townsley, *Can. J. Microbiol.*, **17**, 879 (1971).
40. D. Daba, *J. Phys. A: Gen. Phys.*, **5**, 318 (1972).
41. D. Daba, *J. Phys. D: Appl. Phys.*, **7**, 1458 (1974).
42. T. B. Jones and G. W. Bliss, *J. Appl. Phys.*, **48**, 1412 (1977).
43. H. A. Pohl and R. Pethig, *J. Phys. E: Sci. Instrum.*, **10**, 190 (1977).

Note added in proof:

Since this Chapter was written, an extensive account has been given by Pohl of the application of dielectrophoresis to the characterization and separation of cells.

H. A. Pohl, In *Methods of Cell Separation*, Vol. 1 (N. Catsimpoolas, Ed.), Plenum Press, New York (1977), Ch. 3.

Chapter 7

Biological Membranes and Tissue

In this chapter the dielectric properties of membranes, both natural and artificial, and of living tissue will be described. There is now an increasing interest regarding the dielectric properties of animal and human tissue in connection with the development of the therapeutic use of radio-frequency and microwave power in diathermy, and in understanding the possible hazards associated with microwave radiation. At frequencies below about 10 kHz, the dielectric properties of living tissue are associated mainly with the membranes which enclose the tissue cells, and in particular with the high resistance value of the membrane. The membrane also possesses a relatively high capacitance value, and as the frequency of an impressed electrical field is increased the capacitive reactance effectively 'shorts-out' the membrane resistance so that the effective cell resistance is associated mainly with the cytoplasmic medium within the cell. On death of the cell, or through membrane damage, the membrane loses its high resistance property.

The dielectric properties of membranes will, therefore, be considered in some detail before proceeding to biological tissue. A separate consideration of membranes is also warranted by their great biological importance. Much of modern cell biology is now directed towards a greater understanding of the complex structure and functions of membranes, and dielectric studies have already made important contributions. The unique importance of biological membranes has been poetically summarized by Lewis Thomas[1] as follows:

'It takes a membrane to make sense out of disorder in biology . . .
To stay alive, you have to be able to hold out against
equilibrium, maintain imbalance, bank against entropy, and you
can only transact this business with membranes in our kind of world.'

BIOLOGICAL MEMBRANES

Biological membranes are ultrathin structures composed primarily of lipids and proteins, with smaller amounts of such substances as water, cholesterol, sugar groups, and metal ions. The concept that cellular membranes contain significant amounts of lipid material originated in 1899 when Overton[2] found that lipids readily diffused across cellular membranes. Since lipid molecules are amphiphilic,

207

Table 7.1   The five basic types of biological membrane and their approximate lipid and protein contents on a dry weight basis

Membrane type	Composition
1. Plasma	
e.g. Mouse liver cells	54% lipid, 46% protein
Human erythrocyte	48% lipid, 52% protein
Amoeba	43% lipid, 57% protein
Mitochondrial outer membrane	48% lipid, 52% protein
2. Chloroplast thylakoid	40% lipid, 52% protein, 8% pigments
3. Mitochondrial inner cristae	24% lipid, 76% protein
4. Retinal rods	40% lipid (mainly phospholipid), 60% protein (e.g. opsin)
5. Nerve axon myelin	81% lipid, 19% protein

being hydrophilic at one end and hydrophobic at the other, they tend to line up neatly on aqueous surfaces with the polar (hydrophilic) ends pointing downward in contact with water molecules and the hydrophobic ends sticking upwards into the air. This naturally led to the concept that cell membranes consist of lipid mono-layers. In the 1920s the Dutch scientists Gorter and Grendel extracted the phospholipids from the membranes of human erythrocytes and, using the Langmuir trough technique, they were able to determine the area this extracted lipid could cover in the form of a lipid monolayer. They were able to conclude that there was just sufficient lipid to form a bimolecular lipid leaflet surrounding each erythrocyte cell.[3] Recent measurements using modern techniques produce the same conclusion.[4,5]

The biological membranes can usefully be classified into five basic types according to their specific function. These are the plasma membrane of cells and cytoplasmic organelles, the thylakoid membrane of chloroplasts, the inner cristae membrane of mitochondria, the myelin nerve membrane of axons, and the membrane of retinal rods and cones. These five basic membrane types are listed in Table 7.1, together with their typical content of lipid and protein material.[6,7]

It can be seen that proteins nearly equal or exceed the quantity of lipid in most membranes. The major exception is that of myelin, where proteins constitute less than one-fifth of the membrane material. This difference is most probably due to the fact that myelin is thought to act as an insulator of the neuron for the conduction of nervous impulses, whereas the other membranes have functions such as the active transport of molecules, and energy transduction and conversion which require the participation of protein molecules.

Most membrane lipids can be considered to be based on glycerol which is a simple three-carbon molecule with hydroxyl groups attached to each carbon atom, as shown in Figure 7.1(a). In membrane lipids, two of the hydroxyl groups are replaced by long hydrocarbon-chain fatty acids, and the third hydroxyl group is replaced by a polar group which either carries, or can accept, electric charge. The

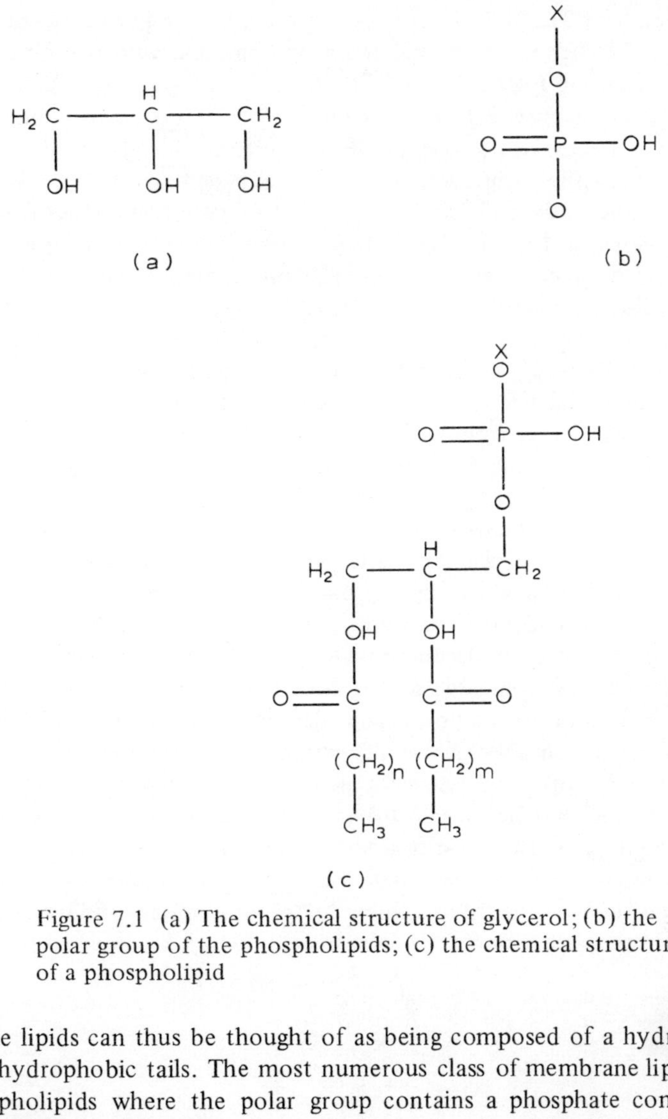

$$H_2C — \underset{\underset{OH}{|}}{\overset{\overset{H}{|}}{C}} — \underset{\underset{OH}{|}}{CH_2}$$
$$\underset{OH}{}$$

(a)

$$X$$
$$|$$
$$O$$
$$|$$
$$O = P — OH$$
$$|$$
$$O$$

(b)

$$X$$
$$O$$
$$\|$$
$$O = P — OH$$
$$|$$
$$O$$
$$|$$
$$H_2C — \overset{\overset{H}{|}}{C} — CH_2$$
$$| \qquad |$$
$$OH \quad OH$$
$$| \qquad |$$
$$O = C \quad C = O$$
$$| \qquad |$$
$$(CH_2)_n \ (CH_2)_m$$
$$| \qquad |$$
$$CH_3 \quad CH_3$$

(c)

Figure 7.1 (a) The chemical structure of glycerol; (b) the polar group of the phospholipids; (c) the chemical structure of a phospholipid

membrane lipids can thus be thought of as being composed of a hydrophilic head and two hydrophobic tails. The most numerous class of membrane lipids is that of the phospholipids where the polar group contains a phosphate complex with a variable group X, as shown in Figure 7.1(b). In lecithin (phosphatidylcholine), for example, the phosphate has a choline $(CH_3)_3N^+ \cdot CH_2 \cdot CH_3 \cdot CH_2^-$ group attached. Other possible X groups include the amino acids, serine, and threonine, amines having a structure similar to choline, sugars, and metal ions such as calcium. The phospholipids from biological membranes often contain both a saturated and an unsaturated hydrocarbon chain, with the unsaturated one nearly always being situated in the centre of the basic glyceryl framework. These fatty acids are often characterized by two numbers for convenience, the first giving the number of carbon—carbon double bonds which quantifies the degree of unsaturation. Examples of this are given for the following fatty acids: palmitic (16.0) stearic

(18:0), linoleic (18:2), linolenic (18:3), arachidonic (20:4) and docosahexanonic (22:6). A typical chemical structure for a phospholipid with two saturated fatty acid chains is shown in Figure 7.1(c).

Despite the complex and highly variable structure and function of the various biological membranes, it is commonly considered that they all possess the bimolecular lipid leaflet construction as their fundamental structure. According to this viewpoint the basic building block of all biological membranes is that of the lipid bilayer shown in Figure 7.2(a), where the lipid molecules are orientated with their polar groups facing outward into aqueous environments, and their hydrophobic hydrocarbon chains point inwards to form the membrane interior. This is the basic concept developed by Gorter and Grendel,[3] and which was later extended by Davson and Danielli[8] to include adsorbed protein layers on the membrane surfaces. The lipid bilayer construction also appears to be the major structural entity of cell membranes from electron microscopy studies (e.g. references 9 and 10), where the cell membranes appear in the form of a triple-layered image consisting of two dark lines on either side of a less dense region. The dark lines are considered to represent the deposition of electron-dense fixative at the polar heads of the lipid molecules, whilst the lighter region is the unstainable and electron-transparent hydrocarbon phase in the interior of the lipid bilayer. The overall thickness of this triple-layered structure is measured to be of the order 5–10 nm, consistent with that of a bimolecular lipid leaflet. Such electron microscopic studies should be viewed with some caution of course, since the technique does require complete dehydration of the cells and the introduction of 'foreign' chemical fixative into the membranes. X-ray diffraction studies (references 11 and 12) on nerve myelin and retinal rod segments also provide strong evidence for the basic lipid bilayer model, but here again the experimental procedures required and the long times required to form a diffraction pattern will tend to mask some of the more subtle features of the membrane structures. Variations of the basic lipid bilayer concept include the 'hydrophobic lipoprotein' model of Sjostrand[13] where the lipid molecules are assumed to form micelles which are covered by a layer of

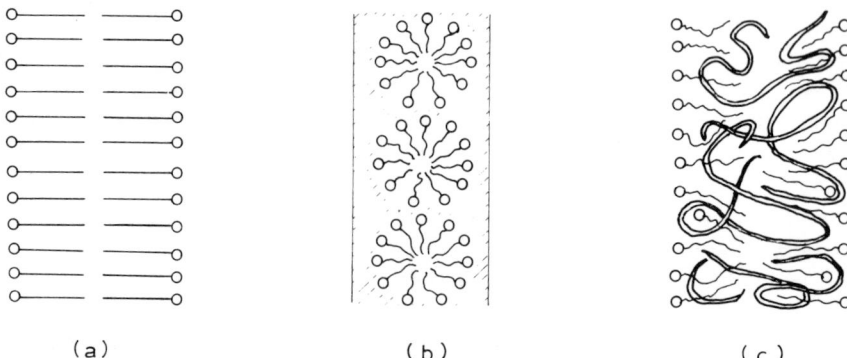

(a) (b) (c)

Figure 7.2 (a) The basic bilayer lipid model for a biological membrane; (b) the spherical lipid micelle membrane model of Lucy;[14] (c) the lipoprotein model proposed by Benson[15]

protein, and the 'spherical lipid micelle' model proposed by Lucy[14] in which globular micelles of lipid are in hexagonal close packing within the membrane, as shown in Figure 7.2(b). For the case of the chloroplast membrane in particular, Benson[15] has proposed that the lipids are bound by hydrophobic regions within the protein molecule structures, to form a two-dimensional lipoprotein aggregate as shown in Figure 7.2(c). A variation of this model has been suggested by Deamer[16] where although the lipids form a bilayer, each phospholipid has one hydrocarbon chain interacting with a hydrophobic bonding site on a membrane-associated protein molecule, and the other chain is directed into the hydrophobic interior of the membrane.

From Table 7.1 it can be seen that three-quarters of the mitochondrial inner membrane is composed of protein material. Green and his colleagues (e.g. references 17 and 18) have investigated alternative membrane models where the proteins, rather than the lipid molecules, are involved in maintaining the basic membrane structure, and they have proposed that in mitochondrial membranes the basic structure consists of a repeating sequence of lipoprotein subunits. The important rôle of the proteins has been emphasized by Korn[19] who has suggested that it is more reasonable for the protein molecules to be considered as the fundamental structural units of membranes, with the lipids being added later. It is now known that proteins are incorporated into the membrane lipid matrix in a number of ways. Some, such as spectrin, are adsorbed or otherwise attached to the membrane surface, molecules like cytochrome-*b* are embedded in the membrane lipid with portions of it projecting into the surrounding medium, and others like glycophorin[20] and (Na^+, K^+)-ATPase[21] extend right across and through the membrane. The actual structure of biological membranes can therefore be considered to be more complex than indicated by Figure 7.2. In seeking for the most basic fundamental structure, it is perhaps difficult to escape from the fact that both on thermodynamic and symmetry considerations, the lipid bilayer leaflet model represents the simplest two-dimensional and most likely irreducibly basic structure.

In Chapter 2, using bovine serum albumin and lysozyme as a basis, the relative permittivity of a protein molecule was estimated to be of the order 2.6–2.7. Fettiplace *et al.*[22] have estimated that the hydrocarbon residues of typical lipids have relative permittivity values in the range 2.2–2.6. We can make our own estimate of the permittivity value with the method described in Chapter 1, using equation (1.44) and Table 1.2. For a close-packed assembly of the hydrocarbon residues of the lipid monopalmitolein (16:1) having a molecular volume[22] of 0.421 nm^2 then the corresponding permittivity value is calculated to be $\epsilon = 2.28$. For monodocosahexanoin (22:6) of molecular volume 0.528 nm^3 then $\epsilon = 2.51$. We can also estimate that the specific resistivity of the lipid–protein matrix will be of the order 10^{12} ohm cm. Assuming the membrane to be of a thin planar structure we can derive an approximate value for the membrane capacitance and resistance on a unit area basis using equations (1.1) where

$$C = \frac{\epsilon_0 \epsilon A}{d} \quad \text{and} \quad R = \frac{d}{\sigma A}$$

These equations will be reasonably adequate for such systems as that of the squid giant axon, since the ratio of the axon diameter to membrane thickness is large. The membrane optical refractive index n can also be calculated using the relationship $n = (\epsilon)^{1/2}$. In the table below the estimates so derived for these physical constants (assuming $d = 6$ nm and $\epsilon = 2.5$) are compared with those values commonly observed for biological membranes.[2 3]

Parameter	Estimate ($d = 6$ nm, $\epsilon = 2.5$)	Natural membranes[2 3]
Capacitance ($\mu F/cm^2$)	0.37	0.5–1.3
Resistance (ohm cm^2)	6×10^5	$10^2 – 10^5$
Refractive index	1.58	~1.6

From this table it can be seen that our simple estimates are not too dissimilar from the actual values measured for cellular membranes. It is known that water molecules are incorporated into membrane structures, and consideration of the effects of such intramembrane water would be sufficient to raise the effective membrane permittivity and so bring the estimated capacitance value into closer agreement with that obtained experimentally. It should also be noted that, because of the reasonable agreement of the capacitance values obtained in the kHz range of frequencies with the estimated value given in the above table, effects associated with orientational polarizations of the membrane protein molecules must be very small. Although this may be taken as evidence that the protein molecules are firmly bound into the membrane structure, a major reason for the apparent lack of dipolar orientational polarizations is associated with the fact that all living cell membranes so far studied have a potential difference across them with the inside of the cell at the negative potential. Depending on the cell this membrane potential has a value between 50 and 100 mV,[2 4] corresponding to there being a field of the order 10^7 V m^{-1} across the membrane and directed towards the cell interior. Proton gradients, giving fields of the order 6×10^7 V m^{-1} across the mitochondrial inner membrane, are also thought to be involved in ATPase activity.

We saw in Chapter 3 that protein molecules can possess dipole moments of the order 300 debye and greater. The Langevin factor $\mu E/kT$ described in Chapter 1 will thus have a value exceeding 2.5 for protein molecules in the membrane field, and this value corresponds to the situation where there is saturation of orientational polarizability. For molecules like water having small dipole moments by comparison, the factor $\mu E/kT$ will have a value of less than 2×10^{-2} and orientational polarizations can take full effect. The fact that the membrane capacitance can be accurately described in terms of its constituent material having a relative permittivity of $\epsilon = 5~6$ indicates that either the membranes have a very low water content, or that the water molecules are irrotationally bound through hydrogen bonding or have formed polarized hydration sheaths around ions and other point charges. For an electrical materials scientist, the fact that the field acting across cell membranes can exceed 10^7 V m^{-1} is of considerable interest. For many dielectric materials (e.g. transformer oil, paraffin, rubber, glass, and plastics) such a field strength is sufficient to cause electrical breakdown effects. In the form of biological membranes, nature has somehow produced a dielectric which although being maintained in a condition very close to electrical breakdown and destruction is capable of orchestrating many electrical and non-electrical events in a very controlled and specific manner.

The membrane field may also have a rôle in orienting protein molecules in the membrane. Optical measurements on intact membranes indicate that some 40 per cent of the protein molecules are in the α-helix configuration, which is a greater proportion than that found for most soluble globular proteins. For example, circular dichroism spectra[7] of intact erythrocyte membrane show a double minimum at wavelengths of 208 and 222 nm, indicating that much of the constituent protein is in the α-helix form. By contrast, the protein β-sheet form would only show a single minimum at 218 nm. In the human erythrocyte membrane,

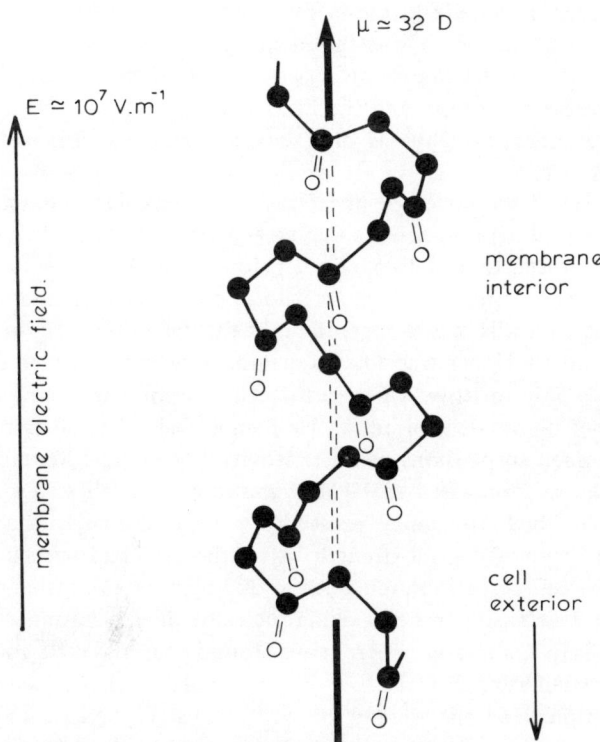

Figure 7.3 A polypeptide α-helix with its resultant dipole moment vector aligned with the membrane field

glycophorin accounts for about 20 per cent of the membrane protein and as already mentioned it extends completely across the membrane. The glycophorin molecule appears to consist of one polypeptide chain of 200 amino-acid residues and has its carboxyl end group protruding into the cytoplasmic interior of the red blood cell, whilst its terminal amino group extends beyond the outer surface of the cell membrane.[25] Outside the cell membrane, there are about 25 oligosaccharide groups attached to the glycophorin molecule. The α-helix polypeptide chain configuration will be stable in the hydrophobic membrane interior since there will be no competition for the hydrogen bonds from water molecules, and it would seem very probable that the glycophorin portion inside the membrane is in the α-helix form. Another protein molecule known to extend right across the membrane is that of bacteriorhodopsin in the purple membrane of the bacterium *Halobacterium halobium*.[26] Each of the bacteriorhodopsin molecules take the form of seven α-helices extending right across the membrane, and up to 12 of these protein molecules group together to form a single purple membrane particle (having 84 transmembrane α-helices) which appears to have the function of being a light-driven proton pump. In Chapter 2 it was shown that as a result of the peptide dipole moment of 3.63 debye, just one turn of a polypeptide α-helix will have a resultant dipole moment parallel to the helix axis of the order 13 debye. We will expect the orientation of such an α-helix in the membrane to be such that its dipole moment will be aligned with the field across the membrane. This is illustrated in Figure 7.3 where the polypeptide backbone of a two-and-a-half turn α-helix is outlined to show just the carbonyl groups, and the helix is so oriented that its resultant dipole moment of the order 32 debye is directed along the membrane field direction towards the cell interior.

Our knowledge of the dielectric properties of biological membranes results from measurements on cellular suspensions that can be traced back to the last century. In 1899 Stewart reported that at frequencies below about 1 kHz red blood cells were poorly conducting,[27] and a few years later Bernstein postulated that the only resistive part of the cells was a membrane at the cell surface.[28] In an important series of experiments Höber was able to demonstrate that the membrane of a red blood cell was highly resistive, whereas the cell interior was of low resistivity and similar to that of dilute electrolytes.[29] He demonstrated that at frequencies of the order 1 kHz red cell suspensions had a resistivity exceeding 1000 ohm cm but that as the frequency was increased to 10 MHz the resistivity fell to a value of around 200 ohm cm. This high frequency resistivity value is similar to that exhibited by NaCl solutions of physiological strength. After the cells had been haemolysed the resistivity of the cell suspension remained at 200 ohm cm even when the frequency was reduced to 1 kHz. Experiments along the same lines were made by Osterhout on the marine kelp *Laminaria*, where it was found that at 1 kHz the resistivity of the plants decreased after their death to a value similar to that of sea water.[30] Such experiments rationalized the already known fact, established by 1876[31], that the resistivity of living tissue was much higher than that of dead tissue. Philippson[32] interpreted his a.c. studies of blood cell suspensions, liver, and muscle in terms of the following equivalent circuit:

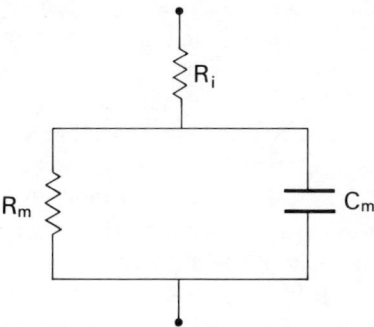

where R_m and C_m represents the membrane resistance and capacitance respectively, and R_i is the resistance of the cell interior. The characteristic of this circuit is that at low frequencies the cell resistance R is given by $R = R_i + R_m$, and as the frequency increases the resistance value tends to $R = R_i$.

Fricke developed a theoretical model to describe the dielectric properties of conducting spheroids, each surrounded by a non-conducting layer (the membrane), dispersed in a conducting medium.[33] This model enabled Fricke to estimate the capacitance of the erythrocyte membrane, which came to a value of about $1 \mu F/cm^2$. This capacitance value has since been found to be almost a physical constant for living cells, bacteria, and organelles (e.g. references 34–39). It is interesting to note that the membrane capacitance value, which indicated that the cell membrane could not exceed more than a few molecular diameters in thickness, was for many years the only clear evidence of the extremely thin structure of biological membranes. As the measurement frequency was reduced below 200 Hz, it was found that the membrane of squid giant axon exhibited an inductive reactance.[40] This surprising, and at first very mystifying result which indicated that the membrane had an inductance of the order 0.2 henry per cm^2, has since been explained in terms of potassium ion flow across the nerve membrane during changes in the membrane conductance.[41]

The first measurement of the membrane resistance seems to have been made by Blinks, who obtained a resistance value of the order 10^4 ohm cm^2 for the *Valonia* cell and 2.5×10^5 ohm cm^2 for the *Nitella* cell.[42] Measurements on other membranes such as that of nerve, muscle, marine eggs, and motoneurones give membrane resistances typically in the range 100 to 10^4 ohm cm^2.[34,43-45] Schwan extended Fricke's theoretical model of a cell suspension to include the membrane resistance components, and was able to deduce from the experimental results that the membrane capacitance is practically independent of frequency.[46] By comparison with the capacitance measurement which now appears to provide reproducible and physically meaningful results, a determination of the membrane resistance is more difficult to achieve.[41,47] Recently Fytak and Terlecki[48] have developed their own theoretical model to describe their measurements on erythrocyte suspensions, and their results for the membrane resistance appear to be lower than that previously determined by others. The concept that the membrane resistance was limited by the conduction of potassium ions through it resulted from electrical measurements across the squid axon[49] and at a later date from measurements on crab axon[50] and frog muscle fibres.[51] It is now known that virtually all animal cell membranes are actively involved in ion transport processes that lead to an asymmetric distribution of ions across the membrane. For example, in most cells the cytoplasm is considerably more rich in potassium ions and poorer in calcium and sodium ion concentration than the extracellular fluid surrounding the cell. There is also an asymmetric distribution of calcium and magnesium ions across mitochondria membranes, there being a much higher concentration of magnesium ions within the mitochondria than in the surrounding cytoplasm (cytosol) of the host cell. A great deal of effort is currently being made to increase our understanding of the ionic 'pumps' involved in ion transport across membranes.

A current concept in membrane biology is that the passage of ions across a membrane is facilitated by a discrete number of localized pores or channels. Electrical noise can be expected to be found in such membranes as a consequence of fluctuations in either, or both, the conductance state and number of these channels. The first studies of membrane noise were made in 1965 on isolated frog sciatic nerve fibres.[52,53] In addition to thermal Nyquist noise the membrane voltage was found to contain a large excess noise which in the range 0.1 Hz to 10 kHz varied with frequency f as $f^{-\alpha}$ where α is close to unity. Compelling evidence indicated that this $1/f$ noise was mainly associated with the passive movement of potassium ions across the membrane. This result was confirmed by later measurements on the same preparation[54,55] and also on giant axons of lobsters.[56] The amplitude $S(f)$ of the current noise spectrum, typically over the frequency range f from 10 Hz to 2 kHz, is observed to follow the relationship

$$S(f) = (A + k|\bar{I}|^{\mathrm{r}})/f$$

where $\bar{I}$ is the average membrane current and A, k, and r are parameters independent of the membrane voltage. The parameter r usually has a value around 2, but for the measurements on lobster axons the parameter r had a mean value of 1.5 and the value for A increased with increasing external potassium concentration. From the above experimentally determined expression for the noise spectrum, it can be seen that $1/f$ noise occurs even when there is no net current flow across the membrane. A detailed discussion of the occurrence of $1/f$ noise in nerve membranes has been given by Verveen and De Felice.[57] Recent work on the squid giant axon has investigated both potassium and sodium ion current noise.[58] A typical $1/f$ noise characteristic for nerve membranes is shown in Figure 7.4.

It is reasonable to believe that quantitative experimental studies of $1/f$ noise on well-characterized membrane areas will lead to a better understanding of ion–ion and ion–channel interactions. Unlike Nyquist noise which is an equilibrium characteristic of a thermodynamic system containing charged particles, $1/f$ noise phenomena occur only in non-equilibrium systems.[59] $1/f$ noise has been found in artificial membranes,[60] in glass microelectrodes,[61] and across small holes separating two different electrolytic solutions.[62] Apart from these examples, noise with a

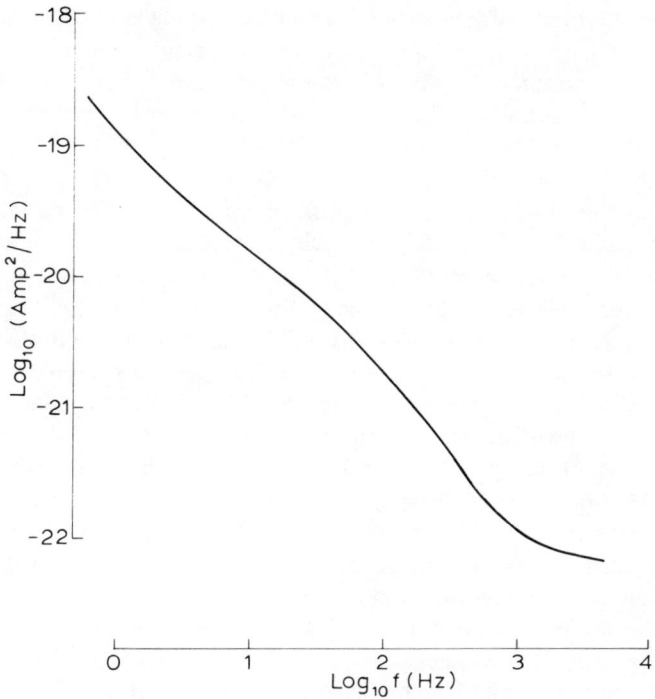

Figure 7.4 A typical $1/f$ noise spectrum for nerve membranes

roughly f^{-1} spectrum is a common phenomenon in many electrical systems, so that there may be an alternative explanation for it other than that associated with membrane resistance fluctuations. In this respect the observation that the $1/f$ noise persists when there is zero current flow across the nerve membranes may be significant. As $1/f$ noise is a non-equilibrium process, then even at zero net current flow there must exist some gradient in the potential of some ion across the membrane. If it can be established that $1/f$ noise exists in the absence of all types of electrochemical gradients, then its origin must lie outside that of transmembrane conduction fluctuations.[59]

Various theoretical models have been proposed to describe the $1/f$ noise mechanism in nerve cell membranes,[63-65] although they do not really help in understanding the underlying molecular mechanisms involved, as they are mostly based on rather mechanistic concepts. Similarly, several models have been proposed[66-68] to describe the enzyme mechanism involved for the active transport of sodium and potassium ions across cell membranes, but while these mechanisms suggest conformational changes they do not provide an insight into how these mechanisms operate at the molecular level. As previously mentioned there is now evidence that the major polypeptide component of (Na^+, K^+)-ATPase spans the membrane. This has led Shamoo[69] to suggest that the ions may be transported in the α-helices of the protein, and Kyte[70] has proposed that the ions may pass through the intermolecular channels formed between adjacent helical polypeptide chains. Recently Chandler *et al.*[71] have expanded these models to propose the relatively simple mechanism whereby the translocation of Na^+ and K^+ across the membrane is brought about by gliding edge dislocations in the protein α-helix structure. Dislocations can occur in an α-helix as a result of the appearance of a broken hydrogen bond. Such a dislocation can be propagated along the helix as a localized wave of slip. Two kinds of dislocation are possible, one having a non-hydrogen-bonded hydrogen atom (H-dislocation), and the other having a non-hydrogen-bonded oxygen atom (O-dislocation). The H-dislocation will have a small associated positive charge and the O-dislocation a small negative charge. Since only one hydrogen bond at a time is disrupted, the 'gliding' of a dislocation along a helix will not be too energetically unfavourable. Chandler *et al.* have shown that if an O- and H-dislocation were to approach each other in an α-helix, the helical region between them assumes a conformation with a larger helix diameter that allows sufficient room for a Na^+ ion (but not K^+ which is too large) to fit inside the helix. This bulge could propagate by glide of the two edge dislocations, the ion being attracted through the membrane by the negatively charged O-dislocation and repelled by the H-dislocation. The appearance of an O-dislocation in an α-helix produces a local deformation where the helix diameter becomes smaller. In this way the intermolecular spacing between close-packed parallel helices would become larger and able to accommodate a K^+ ion, and such a K^+ ion could be attracted through the membrane by negatively charged O-dislocations.

Whatever the molecular mechanism involved in the translocation of membrane ions, it is most probable that the passage of the ion through a channel involves the surmounting of several energy barriers, as depicted in Figure 7.5. By analogy with Figure 1.10, a system of charges hopping back and forth over a distribution of potential barriers will be dielectrically indistinguishable from a set of relaxing dipoles. We can imagine that in a membrane there will be a set of channels for ionic conduction consisting of potential energy barriers of the form shown in Figure 7.5.

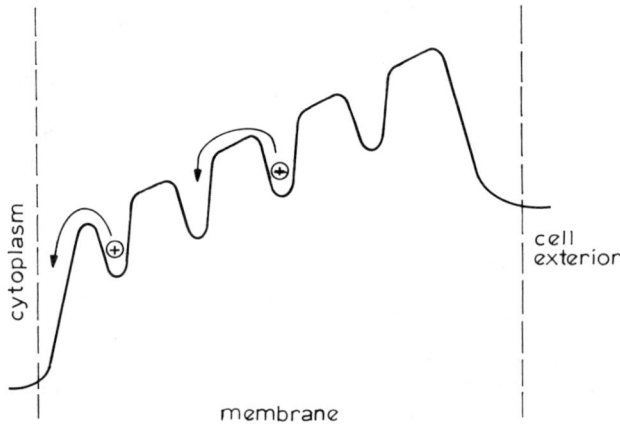

Figure 7.5 The activated hopping of ions through a membrane

As described in Chapter 1, a distribution of energy barrier heights leads to a distribution of polarization relaxation times. It can be shown[72] that for a system of charge carriers hopping over a broad distribution of energy barrier heights, the resulting conductivity will vary with the frequency f as f^{α} with $\alpha \simeq 1.0$. In fact this type of conductivity is characteristic of any relaxation process with a wide distribution of relaxation times τ, with τ an exponential function of a random variable ξ

$$\tau = \tau_0 \exp(\xi)$$

Because of the exponential dependence of τ on ξ, then even a distribution of ξ uniform over only a relatively narrow range will cause the approximate $f^{1.0}$ behaviour to extend over a large frequency range. If ξ can vary in the region between ξ_1 and ξ_2 then the conductivity will vary approximately linearly with frequency in the frequency range

$$\exp(-\xi_2) < 2\pi f \tau_0 < \exp(-\xi_1)$$

For hopping charge carriers such as electrons or ions, the random variable ξ corresponds to the energy barrier height (E/kT). Such a hopping charge mechanism could possibly be the basis for the existence of $1/f$ noise in membranes. In Chapter 4 the dielectric properties of electrolyte solutions are described, and in the so-called 'kinetic polarization deficiency' model the concept is envisaged whereby a migrating ion is able to induce dipolar rotations of the surrounding solvent molecules. In the model proposed by Chandler et al. ions pass through or alongside α-helices, and it is almost certain that during the passage of such ions there will be perturbations in the orientations of the peptide unit dipole moments in the α-helices. Such induced orientational polarizations of the α-helix dipole moments could also contribute to the observed electrical noise in membranes.

We have only considered what may be called the passive properties of biological membranes. The electrical theories and techniques used in the investigation of nerve impulses, and other electrical excitations in excitable nerve and muscle cell membranes, can be found in several books[41,73-75] and will not be described here. Finally, by way of summarizing the basic electrical properties of biological membranes, some resistance and capacitance values obtained for a selection of cell membranes are tabulated in Table 7.2.

Table 7.2   Some capacitance and resistance values for a variety of cell membranes

System	Membrane capacitance (μF/cm^2)	Membrane resistance (ohm cm^2)	Cytoplasm resistivity (ohm cm)	Reference
Starfish egg	0.5	3×10^3	–	44
Crab (leg nerve)	1.1	8×10^3	90	76
Lobster (leg nerve)	1.3	2.3×10^4	60	77
Frog (sciatic nerve)	0.55	–	560	78
Cat (sciatic nerve)	0.65	–	720	78
Squid (stellar nerve)	1.1	1.5×10^3	30	79
Lymphocyte (turtle)	0.8	–	140	80
Leucocyte (rabbit)	1.0	–	140	80

ARTIFICIAL MEMBRANES

Because of their inherent complexity, it is obviously of advantage to be able to study the basic properties of biological membranes using a much simpler model. Two such model systems exist, namely bilayer lipid membrane (BLM) and lipid microvesicles (LMV). Both these models use amphipathic lipids to form membranes of bimolecular thickness separating two aqueous solutions, and as such they can be used to investigate the dielectric, permeability, and electrical excitability characteristics of the fundamental bilayer lipid leaflet structure of biological membranes. Apart from their interest to membrane biologists these artificial membrane systems are of value to physical chemists studying colloidal and interfacial phenomena, as well as the basic properties of amphipathic compounds.

Bimolecular lipid membranes (BLM)

The formation of these membranes, also referred to as bilayer lipid membranes or black lipid membranes, is not difficult. The lipids that have been used to form a stable BLM include many of the naturally occurring phospholipids and some mono- and diglycerides. Gross lipid extracts from brains, eggs, yeast, erythrocytes, bacteria, mitochondria, and chloroplasts have also been used. Whilst these may produce more 'realistic' biological membrane models, they can cause problems if precise chemical and physical characterization of the lipid bilayers are required. For this reason synthetic lipids and materials like highly purified egg lecithin have been used by some experimenters. Oxidized cholesterol, interface-active agents (surfactants), proteins, and carotenoid pigments have also been incorporated into the BLM structures. The formation, chemical composition, physical characteristics, and stability of BLM preparations have been reviewed by Fettiplace et al.[22] and Tien[81] and will not be described here.

A wide variety of experimental arrangements have been used for studying the electrical properties of BLM, but the most basic type is the vertical film arrangement. The principal component is a hollow PTFE (teflon) bucket in which a small part of the wall has been thinned to about 0.2 mm and a smooth-edged hole of diameter around 0.4 mm formed through it. This bucket is placed in a glass vessel so that the hole in the PTFE wall can be immersed in aqueous solution and illuminated for inspection through a microscope. BLM films are formed either by passing a small brush loaded with lipid solution (formed in a suitable solvent such as hexane or decane) across the hole, or by introducing lipid into the hole from a pipette. To begin with the lipid film in the hole thins as a result of gravitational forces in association with capillary action at the PTFE border. When the film approaches molecular dimensions in thickness, London–van der Waals forces accelerate the thinning process until these forces are opposed at equilibrium by the steric repulsion of the hydrocarbon chains in the two opposing lipid monolayers. In its stable bimolecular form there is nearly a $180°$ phase difference between light reflected from the front and back face of the film, to give almost a complete lack of light reflection from the film. This has led to the term 'black lipid membrane'. It is an exciting experience to make, after several unsuccessful attempts, one's first

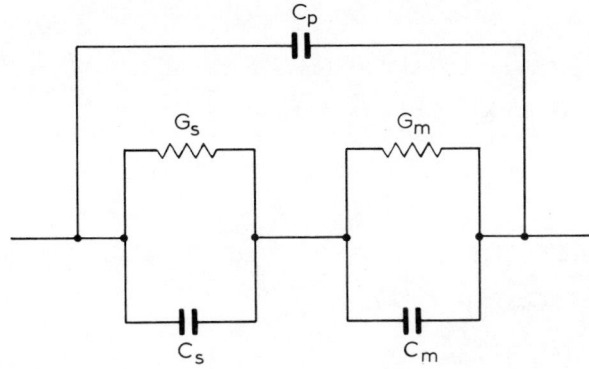

Figure 7.6 The equivalent a.c. circuit of an experimental BLM, where G_m is the conductance of the membrane and C_m is its capacitance. G_s and C_s is the conductance and capacitance of the surrounding aqueous solution, respectively, and C_p is the stray capacitance of the experimental circuit

stable BLM — to see the silvery white lipid film thin to produce swirling pink and green interference patterns until finally the awaited 'black spot' appears and grows to cover the entire hole apart from a bright circular rim at the edge. Similar phenomena excited the interest of Hooke more than 300 years ago, and a little later that of Newton, in studies of the formation of soap films. The electrical properties of these lipid bilayer films can be monitored by electrodes placed in the aqueous compartments on each side of the film. For a.c. measurements the electrodes may be of platinum—black, for example, but for d.c. studies reversible electrodes should be used, such as silver—silver chloride, connected via saturated KCl salt bridges to the aqueous solutions. The electrodes should be of as large a surface area as possible so as to reduce their impedance. A detailed account of the various electrical measurement procedures used in the study of BLM films is given in the two reviews already mentioned.[22,81] The electrical properties that have been measured for BLM films include capacitance, resistivity, the dielectric breakdown voltage, and the membrane potential.

The equivalent a.c. circuit for electrical measurements on BLM films is shown in Figure 7.6, where G_m, G_s and C_m, C_s are the conductance and capacitance of the BLM films and of the aqueous solution, respectively, and C_p is the stray capacitance appearing in parallel across the measurement assembly. Standard circuit analysis shows that, as a function of frequency f, the real and imaginary components C' and C'' of the circuit's complex capacitance C^*

$$C^* = C' - jC''$$

$$= C' - \frac{jG}{2\pi f}$$

are given by

$$C' = C_1 + C_p + (C_2 - C_1)/(1 + \omega^2 \tau_0^2) \tag{7.1}$$

$$C'' = [(C_2 - C_1)\omega\tau_0]/(1 + \omega^2 \tau_0^2) + G_1/\omega \tag{7.2}$$

where

$$\omega = 2\pi f,$$

$$C_1 = C_s C_m/(C_s + C_m)$$

$$C_2 = (C_s G_m^2 + C_m G_s^2)/(G_m + G_s)^2$$

$$G_1 = G_m G_s/(G_m + G_s)$$

and

$$\tau_0 = \frac{1}{2\pi f_0}$$

$$= (C_m + C_s)/(G_m + G_s)$$

The usual experimental arrangement is to have $C_m \gg C_s$ and $G_s \gg G_m$. Equations (7.1) and (7.2) can be compared directly with equations (1.20) and (1.21) describing the components of the complex permittivity ϵ^*. As the factor G_1/ω is mostly always negligibly small, and since the stray capacitance C_p is largely frequency independent, then a plot of C' versus C'' for various frequency values should produce a semicircular arc of the form of Figure 7.7, with C'' tending to zero at low and high frequencies and reaching a maximum value at the characteristic frequency f_0. In a typical experiment on a BLM separating two NaCl solutions, the various components of the equivalent circuit of Figure 7.6 will have the following approximate values; $G_s = 10^{-3}$ mho, $G_m = 10^{-10}$ mho, $C_s = 10$ pF, $C_p = 25$ pF and $C_m = 4 \times 10^3$ pF. The variation of C' and the conductance G ($G = 2\pi f C''$) with frequency f, as determined from equations (7.1) and (7.2), is shown in Figure 7.7(a). It can be seen that at the lower frequencies, the value for C' corresponds to the membrane capacitance value of 4000 pF, and as the frequency is increased C' falls to a high frequency limiting value equal to the sum $C_s + C_p$. At the higher frequencies the conductance value approaches that corresponding to the electrolyte solution, just as if the lipid film had not been present. This is the same effect as that described earlier in this chapter for measurements on living cells, where at the higher frequencies the effect of the cell membrane on the resultant resistivity became negligible. The 'semicircular' variation of C' with C'' is shown in Figure 7.7(b) using the same symbols used in Figure 7.7(a) for each frequency value. The maximum value for C'' occurs at the characteristic frequency f_0 ($\sim$40 kHz). Both the characteristic frequency, f_0, and the limiting high frequency conductance increase in value as the electrolytic strength (and hence conductivity) of the aqueous solution either side of the BLM increases. When weak electrolyte solutions are used, an extrapolation of the results to lower frequencies is often the only way of obtaining an accurate value for the membrane capacitance. Hanai et al.[82] have presented their

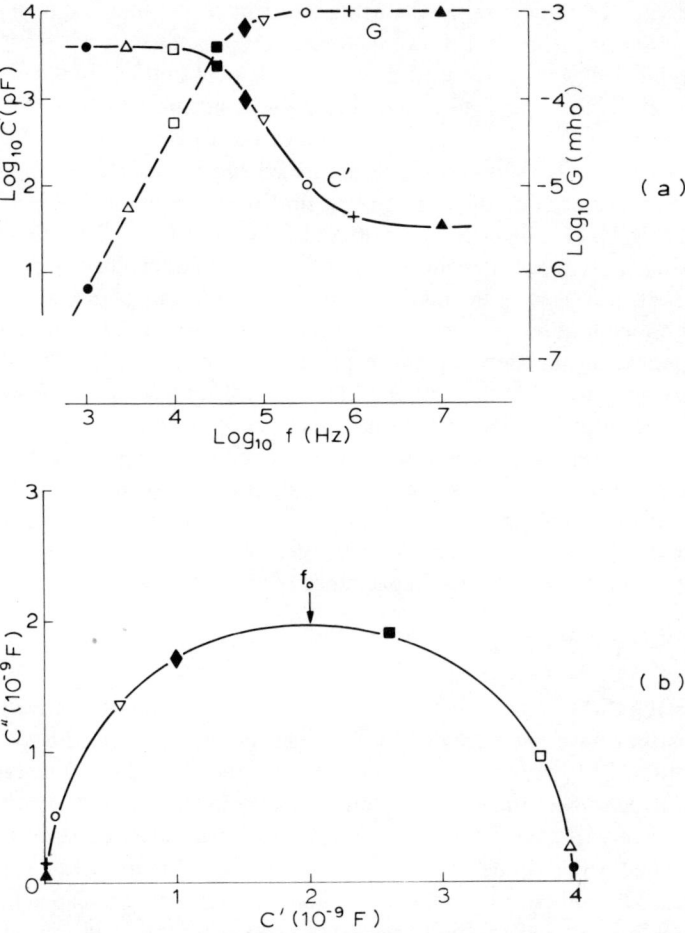

Figure 7.7 (a) The variations of C' and G with frequency for the equivalent circuit of Figure 7.6 with $G_s = 10^{-3}$ mho, $G_m = 10^{-10}$ mho, $C_s = 10$ pF, $C_m = 4 \times 10^3$ pF and $C_p = 25$ pF. (b) The variation of C' with C'' using the same symbols as in (a) for the frequency values. The characteristic frequency f_0 is indicated

results for a.c. measurements on egg yolk lecithin films in the same form as Figure 7.7, and semicircular arc plots of membrane resistance versus membrane reactance have often been used to present the measurements made on living cells.[41] In the same way that Schwan[46] was able to show that the capacitance of a living cell membrane is practically independent of frequency, so too several workers[82-84] have shown that the capacitance of BLM films is frequency invariant up to 10 MHz. In this respect it would have been of great interest to investigate these films for frequencies up to 20 GHz, where dispersions due to orientational polarizations of membrane-bound water may have been detected. By contrast, Coster and Simons[85] have observed anomalous dielectric dispersions in lipid films which were suggestive of the presence of thin water layers being adjacent to the film and having dielectric properties different from that of normal bulk water.

In Chapter 5 it was shown how a diffuse electrical double layer can form at the surface of an object immersed in an electrolyte solution, and that the weaker the ionic concentration of the electrolyte becomes the further does the diffuse double layer extend from the surface of the immersed object. As a result of ions sorbed onto the film surface, or of ionic groups that occur in lipids like lecithin, such a diffuse double layer will occur on both sides of BLM films. The effective capacitance of such a BLM will depend not only on the polarizability of the membrane material itself, but also on the distribution of charge in the diffuse electrical double layer at the membrane surfaces. The effective thickness of the membrane, as far as the charge distribution defining the capacitance is concerned, will not be simply given by the physical thickness of the film — it will tend to be of greater thickness. Everitt and Haydon[86] have shown that electrical double layer effects can lead to greatly reduced values for the membrane capacitance, especially for very weak electrolyte solutions and low surface charge densities, and this does in fact appear to have been observed for some BLM films.[87,88] The effect of the lipid polar head groups on the measured capacitance has been shown to be negligible[89] and for most practical purposes the BLM capacitance is accurately given by

$$C = \frac{\epsilon_m \epsilon_0}{d} \, \text{F m}^{-2}$$

where d is the thickness of the hydrocarbon part of the film of relative permittivity ϵ_m. Values that have been obtained for the capacitance of various BLM[22,81] range from 0.3 to 1.3 μF/cm^2 and so cover the same range as that observed for many natural biological membranes. The main difficulty in the accurate determination of the capacitance is the measurement of the precise area of the BLM. Woo and Wei[90] have developed a theoretical model, based on the considerations of surface charge and surface recombination or binding of ions, to describe the capacitances of the diffuse double layer and the polar head region as a function of ionic strength, pH, and external voltage bias. The capacitances of oxidized cholesterol lipid membranes were then measured under various conditions, and were found to remain constant to within 5 per cent. The experimental results were in fairly good agreement with the theory. Their theory has shown that large capacitances, of the order $10–100 \, \mu$F/cm^2, occur at lipid film interfaces as a result of both a dipole layer at the film surface and of the diffuse double layer. This interface capacitance, on either side of the membrane, is very sensitive to environmental conditions and it acts as a guard to maintain the stability of the membrane capacitance itself. In this way different biomembranes could have a variety of environmental conditions, but if they have the essential bilayer lipid construction they will all have the same capacitance of the order 1 μF/cm^2 and good stability under changing physiological conditions. This highly stable character results from the guarding and protection action of the interface capacitances. The membrane capacitance value itself is largely determined by the central hydrocarbon region of the membrane. This work by Woo and Wei has brought attention to a previously largely overlooked aspect of membrane biophysics and it could have significant implications.

The measurement of BLM resistivity does not produce so reproducible results as those for the capacitance. This is possibly associated with the indeterminate effects of impurities in the lipid material as well as from surface current leakage effects. The resistance values obtained for BLM fall in the range $10^3 - 10^{10}$ ohm cm^2 and so are considerably greater than that observed for cell membranes.[81] Electrical breakdown in BLM has been observed to occur for applied fields ranging from 2×10^7 volts/metre to 10^8 volts/metre[81] and a model has been proposed which describes the electrical breakdown in BLM in terms of electromechanical instability.[91] Wei and Woo,[92] on the other hand, have proposed a breakdown phenomenon involving avalanche multiplication of ions that have been released from traps at the membrane surface.

Lipid microvesicles (LMV)

The LMV, also known as liposomes or spherules, provide a second model system for biological membranes. LMV are formed when materials like phospholipids are allowed to swell in an electrolyte solution at room temperature, and at the same time given gentle mechanical agitation. The lipid micelles formed are fragmented by the agitation to produce microvesicles of diameter ranging from 5 to 50 μm, and they are composed of 'closed' lipid bilayer sheets which contain aqueous solution. If ultrasonication is used instead of mechanical agitation then much smaller micro-vesicles can be produced having diameters of 50 nm and less. Unlike the BLM, organic solvents are not required for the production of LMV. The various methods of producing LMV have been described in several publications[93-97] and their use as models for biological membranes has been reviewed by Sessa and Weissmann.[98] As for BLM, a wide variety of both natural and synthetic lipids have been used to form LMV. Their membranes are more permeable to water than to electrolytes by a factor of the order 10^8 and so like natural membranes they are osmotically active.

Because of their small size a direct measurement of their electrical properties is not possible. However, from chlorine ion permeability measurements the membrane resistance value has been estimated at 10^7 ohm cm^2 for phospholipid LMV. *Schwan et al.*[99] have measured the capacitance of a suspension of microvesicles in the frequency range 1 kHz to 100 MHz. Two dielectric dispersions were observed, the lower frequency dispersion was assigned to relaxations of the counter-ions in the electrical double layer around each vesicle, and the dispersion at around 1 MHz was considered to result from an interfacial Maxwell–Wagner mechanism. The relative permittivity of the vesicle membrane was estimated to have a value of 10 or greater, and the water contained within the vesicle was considered to be dielectric-ally indistinguishable from normal bulk liquid water.

BIOLOGICAL TISSUE AND FLUIDS

A full understanding of the electrical properties of biological tissue and fluids requires knowledge of the concepts outlined in this book so far. This can be appreciated from inspection of Figure 7.8 which shows the frequency variation of the relative permittivity typically obtained for biological tissue. Figure 7.8 is

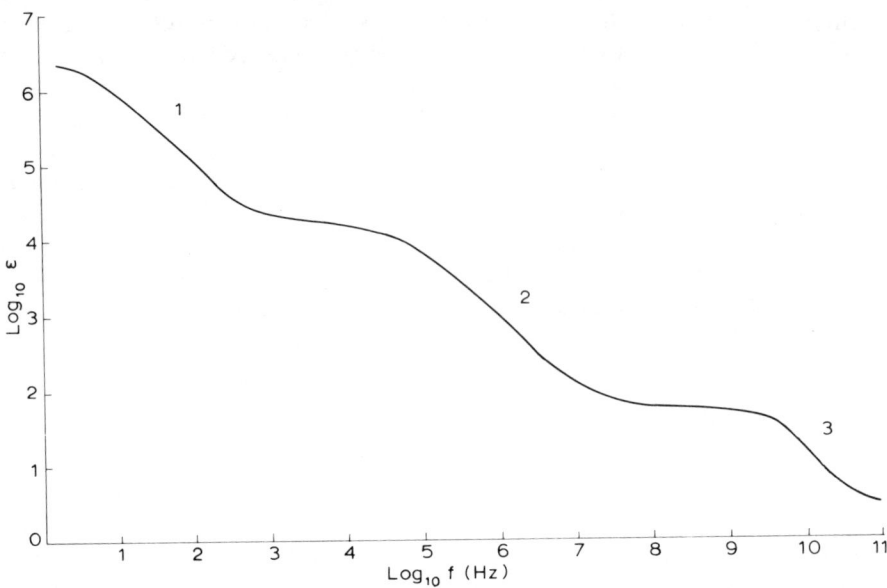

Figure 7.8 The frequency variation of the permittivity ϵ of a typical biological tissue

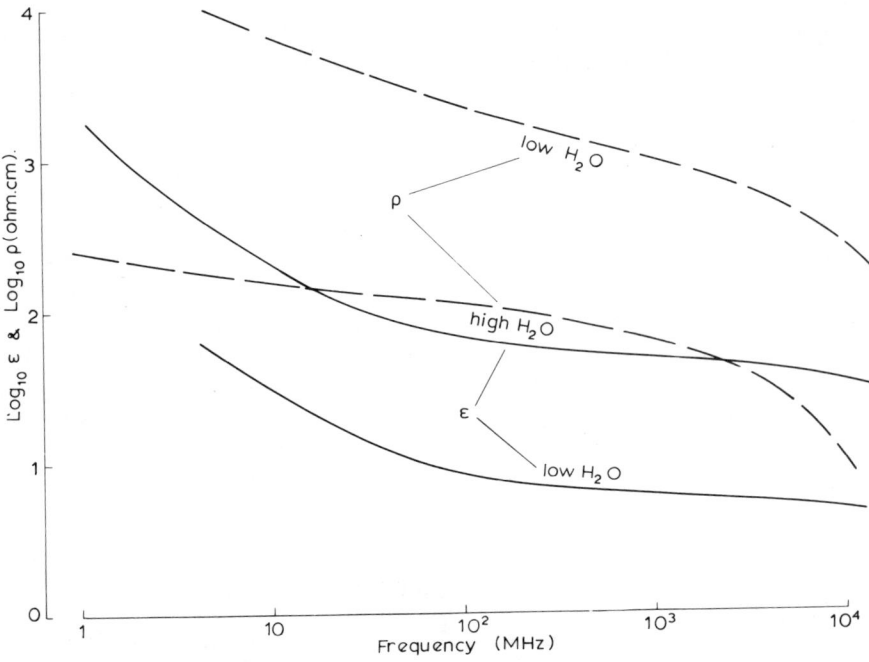

Figure 7.9 The typical frequency variation of the permittivity ϵ and resistivity ρ for tissues of low and high water content

characterized by three regions where the permittivity changes rapidly with frequency. The low frequency region 1 is now generally considered to reflect dielectric dispersions associated with interfacial phenomena, and in particular with ionic conduction and relaxation effects at membrane surfaces, as described in Chapter 4. The intermediate frequency region 2 is associated with the capacitance of cell membranes already described in this chapter, where the effect of the membrane capacitance falls as the frequency increases until at around 100 MHz only the intra- and extracellular fluids dominate the dielectric properties. The relatively steady permittivity value in the range 100–3000 MHz is essentially governed by the water content of the tissue. In turn, the properties of this water are influenced by its content of solvated ions, proteins, and other biomolecules and also the way in which it interacts with biological surfaces. As described in Chapter 3, at frequencies of the order 100 MHz and above, solvated biomacromolecules such as proteins exhibit negligible orientational polarization properties. Some 75 per cent of all the proteins in biological tissue are on average dissolved in the aqueous interior of cells, and the protein content of most tissue is at least 20 per cent by volume. Such proteins and other biomolecules therefore represent a significant amount of dielectric 'voids' in the tissue fluid and as such the fluid's permittivity will be expected to be less than that of normal bulk liquid water. The ionic content of the tissue fluid, equivalent to approximately 0.9 per cent saline

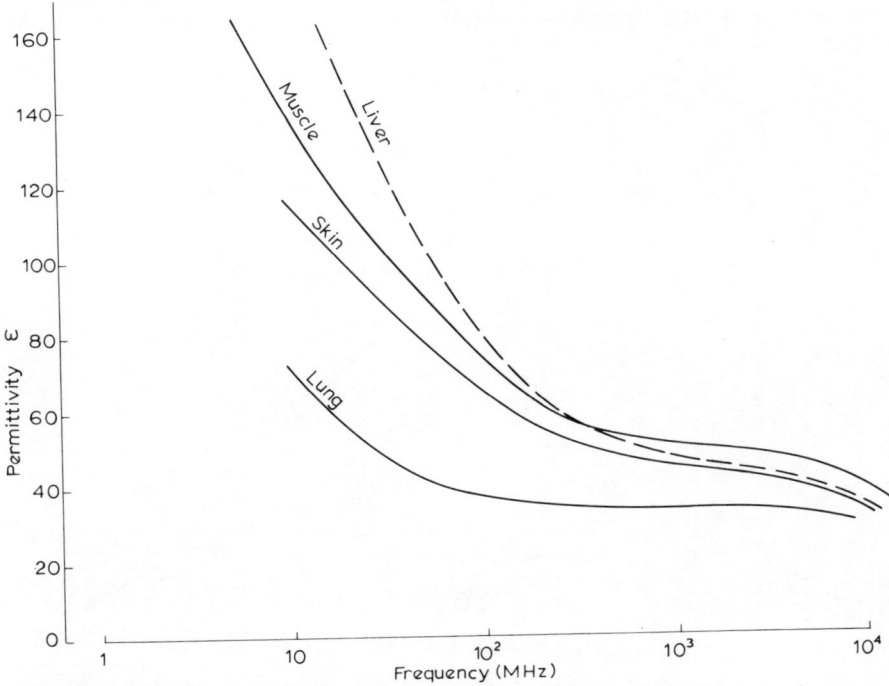

Figure 7.10 The frequency dependency of the permittivity ϵ for several tissues of high water content

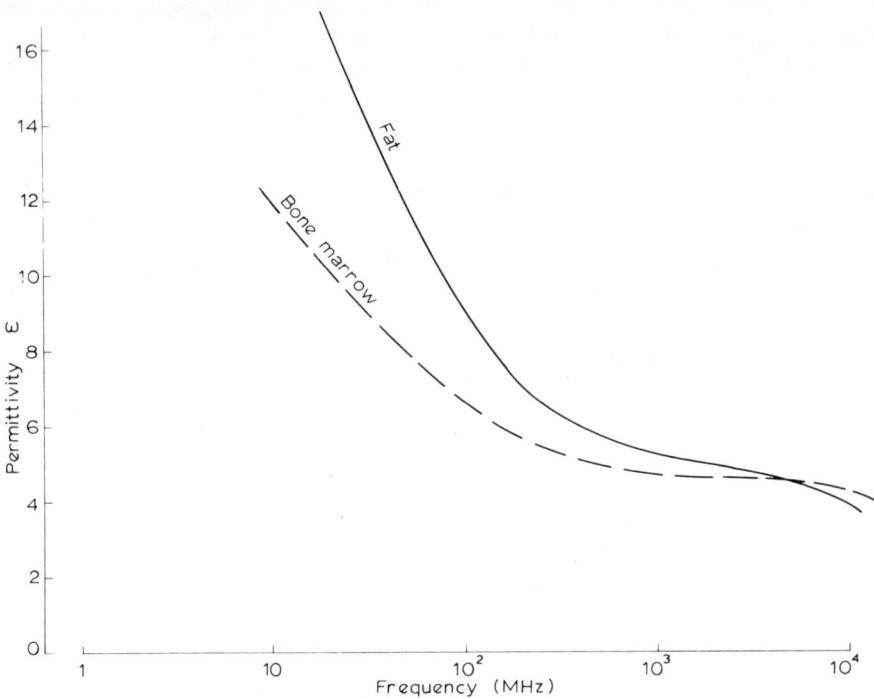

Figure 7.11 The frequency dependency of the permittivity ϵ for fat and bone marrow

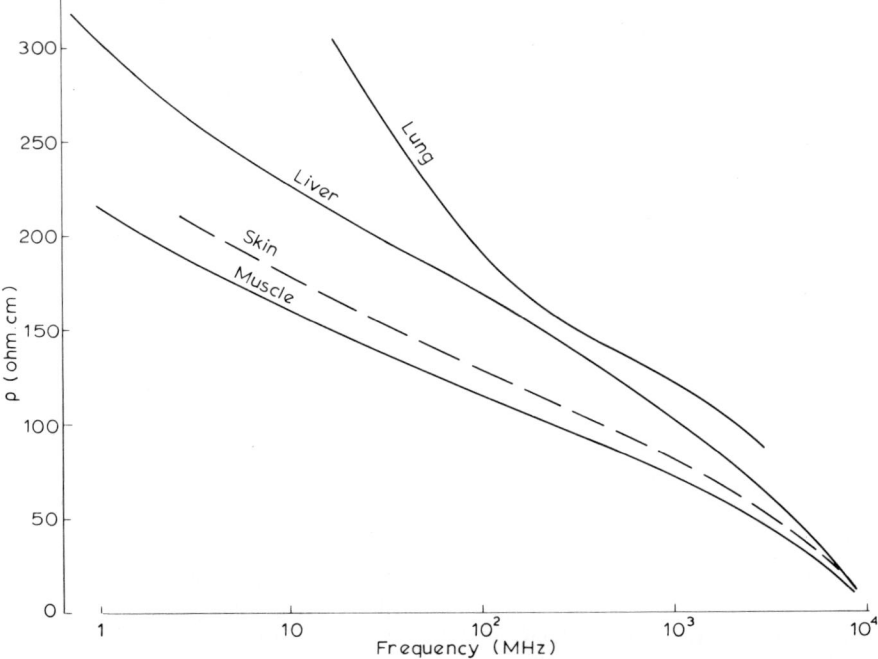

Figure 7.12 The frequency dependency of the resistivity ρ for several tissues of high water content

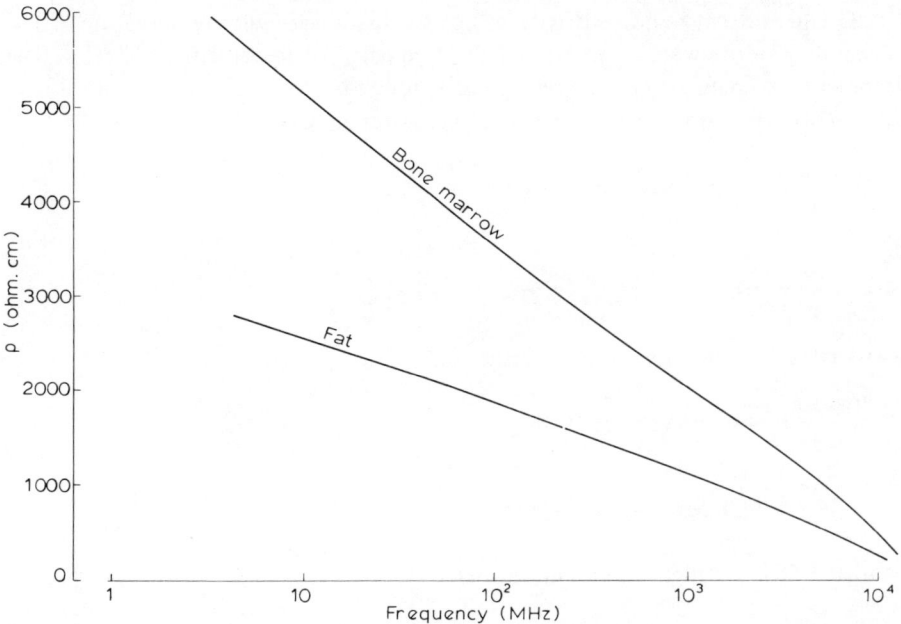

Figure 7.13 The frequency dependency of the resistivity ρ for fat and bone marrow

solution, will also result in a dielectric decrement as described in Chapter 4. At 500 MHz the relative permittivity of living tissue has been found to be given reasonably accurately by the expression

$$\epsilon(\text{tissue}) = 78 - W$$

where W is the weight (gram) content of the protein or other solid matter in 100 ml of the tissue.[46] The high frequency region 3 of Figure 7.8 represents the dielectric dispersion associated with the orientational polarizability of the water molecules, as described in Chapter 4.

The first systematic investigation of the dielectric properties of mammalian tissue appears to have been that reported by Osswald in 1937[100] and by Rajewsky in 1938.[101] Such studies have been greatly extended by several other investigators (see, for example, references 102—108). As might be expected, tissues such as those of liver, muscle, and skin having a high water content exhibit larger permittivities than materials such as fat and bone marrow which have very low water contents by comparison. Similarly, the lower the water content of the tissue the greater is its resistivity. This is demonstrated in Figure 7.9 which shows the typical frequency variations of the permittivity and resistivity of low and high water content tissue. From the work cited above, it is possible to determine (sometimes by extrapolation), the average dielectric properties of several tissues for the frequency range 1 MHz to 10 GHz. These properties are shown in Figures 7.10, 7.11, 7.12, and 7.13 and relate to the normal physiological temperature of 37 °C.

The permittivity and resistivity values for tissue vary with temperature in a way depending on its water content and the frequency of measurement.[101,108] These temperature changes can be summarized as follows.

At 50 MHz for tissue of both low and high water contents

$$\frac{\Delta\rho}{\rho} \simeq -2 \text{ per cent per } °C \text{ rise}$$

$$\frac{\Delta\epsilon}{\epsilon} \simeq +0.5 \text{ per cent per } °C \text{ rise}$$

At 1 GHz for low water content tissue

$$\frac{\Delta\rho}{\rho} \simeq -4 \text{ per cent per } °C \text{ rise}$$

$$\frac{\Delta\epsilon}{\epsilon} \simeq +1.1 \text{ per cent per } °C \text{ rise}$$

and at 1 GHz for high water content tissue

$$\frac{\Delta\rho}{\rho} \simeq -1.3 \text{ per cent per } °C \text{ rise}$$

$$\frac{\Delta\epsilon}{\epsilon} \simeq -0.4 \text{ per cent per } °C \text{ rise}$$

Most of this work has been performed on excised tissue, although some measurements in the range 10 Hz to 10 kHz have been *in vivo* by inserting electrodes into various tissues of live dogs.[109] After it has been excised the dielectric properties of tissue changes with time and within about 24 hours the permittivity and resistivity values decrease, especially at the low frequencies. Referring to Figure 7.8, the permittivity in the frequency region 1 begins to decrease quite rapidly, followed by a reduction of the permittivity in frequency region 2. This is consistent with the breaking down of the cellular membrane structures in the tissue, and along with it the disappearance of the Maxwell–Wagner type interfacial effects responsible for the extremely high permittivity values at the lower frequencies. The membrane capacitance effect responsible for the permittivity variation around 1 MHz will also disappear. By far the greatest changes with time after excision occur for the low frequency permittivity values. This change in the dielectric properties of dying tissue has been used to develop a non-destructive test to determine the freshness of fish.[110] One obvious cause for there being a difference in the properties of living and of excised tissue is the presence of blood. The dielectric properties of blood have been measured by several laboratories[100,103-107] and the results have been very consistent from laboratory to laboratory. The dielectric properties of blood at 37 °C are outlined in Figure 7.14.

Another biological fluid that has received extensive investigations of its dielectric properties is cow milk, and the measurements previous to 1956 have been sum-

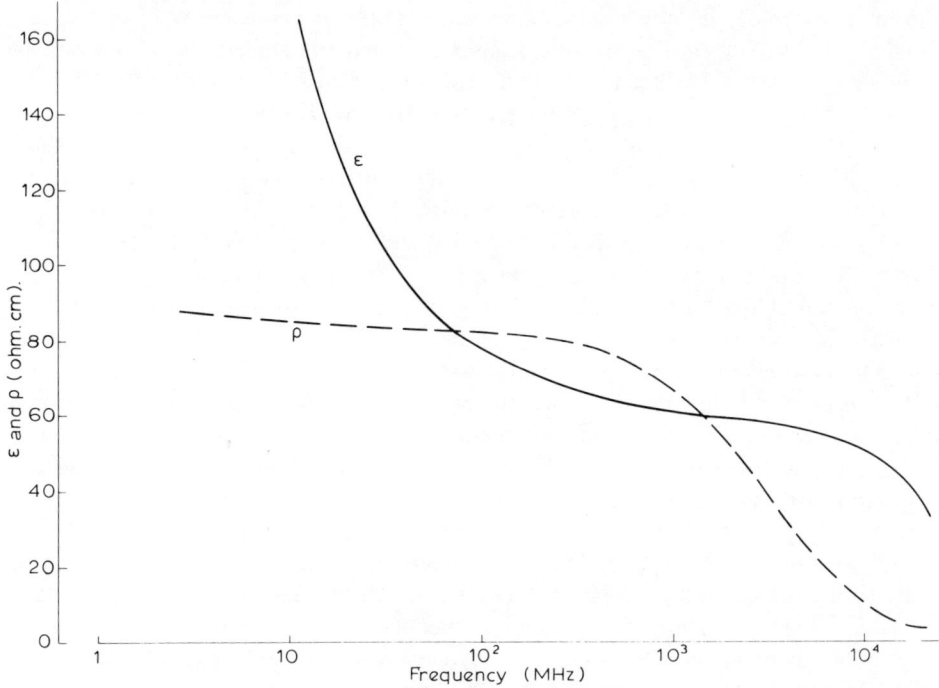

Figure 7.14 The frequency dependency of the permittivity ϵ and resistivity ρ of blood at 37 °C

marized by Schulz.[111] Milk from normal healthy cows generally has a resistivity in the range 180–250 ohm cm at the kHz range of frequencies. Homogenization has no effect, and souring decreases its resistivity by only a small amount, equivalent to a decrease of the order 1 ohm cm per 0.01 per cent increase in lactic acid content. McPhillips and Snow, by monitoring the conductivity, were able to follow acid production in milk by *Streptococcus lactis* as a function of time.[112] The resistivity of milk is largely governed by the electrolytes dissolved in it, and normal milk contains about 0.14 per cent potassium, 0.12 per cent chloride and calcium, and 0.06 per cent sodium. The most important factor affecting the chloride content, and so the resistivity of milk, is mastitis. This condition is caused by numerous types of bacteria, some of which, like coliforms, *Pseudomonas*, and *Salmonella*, are Gram-negative and so are resistant to many common antibiotics. The onset of mastitis results in a fall in lactose content of the milk, and extra sodium chloride is secreted into the milk to maintain the osmotic pressure which is always equal to that of the cow's blood. The lactose content (4.7 per cent) in normal milk exerts an osmotic pressure of 3.30 atmospheres and the chloride content one of 1.33 atmospheres,[113] so that a reduction of 44 per cent of the lactose content requires a chloride increase of 0.1 per cent to maintain the normal osmotic pressure. Such a chloride increase results in a significant reduction in the milk resistivity. Linzell *et al.*[114] have shown that the monitoring of the resistivity of the milk in each of the four quarters of the cow's udder provides a simple and direct way of screening a dairy herd for subclinical mastitis. Considering that the annual cost of mastitis to the dairy industry has been estimated to be in excess of 500 million dollars for the USA alone, it is surprising that this promising technique has not been incorporated into regular farm husbandry. In fact the study of the impedance changes in media that are brought about by microbial metabolism has been conducted by several workers over many years. In 1899 Stewart[115] and then in 1912 Oker-Blom[116] demonstrated that as defibrinated blood was allowed to putrify over a period of 25 days, its conductivity increased by at least ten times. Later Parsons *et al.*[117,118] demonstrated that the metabolic activity of clostridia incubated anaerobically in gelatin and nutrient broths could be monitored by conductivity measurements, which could also be correlated directly with the production of ammonia. In 1938, Allison *et al.*[119] demonstrated that the proteolysis by bacteria could be monitored by electrical conductivity measurements. Recently there has been a greatly increased interest in such studies and in developing impedance measurement systems for the monitoring of enzyme activity,[120] blood coagulation,[121] and bacterial activity.[122-126]

One of the oldest applications of electromagnetic waves in medicine is that of diathermy used to achieve heating in tissue beneath the skin and subcutaneous fatty layers.[127] This can produce therapeutic benefits as a result of local increases in metabolic activity and in blood flow increase from dilation of the blood vessels. This in turn can lead to stimulation of healing and body defence mechanisms. The treatment of cancer is also being approached using microwaves to heat tumours selectively whilst the rest of the body is maintained in a hypothermic condition some 25 °C below normal body temperature, and the warmed tumour, with its high metabolic rate, is then attacked with anti-cancer drugs.[128] Other possible uses of electromagnetic radiation in medicine include the rewarming of refrigerated blood, the thawing of frozen human organs, rapidly reversing a patient's hypothermic state in association with open heart surgery, and diagnostic applications including the monitoring of cardiac ventricular volume changes.[129] For a full understanding of the way in which electromagnetic radiation can heat biological tissue, consideration has to be given to the propagation properties of the radiation. This is achieved by solving the wave equation

$$\nabla \cdot \nabla E + \omega^2 \mu \epsilon \left(1 + \frac{i}{\omega \rho \epsilon} \right) E = 0$$

which describes the way in which the E-field component of a plane e.m. wave with harmonic time dependence travels through a medium of magnetic permeability μ, permittivity ϵ, and resistivity ρ.[130] For a wave travelling in the z-direction, the solution of this equation is

$$E = E_0 \, e^{i(kz - \omega t)}$$

where the propagation constant k is given by

$$k = \alpha + i\beta$$

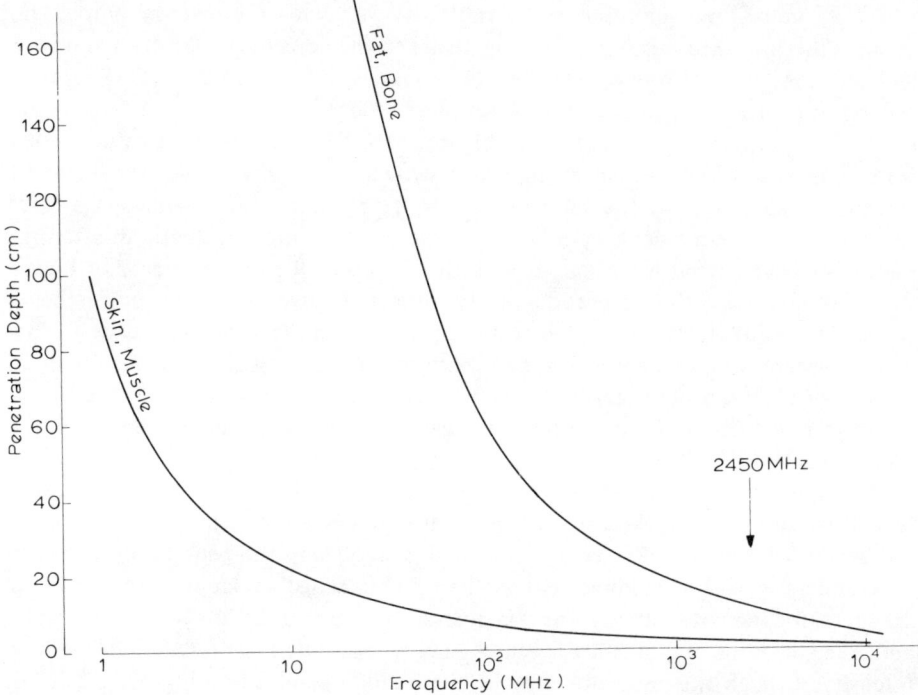

Figure 7.15 The variation of the penetration depth (corresponding to 86.5 per cent power reduction of a propagating e.m. wave) with frequency for tissues with low water content (fat, bone) and high water content (skin, muscle)

The wavelength λ of the propagating wave is given by $\lambda = 2\pi/\beta$ and is considerably reduced from the value corresponding to transmission through air as a result of the high permittivity of the tissue. The factor α determines the extent to which the wave is attenuated as a result of absorption of the wave power density as the wave penetrates into the tissue. This absorption of power can be quantified by defining a depth of penetration corresponding to the distance that the propagating wave travels before the power density is attenuated by a factor of e^{-2} (corresponding to a power reduction of 86.5 per cent). The frequency variation of such a penetration depth for tissues of high and low water content is shown in Figure 7.15, and has been derived from the work of Johnson and Guy.[129] Other tissues with intermediate water contents (e.g. brain, lung, kidney, liver) will have penetration depth values between the extreme values shown in Figure 7.15.

Human bodies are not composed of homogeneous tissue, but have various interfaces between tissues of differing permittivity and resistivity. As a result there will be reflections of incident e.m. radiation at such interfaces, characterized by the reflection coefficient R given by

$$R = re^{i\phi} = \frac{\sqrt{\epsilon_1^* - \epsilon_2^*}}{\sqrt{\epsilon_1^* + \epsilon_2^*}}$$

where ϵ_1^* and ϵ_2^* are the complex permittivity values of the material before and beyond the interface, respectively, r^2 is equal to the percentage of reflected power and ϕ is the phase difference between the incident and reflected waves. Values of reflection coefficient calculated by Johnson and Guy,[129] for various interfaces are shown as a function of frequency in Figure 7.16. When a wave in a tissue of low water content is incident on an interface with a tissue of high water content of thickness greater than the penetration depth, the reflected wave is almost 180° out of phase with the incident wave. This produces a standing wave with an intensity minimum near to the interface. This is the situation for the interfaces of Figure 7.16. By contrast, the amplitude of the reflected wave caused by power being transmitted initially through high water content tissue and meeting tissue of low water content will be almost in phase with the incident wave. This results in an intensity maximum near the interface. Extensive calculations of microwave power absorption distributions in various biological tissues have been presented in the literature.[108,129,131]

A frequency of 2450 MHz is commonly used for diathermy, probably because it is able to utilize well-established microwave power equipment and techniques. Inspection of Figure 7.15 reveals that at this frequency the penetration depth in skin and muscle, for example, is less than 2 cm, whilst in fat it is of the order 10 cm. Subcutaneous fat may have a thickness of up to 2.5 cm for some people, and deep heating of the body requires energy transmission of reasonable power intensity through this fat to the muscle layers and beyond. From Figure 7.16 it can

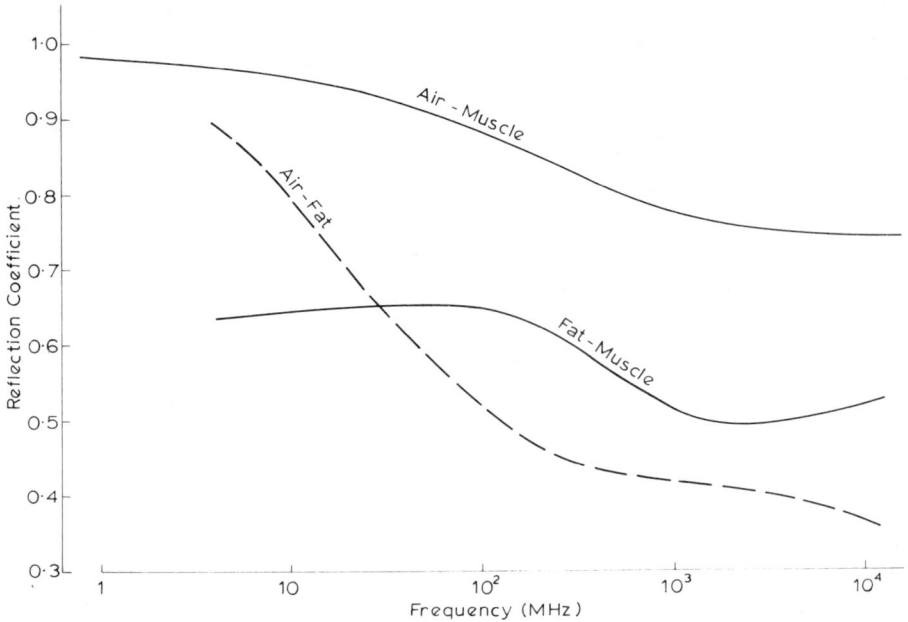

Figure 7.16 Frequency variation of the reflection coefficient of e.m. radiation incident at several interfaces. For all cases shown here the reflected wave is about 175° out of phase with the incident wave

be seen that at the fat—muscle interfaces more than half the incident power is reflected, and considerable power will also be lost in passing across the air—body interface. These considerations suggest that a frequency rather lower than 2450 MHz would be more useful for diathermy if more than just surface heating is required, since the depth of penetration increases as the frequency is lowered. The possibility of creating standing waves producing excessively hot local regions at fat—muscle interfaces would also be reduced. Such factors are most probably responsible for the good reports of the use of 13.56 MHz diathermy for cancer therapy, where deep body tumours are literally cooked to death by having their temperatures raised above the thermal death point for tissue of around 46 °C, whilst at the same time the more rapid blood flow through the surrounding healthy tissue prevents excessive heat being induced in the healthy tissue.[132] The form of Figure 7.15, which shows that the depth of penetration continues to increase rapidly with decreasing frequency, is somewhat at variance with the conclusion expressed by Schwan that comparatively little can be gained by using a frequency of less than about 900 MHz, since there is a less pronounced dependence of penetration depth below this frequency.[133] Schwan does, however, suggest that a frequency lower than 2450 MHz should be used for diathermy, and he has given a valuable review of the subject of radiowave, microwave, and ultrasonic diathermy (in reference 127).

BIOLOGICAL EFFECTS OF e.m. FIELDS

Apart from beneficial effects, electromagnetic radiation can also produce harmful results, either as a result of thermal damage or from, as yet mostly ill-defined, non-thermal effects. The extent of possible damage can range from the extreme example of the Californian woman who inadvertently killed her poodle dog by trying to dry it in a microwave oven, to reports from the Soviet Union of blood disorders in electrical power station workers. The question of possible non-thermal effects due to e.m. radiation is a controversial one, and officially in the UK and the USA only thermogenic effects have been taken into account in defining the permissible levels of exposures to microwaves.[134,135] These limits are, for the frequency range 10 MHz to 1000 GHz, a maximum power density of 10 mW/cm^2 and exposures should not exceed 1 mW/cm^2 for a continuous period of less than 0.1 hour. For the Soviet Union the corresponding limit is 0.01 mW/cm^2 for one working day, and the wearing of protective glasses is mandatory for exposures not to exceed 1 mW/cm^2 for 20 minutes or 0.1 mW/cm^2 for 2 hours. These more strict controls reflect experimental evidence that has been obtained in the USSR to show that e.m. radiation can cause changes in the nervous and cardiovascular systems.

Concern has been felt by some that the advent of modern electrical technology has introduced unnatural electrical and magnetic fields of harmful magnitude into our everyday environment. The natural electric field strength at the surface of the Earth is of the order 150 volts/metre, which may increase up to a value of 10^4 volts/metre during thunderstorm activity. The geomagnetic static field has a maximum value of about 5×10^{-5} tesla (0.5 gauss). The amplitude of these natural electric and magnetic fields falls rapidly with increasing frequency, until at 50 Hz the natural a.c. electric field background is around 10^{-4} volts/metre and that for the magnetic field is only 10^{-12} tesla.[136] By contrast modern electrical technology has introduced into our environment unnatural 50 and 60 Hz fields. For example, when measured at a distance of 30 cm, electric fields of 2 volts/metre are generated by electric light bulbs and an electric blanket produces a field of 250 volts/ metre,[137] and hair dryers and power feeder cables can produce magnetic fields of 2.5×10^{-3} tesla (25 gauss).[138] Electric fields in excess of 10^4 volts/metre can be experienced beneath overhead 525 kV power cables,[139] and 60 Hz fields of the order 4×10^5 volts/metre have been measured at the surface of the heads of linemen working close to an energized 200 kV conductor.[140] Such high fields result from the conducting body of the linemen distorting the field around the conductor, so that at certain regions around their bodies the local field exceeds that which would have occurred in free space. These high fields will not be uniform, and it is possible that dielectrophoretic forces acting within the body could produce physiological disorders. The dielectrophoretic forces acting on golden eagles flying near to such overhead power lines would seem to be of significant effect, as mentioned in Chapter 6. The capacitance of the average human body has been estimated to be of the order 100 pF,[141] so from the relationship

$$I = 2\pi f C V_0$$

where V_0 is the open-circuit potential (height x field), we can estimate the current induced in linemen, of height 1.75 metre, subjected to an average 60 Hz field of 10^5 volts/metre to be 6.6 mA. To calculate the equivalent energy $(J^2 \rho)$ this represents in terms of the possible heating effect, we need to have values for the current density J and the body resistivity ρ. We can take the body's average cross-sectional area to be 0.125 m² and its resistivity as 5 ohm m, and these values lead to the estimate that the linemen absorb about 1.4×10^{-2} W/m³ of thermal energy. The average human adult has a resting metabolic rate of 17 cal min^{-1} kg^{-1}, equivalent to 1.1×10^3 W/m³ for an average man of weight 75 kg. This would indicate that although the linemen appear to be subjected to very high field stresses, the resulting heating effect is negligible in comparison with their normal metabolic rates. Also, when it is considered that fields of the order 10^7 volts/metre occur across cell membranes, and that the minimum current density in tissue which can cause fibrillation of the heart muscle has been estimated[142] at between 1 and 10 A/m², then it does not seem likely that dangerous current densities could be developed in the human body as a result of such low frequency fields. This is certainly the conclusion expressed by Schwan.[142] There are others, however, who consider that this is a rather optimistic viewpoint. The hazards of electromagnetic radiation to the human race and especially those associated with microwaves, have been clearly detailed in a recent thought-provoking publication entitled *The Zapping of America*.[143]

In 1966 Asanova and Rakov conducted a study of the personnel employed at a 500 kV electrical switchgear substation in the USSR, where each person on average was subjected to fields of the order 10^4 volts/metre for 5 hours each day.[144]

Definite changes were found in the blood cell counts and cell morphology, and also in the functioning of the central nervous systems of a significant number of the workers. Unfortunately no controls were used in these studies. A nine-year study has been made in the USA on 10 men who were exposed to fields of around 7×10^4 volts/metre whilst performing maintenance work on overhead power cables.[145] Extensive medical examinations were made and no significant changes of any kind could be found in the men's blood counts and blood chemistry. A laboratory study on the effects of 3-hour exposures of 50 Hz fields up to 2×10^4 volts/metre has been made in Germany, and a slight increase in leucocytes, neutrophils, and reticulocytes was found in people exposed to such fields as compared with the controls who had not.[146] Also, on exposing mice and rats to a 50 Hz, 10^5 volts/metre electric field, significant increases in the white blood cell counts, and a decrease in the lymphocyte fractions, were found.[147] No conclusions can really be drawn from these various reports, apart from the indication that more carefully controlled studies, and attempts to reproduce the results of others, should be made. The human body is able to regulate the white blood cell populations more rapidly than those of the red blood cells, and this is perhaps significant when considering why only changes in the white blood cell counts have been found. Rats that have been exposed for 30 days to 60 Hz, 1.5×10^4 volts/metre fields, have been found to lose weight, whilst their blood gained in albumin content and suffered a lowering of their γ-globulin and corticosterone levels.[148]

Non-thermal effects have been reported to occur[149-152] on the central nervous systems for incident microwave power levels below 10 mW/cm^2 and changes in behavioural patterns have been observed in rats exposed to low level microwave power.[153] Recently Bawin et al.[154] have demonstrated that the central nervous system reacts in a frequency-selective way to external fields. A carrier signal at 147 MHz of power level 1 mW/cm^2 was modulated at nine separate frequencies in the range 0.5–35 Hz, and irradiated into the brains of chicks in vitro. A 20 per cent greater efflux of calcium ions was observed for modulation frequencies of 11 Hz and 16 Hz, a smaller increase was observed for the 6 Hz and 9 Hz signals, and no significant effects were observed for the other modulation frequencies. The possible significance of the definite effects of the 11 and 16 Hz frequencies is almost certainly related to the fact that these two frequencies are constituents of the electroencephalogram (e.e.g.) patterns of the aroused chick brain. The physiological and behavioural effects of electric fields have been investigated on both human and non-human subjects. Two independent studies (references 136, p. 81; 155) suggest that when humans are exposed to low frequency (2–12 Hz) electric fields of the order 1–4 volts/metre, a significant increase occurs in their reaction times, and that this tendency is frequency selective. Experiments extending over 4 weeks have been conducted to show that the circadian rhythm period of humans is decreased by up to 2½ hours if they are subjected to 10 Hz square wave fields of 2.5 volts/metre.[156] This lends support to the viewpoint expressed by Brown[157] that the circadian 'clock' of an organism results from its interactions with the Earth's electromagnetic environment. A related effect has been observed by Marron et al.[158] who found a significant delay in the mitotic cycle of the slime mould Physarum polycephalum after they had been subjected to electric and magnetic fields of 60 or 75 Hz at levels of 0.7 volts/metre and 2 gauss. By contrast to these positive results no effects have been found in the growth and development of chicks[159] and mice[160] after their exposure for prolonged periods to low frequency electric and magnetic fields.

The application of electrical fields has been found to have remarkable effects on the rate of bone growth, and the regeneration of nervous tissue and amputated limbs of frogs and dogs. It is now known that bone, whose major structural elements are collagen and the crystalline mineral apatite, has piezoelectric properties. This property, which manifests itself as a generation of electric potential according to the mechanical stress applied, has been related to the growth and development of bone and its ability to heal after fracture.[161] Definite increases in the rate of bone growth as a result of the application of either d.c. electric currents or low frequency electric fields has been observed by several investigators.[161-165] Bassett et al.[166] have used pulsed magnetic fields to induce electric fields in the surgically fractured legs of beagle dogs. Bones that had healed under the influence of 1 Hz pulses showed no greater strength than the untreated fractures, whereas those healed under 65 Hz pulses had a significantly greater strength, and the structure in the callus of the exposed bone was more highly organized. Rinaldi et al.[167] placed electrodes on either side of a hole that had been drilled in cortical bone, and a small current of the order 4 μA was passed across the bone defect. An improved healing rate was detected and an increase of thymidine was observed, showing that an increase in the mitosis of the osteogenic cells had resulted from this electrical stimulation.[167] If the leg of a frog is amputated, it does not normally regenerate. However, it has been found that the application of a direct current of magnitude 0.1 A along the stump can result in true regeneration of the limb.[168] An enhancement of neurite growth is considered to lead to this result, or alternatively, electrophoretic effects cause the transfer of 'information-bearing' molecules. The applied current could also produce electrical patterns similar to those which existed in the developing embryonic state. Even more remarkable results have demonstrated that substantial regeneration of rat limbs can result from electric currents of the order 5×10^{-9} A produced by surgically implanted bimetallic elements. An organized growth of muscle, bone, and connective tissues was observed for 8 of the 22 rats investigated.[169,170] Sisken and Smith[171] have found that if non-uniform fields, producing direct-current densities of 11.5 A/m^2, are applied for 12 hours *in vitro* to the trigeminal ganglia of chick embryo, then an increased rate of growth is noted for this nervous tissue. Time-lapse photography showed that cell migration and mitotic behaviour were stimulated by the field, and the nerve fibre cells grew elongated toward the negative electrode at the rate of 2.4 mm per day. These workers consider that apart from direct stimulation, this effect may result from a change in calcium ion concentration causing changes in the functioning of the cell membranes. It is of interest to note that the application of non-uniform fields was found to be necessary, highly suggestive of the fact that dielectrophoretic effects were involved.

Apart from the work on bone, most of the reports of the non-thermal effects of electrical fields have not been corroborated by other independent laboratories, in the sense that the experiments have either not been attempted elsewhere or else the experimental conditions were not the same. For example, the investigations in the USSR and the USA of the effects of high field stresses in people working near electrical installations were conducted for differing lengths of exposure time, for a different number of people and at a differing level of voltage stress. In this way the negative results obtained in the USA cannot be taken to disprove the conclusions reached by the Soviet workers. At present we have no clear concepts of the possible factors that could be involved in the non-thermal effects observed for various biological systems. Dielectrophoresis may play a role, and in this respect the pearl-chain formation effect described in Chapter 6 may be of significance, especially in the work on regenerating nervous tissue and limbs where it may produce an ordering effect. At the higher power levels dielectric saturation effects may be present. Schwan has suggested[172] that as a result of such a saturation, macromolecular polar side chains may be forced to orient themselves in the direction of the applied electric field causing hydrogen bonds to break. This could result in enzyme conformational changes and in changes in the structure of their hydration layers, which in turn could lead to subtle biochemical effects. Large electric field stresses could also result in changes in mechanisms based on solid-state physical processes. The piezoelectric properties of collagen is one such example, and there may be many others. The possible solid-state physical concepts that could be responsible for some of the subtle and sensitive properties of biological systems forms the subject matter of the next two chapters.

References

1. L. Thomas, *The Lives of a Cell*, p. 170. The Viking Press, New York (1974).
2. E. Overton, *Vierteljschr. Naturforsch. Gessel. (Zurich)*, **40**, 149 (1899).
3. E. Gorter and F. Grendel, *J. Exp. Med.*, **41**, 439 (1925).
4. R. S. Bar, D. W. Deamer, and D. G. Cornwell, *Science*, **153**, 1010 (1966).
5. D. M. Engleman, *Nature*, **223**, 1279 (1969).
6. *Chemistry of the Cell Interface*, Part A, p. 233. Ed. H. D. Brown, Academic Press, Inc., New York (1971).
7. S. J. Singer, in *Cell Membranes*, p. 36. Eds. G. Weissmann and R. Claiborne, HP Publishing Co., Inc., New York (1975).
8. H. Davson and J. F. Danielli, *J. Cell. Physiol.*, **5**, 495 (1935).
9. H. Fernandez-Moran and J. B. Finean, *J. Biophys. Biochem. Cytol.*, **3**, 725 (1957).
10. J. F. Robertson, in *Principles of Bimolecular Organisation*. Ciba Symposium, Ed. G. E. W. Wolstenhome and M. O'Connor, Churchill Press (1966).
11. A. E. Blaurock and M. H. F. Wilkins, *Nature*, **223**, 906 (1969).
12. D. L. D. Caspar and D. A. Kirschner, *Nature*, **231**, 46 (1971).
13. F. S. Sjostrand, *J. Ultrastruct. Res.*, **9**, 340 (1963).
14. J. A. Lucy, *Nature*, **227**, 815 (1970).
15. A. A. Benson, *J. Am. Oil Chem. Soc.*, **43**, 265 (1966).
16. D. W. Deamer, *Bioenergetics*, **1**, 237 (1970).
17. D. E. Green and J. F. Perdue, *Proc. Nat. Acad. Sci. USA*, **55**, 1295 (1966).
18. J. Vanderkooi and D. E. Green, *Proc. Nat. Acad. Sci. USA*, **66**, 615 (1970).
19. E. D. Korn, *Ann. Rev. Biochem.*, **38**, 263 (1969).
20. V. T. Marchese, in reference 6, pp. 45–53.

240

21. J. Kyte, *J. Biol. Chem.*, **250**, 7443 (1975).
22. R. Fettiplace, L. G. M. Gordon, S. B. Hladky, J. Requena, H. P. Zingsheim, and D. A. Haydon, in *Methods in Membrane Biology*, Vol. 4, pp. 1–75. Ed. E. D. Korn, Plenum Press, New York (1975).
23. H. T. Tien and A. L. Diana, *Chem. Phys. Lipids*, **2**, 55 (1968).
24. A. L. Hodgkin, *Biol. Rev.*, **26**, 339 (1951).
25. V. T. Marchese, reference 7, pp. 45–53.
26. R. Henderson, *Ann. Rev. Biophys. Bioeng.*, **6**, 87 (1977).
27. G. A. Stewart, *J. Physiol.*, **24**, 356 (1899).
28. J. Bernstein, *Arch. ges. Physiol.*, **92**, 521 (1902).
29. R. Höber, *Arch. ges. Physiol.*, **133**, 237 (1910); **148**, 189 (1912); **150**, 15 (1913).
30. W. J. V. Osterhout, *Biol. Rev.*, **6**, 369 (1931).
31. F. Kohlrausch, *Nach. K. Ges. der Wiss. (Göttingen)*, **1876**, 213.
32. M. Philippson, *Bull. Acad. Roy. Belgique Cl. Sci.*, **7**, 387 (1921).
33. H. Fricke, *Phys. Rev.*, **26**, 678 (1925).
34. K. S. Cole, *Tabulae Biologicae (Cell)*, **19**, 24 (1942).
35. H. Fricke, *Nature*, **172**, 731 (1953).
36. H. Fricke, H. P. Schwan, K. Li, and V. Bryson, *Nature*, **177**, 134 (1956).
37. H. Pauly, L. Packer, and H. P. Schwan, *J. Biophys. Biochem. Cytol.*, **7**, 589 (1960).
38. H. P. Schwan and H. J. Morowitz, *Biophys. J.*, **2** 395 (1962).
39. K. Anami, T. Harai, and N. Koizumu, *J. Membrane Biol.*, **28**, 169 (1976).
40. K. S. Cole and R. F. Baker, *J. Gen. Physiol.*, **24**, 771 (1941).
41. K. S. Cole, *Membranes, Ions and Impulses*, Univ. of Calif. Press (1972).
42. L. R. Blinks, *J. Gen. Physiol.*, **13**, 361 (1930); **13**, 495 (1930).
43. T. Araki and T. Otani, *J. Neurophysiol.*, **18**, 472 (1955).
44. A. Tyler, A. Monroy, C. Y. Kao, and H. Grundfest, *Biol. Bull.*, **11**, 153 (1956).
45. C. Y. Kao, *J. Gen. Physiol.*, **40**, 107 (1956).
46. H. P. Schwan, *Advan. Biol. Med. Phys.*, **5**, 147 (1957).
47. H. P. Schwan, *Proc. IRE*, **47**, 1841 (1959).
48. J. Fytak and J. Terlecki, *Biofizika*, **18**, 873 (1973).
49. K. S. Cole and H. J. Curtis, *J. Gen. Physiol.*, **24**, 551 (1941).
50. A. L. Hodgkin, *J. Physiol.*, **106**, 305 (1947).
51. B. Katz, *Proc. Roy. Soc. (London)*, **B135**, 506 (1948).
52. H. E. Derksen, *Acta Physiol. Pharmacol. néerl.*, **13**, 373 (1965).
53. A. A. Verveen and H. E. Derksen, *Kybernatik*, **2**, 152 (1965).
54. A. A. Verveen and H. E. Derksen, *Proc. IEEE*, **56**, 906 (1968).
55. E. Siebenga and A. A. Verveen, *Pflügers Arch. ges. physiol.*, **318**, 267 (1970).
56. D. J. M. Poussart, *Biophys. J.*, **11**, 21 (1971).
57. A. A. Verveen and L. J. DeFelice, *Prog. Biophys. Molec. Biol.*, **28**, 189 (1974).
58. F. Conti, L. J. De Felice, and E. Wanke, *J. Physiol. Lond.*, **248**, 45 (1975).
59. M. B. Weissman, *Biophys. J.*, **16**, 1105 (1976).
60. L. J. De Felice and J. P. L. M. Michalides, *J. Membrane Biol.*, **9**, 261 (1972).
61. L. J. De Felice and D. R. Firth, *IEEE Trans. Bio-med. Eng.*, **18**, 339 (1970).
62. F. N. Hooge and J. L. M. Gaal, *Philips Res. Rep.*, **26**, 77 (1970).
63. I. Lündstrom and D. McQueen, *J. Theoret. Biol.*, **45**, 405 (1974).
64. M. B. Weissman, *Phys. Rev. Lett.*, **35**, 689 (1975).
65. J. R. Clay and M. F. Schlesinger, *Biophys. J.*, **16**, 121 (1976).
66. A. G. Lowe, *Nature*, **219**, 934 (1968).
67. R. L. Post, S. Kume, T. Totin, B. Orcutt, and A. K. Sen, *J. Gen. Physiol.*, **54**, 3065 (1969).

68. W. D. Stein, W. R. Lieb, S. J. D. Karlish, and Y. Eliam, *Proc. Nat. Acad. Sci. USA*, **70**, 275 (1973).
69. A. E. Shamoo, *Ann. N.Y. Acad. Sci.*, **242**, 389 (1974).
70. J. Kyte, *J. Biol. Chem.*, **250**, 7443 (1975).
71. H. D. Chandler, H. R. Hepburn, and C. J. Woolf, *S. Afr. J. Sci.*, **73**, 58 (1977).
72. T. J. Lewis and R. Pethig, in *Excited States of Biological Molecules*, pp. 342–52. Ed. J. B. Birks, Wiley-Interscience, Chichester, New York (1975).
73. *Biophysics and Physiology of Excitable Membranes*, Ed. W. J. Adelman, Van Nostrand Reinhold Co., New York (1971).
74. L. A. Geddes, *Electrodes and the Measurement of Bioelectric Events*, Wiley-Interscience, New York (1972).
75. J. J. B. Jack, D. Noble, and R. W. Tsien, *Electric Current Flow in Excitable Cells*, Clarendon Press, Oxford (1975).
76. A. L. Hodgkin, *J. Physiol.*, **106**, 305 (1947).
77. A. L. Hodgkin and W. A. H. Rushton, *Proc. Roy. Soc. (London) B*, **133**, 444 (1946).
78. K. S. Cole and H. J. Curtis, *Cold Spring Harbor Symp. Quant. Biol.*, **4**, 73 (1936).
79. K. S. Cole and G. Marmont, *Fed. Proc.*, **1**, 15 (1942).
80. H. Fricke and H. J. Curtis, *Nature*, **133**, 651 (1934); **135** 436 (1935).
81. H. T. Tien, *Bilayer Lipid Membranes (BLM)*, Marcel Dekker, Inc., New York (1974).
82. T. Hanai, D. A. Haydon, and J. Taylor, *Proc. Roy. Soc. (London) A*, **281**, 377 (1964).
83. J. Taylor and D. A. Haydon, *Disc. Faraday Soc.*, **42**, 51 (1966).
84. S. H. White, *Biophys. J.*, **10**, 1127 (1970).
85. H. G. L. Coster and R. Simons, *Biochim. Biophys. Acta*, **203**, 17 (1970).
86. C. T. Everitt and D. A. Haydon, *J. Theoret. Biol.*, **22**, 20 (1969).
87. D. Rosen and A. M. Sutton, *Biochim. Biophys. Acta*, **163**, 226 (1968).
88. S. H. White, *Biochim. Biophys. Acta*, **323**, 343 (1973).
89. T. Hanai, D. A. Haydon, and J. Taylor, *J. Theoret. Biol.*, **9**, 278 (1965).
90. B. Y. Woo and L. Y. Wei, *J. Biol. Phys.*, **1**, 36 (1973).
91. J. M. Crowley, *Biophys. J.*, **13**, 711 (1973).
92. L. Y. Wei and B. Y. Woo, *J. Biol. Phys.*, **1**, 50 (1973).
93. A. D. Bangham, in *Progress in Biophysics and Molecular Biology*, Vol. 18, p. 29. Pergamon Press, New York (1968).
94. A. D. Bangham, M. M. Standish, and J. C. Watkins, *J. Molec. Biol.*, **13**, 238 (1965).
95. D. Papahadjopoulos and J. C. Watkins, *Biochim. Biophys. Acta*, **135**, 639 (1967).
96. C. H. Huang, *Biochemistry*, **8**, 344 (1969).
97. J. P. Reeves and R. M. Dowben, *J. Cell. Physiol.*, **73**, 49 (1969).
98. G. Sessa and G. Weissmann, *J. Lipid Res.*, **9**, 310 (1968).
99. H. P. Schwan, S. Takashima, V. K. Miyamoto, and W. Stoeckenius, *Biophys. J.*, **10**, 1102 (1970).
100. K. Osswald, *Hochfreq. techn. Electroak.*, **49**, 40 (1937).
101. B. Rajewsky, *Ultrakurzwellen, Ergebnisse der Biophysikalischen Forschung*, Bild 1, Georg Thieme, Leipzig (1938).
102. B. Rajewsky and H. P. Schwan, *Naturwissenschaften*, **10**, 315 (1948).
103. H. F. Cook, *Brit. J. Appl. Phys.*, **2**, 295 (1951); **3**, 1, 249 (1952).
104. H. F. Cook, *Nature*, **168**, 247 (1951).
105. T. S. England, *Nature*, **166**, 480 (1950).
106. J. F. Herrick and F. H. Krusen, *Electr. Engng.*, **72**, 239 (1953).

242

107. H. P. Schwan and K. Li, *Proc. IRE*, **41**, 1735 (1953); **44**, 1572 (1956).
108. H. P. Schwan and G. M. Piersol, *Am. J. Phys. Med.*, **33**, 371 (1954); **34**, 425 (1955).
109. H. P. Schwan and C. F. Kay, *Ann. N.Y. Acad. Sci., Circulation Research*, **4**, 664 (1956).
110. A. C. Jason and A. Lees, U.K. Dept. of Trade and Industry, Torry Research Station Report T71/7 (1971).
111. M. E. Schulz, *Milchw. Forschbericht*, **8**, 641 (1956).
112. J. McPhillips and N. Snow, *Aust. J. Dairy Technol.*, **3**, 192 (1958).
113. J. H. Coste and E. T. Shelborn, *Analyst*, **44**, 158 (1919).
114. J. L. Linzell, M. Peaker, and J. G. Rowell, *J. Agric. Sci. Camb.*, **83**, 309 (1974).
115. G. A. Stewart, *J. Exp. Med.*, **4**, 235 (1899).
116. M. Oker-Blom, *Centralbl. Bakteriol. Abt.*, **65**, 382 (1912).
117. L. B. Parsons and W. S. Sturges, *J. Bacteriol.*, **11**, 117 (1926); **12**, 267 (1926).
118. L. B. Parsons, E. T. Drake, and W. S. Sturges, *J. Bacteriol.*, **51**, 166 (1929).
119. J. B. Allison, J. A. Anderson, and W. H. Cole, *J. Bacteriol.*, **36**, 571 (1938).
120. A. J. Lawrence and G. R. Morres, *Eur. J. Biochem.*, **24**, 538 (1972).
121. A. Ur, *Nature*, **226**, 269 (1970).
122. A. Ur and D. Brown, *Biomed. Eng.*, **9**, 203 (1974).
123. A. Ur and D. Bloom, in *New Approaches to the Identification of Microorganisms*, pp. 61–71. Eds. C. Heden and T. Illeni, J. Wiley, New York (1975).
124. P. Cady, *ibid.*, pp. 73–99.
125. T. G. Wheeler and M. C. Goldschmidt, *J. Clin. Microbiol.*, **1**, 25 (1975).
126. W. K. Hadley and G. Senyk, *Microbiology*, 1975, 12.
127. *Therapeutic Heat and Cold*, Ed. S. Licht, Waverley Press, Baltimore (1972).
128. R. P. Zimmer, H. A. Ecker, and V. P. Popovic, *IEEE Trans. Microwave Theory Tech.*, **MTT–19**, 238 (1971).
129. C. C. Johnson and A. W. Guy, *Proc. IEEE*, **60**, 692 (1972).
130. J. A. Stratton, *Electromagnetic Theory*, McGraw-Hill, New York (1941).
131. A. W. Guy, *IEEE Trans. Microwave Theory Tech.*, **MTT–19**, 214 (1971).
132. H. H. LeVeen, S. Wapnick, V. Piccone, G. Falk, and N. Ahmed, *J. Am. Med. Assoc.*, **235**, 2198 (1976).
133. H. P. Schwan, reference 27, page 94.
134. *Safety Precautions Relating to Intense Radiofrequency Radiation*, HMSO, S.O. Code 43–182 (1960).
135. *Safety Levels and/or Tolerances with Respect to Personnel*, American National Standards Institute, USASI, 95 (1966).
136. H. L. Konig, in *ELF and VLF Electromagnetic Field Effects*, p. 14. Ed. M. Persinger, Plenum Press, New York (1974).
137. A. R. Valentino, *Proc. IEEE. Int. Electromagnetic Compatibility Symp.*, p. 265. (1972).
138. W. K. Hartell, D. A. Miller, and M. M. Abromavage, *ibid.*, p. 125.
139. T. D. Bracken, *IEEE Trans. Pow. App. Sys.*, **PAS–95**, 494 (1976).
140. H. C. Barnes, A. J. McElroy, and J. H. Charkow, *IEEE Trans. Pow. App. Sys.*, **PAS–86**, 482 (1967).
141. D. W. Deno and L. E. Zaffanella, *Transmission Line Reference Book 345 KV and Above*, Electric Power Research Institute, Palo Alto, California (1975).
142. H. P. Schwan, Naval Weapons Laboratory Rept. NWL TW 2713 (1972).
143. P. Brodeur, *The Zapping of America*, W. W. Norton, New York (1977).
144. T. P. Asanova and A. I. Rakov (English Translation by G. Knickerbocker, *IEEE Power Eng. Soc. Special Publ. No. 10*) (1975).

145. W. B. Kowenhoven, O. R. Langworthy, M. L. Singewald, and G. G. Knicker-bocker, *IEEE Trans. Power App. Sys.*, **PAS−86**, 506 (1967).
146. R. Hauf, *Sonderdruck aus Elektrotechnische Zeitung-b*, **26**, 381 (1974).
147. D. Blanchi, *Arch. Fisiol.*, **70**, 30 (1973).
148. A. A. Marino, T. J. Berger, J. T. Mitchell, B. A. Duhacek, and R. O. Becker, *Ann. N.Y. Acad. Sci.*, **238**, 436 (1974).
149. A. S. Presman, I. Yu. Kamensky, and A. N. A. Levitina, *Archiev. Mod. Biol. (USSR)*, **51**, 84 (1961).
150. Yu. A. Kholodov, *Bull. Exp. Biol. Med. (USSR)*, **57**, 98 (1964).
151. A. H. Frey, *J. Appl. Physiol.*, **23**, 984 (1967).
152. A. H. Frey, *IEEE Trans. Microwave Theory Tech.*, **MTT−19**, 153 (1971).
153. S. F. Korbel and W. D. Thompson, *Psychological Rep.*, **17**, 595 (1965).
154. S. M. Bawin, L. K. Kaczmarek, and W. R. Adey, *Ann. N.Y. Acad. Sci.*, **247**, 74 (1975).
155. J. R. Hamer, *Comm. Behav. Biol. A*, **2**, 217 (1968).
156. R. Wever, *Naturwissenschaften*, **55**, 29 (1968).
157. F. A. Brown, *Am. Sci.*, **60**, 756 (1972).
158. M. T. Marron, E. M. Goodman, and B. Greenbaum, *Nature*, **254**, 66 (1975).
159. W. F. Krueger, A. J. Giarola, and H. W. Woodall, *Biomed. Sci. Instrum.*, **9**, 183 (1972).
160. G. G. Knickerbocker, W. B. Kouwenhoven, and H. C. Barnes, *IEEE Trans. Power App. Sys.*, **PAS−86**, 498 (1967).
161. J. H. McElhaney, R. Stalnaker, and R. Bullard, *J. Biomech.*, **1**, 47 (1968).
162. L. S. Lavine, I. Lustrin, R. Rinaldi, and M. Shamos, *Science*, **175**, 1118 (1972).
163. L. S. Lavine, I. Lustrin, M. Shamos, R. Rinaldi, and A. R. Liboff, *Ann. N.Y. Acad. Sci.*, **238**, 552 (1974).
164. L. A. Norton, *ibid.*, p. 466.
165. L. Klapper and R. E. Stallard, *ibid.*, p. 530.
166. C. A. L. Bassett, R. J. Pawluk, and A. A. Pilla, *ibid.*, p. 242.
167. R. Rinaldi, M. Shamos, and L. Lavine, *ibid.*, p. 307.
168. S. D. Smith, *ibid.*, p. 500.
169. R. O. Becker, *Nature*, **235**, 109 (1972).
170. R. O. Becker and J. A. Spadaro, *Bull. N.Y. Acad. Med.* 2nd Series **48**, 627 (1972).
171. B. F. Sisken and S. D. Smith, *J. Embryol. Exp. Morphol.*, **33**, 29 (1975).
172. H. P. Schwan, in *Proc. 2nd Tri-Serv. Conf. on Biol. Effects of Microwave Energy*, p. 33. Rome, N.Y. (1958).

Chapter 8

Electrons, Energy Levels, and Energy Bands

INTRODUCTION

Although the significance of electron and energy transfer in ordered biological structures had been recognized earlier,[1,2] the present interest in the electronic properties of biological materials can basically be considered to arise from the impact of the Karanyi Memorial Lecture, 'The Study of Energy Levels in Biochemistry', given by Szent-Györgyi in 1941.[3] Szent-Györgyi concluded his lecture as follows:

'By means of our active substances we can produce the most astounding biological reactions, but we fail wherever a real explanation of molecular mechanisms is wanted. It looks as if some basic fact about life were still missing, without which any real understanding is impossible. It may be that the knowledge of common energy-levels will start a new period in biochemistry, taking this science into the realm of quantum mechanics.'

The concept of treating materials like proteins as electronically active solids was also considered by one of Szent-Györgyi's early students, K. Laki, who proposed that the proteins, rather than being considered as inhomogeneous solids composed essentially of H, C, N, O, and S atoms, might more usefully be approximated as homogeneous particles composed of tightly packed hydrogen atoms. This line of thought led Laki to suggest that the effects of salts in altering the isoelectric point of casein resulted from a modification of electronic energy bands associated with the continuum of hydrogen bonds in the protein structure.[4] Szent-Györgyi himself has continued to emphasize the concept that electronic conduction and charge-transfer effects may be responsible for the activity and subtlety of many biological functions in his trilogy *Bioenergetics*, *Introduction to a Submolecular Biology*, and *Bioelectronics*, and more recently has drawn attention to its possible relevance in cancer research.[5,6]

Condensed solid matter exhibits an extremely wide range of electronic con-

244

ductivity. For example, the pen used to write the rough draft of this chapter is composed of metal of conductivity around 10^8 mho/m and of plastic (polyethylene) having a bulk conductivity of the order 10^{-17} mho/m. This conductivity range, extending over 25 decades of magnitude at room temperature, would be very much greater at lower temperatures and can truly be said to be of astronomical proportions (the extent of the universe is around 10^{26} m). When the concept of superconductivity is also remembered, then the phenomenon of electronic conductivity seems even more remarkable. It must surely represent the most variable of all the properties exhibited by solid matter, and it would be surprising perhaps to find that electronic conduction effects have not been incorporated into the functioning of some biological mechanisms.

The extent of electronic conduction in a material is basically dependent on two quantities: the number of mobile electrons and their effective mobility. It does not follow that poorly conducting materials have low charge carrier mobilities. For example diamond (an insulator), and germanium and silicon (semiconductors) have electron mobilities more than 25 times greater than that for metals such as copper and silver. It is also of interest to note that the electron mobilities in the two carbon-based solids graphite and diamond are of similar magnitude at around 0.15 m^2/V s, whereas there is a factor of at least 10^{20} difference in their electronic conductivities. This shows that the number of mobile electrons available for conduction is the significant quantity. All materials contain electrons. What is it that influences the extent to which these electrons are available for electronic conduction? We shall see that the answer to this problem is associated with the existence of energy bands.

Most of the electron energy levels in solids are not randomly distributed in their energy values, instead they are grouped together into 'bands' of allowed energies. Convincing evidence for the existence of such energy bands can be derived from X-ray emission and absorption spectra. If solids are bombarded by high-energy electrons, sufficient energy can be given to electrons occupying the lowest lying electron states (orbitals) of the constituent atoms of the solid to ionize or remove them from their parent atoms. Electrons occupying higher energy states are then able to lower their energy by making transitions to these vacated states, and the energy released is given off as soft X-ray emission. This emission mostly takes the form of narrow discrete lines, except for one or two lines that are considerably broadened. This broadening means that some of the energy levels of the more energetic electrons are extended over a band of energies. The extent of broadening of the X-ray emission spectra is related to the width of such bands in so far as they are occupied by electrons. This is shown in Figure 8.1(a). It is found that the most energetic, valence, electrons in aluminium occupy energy levels extending over a range of about 12 eV, and for sodium the range is 3 eV. In ionic crystals such as sodium chloride the existence of two energy bands is indicated, corresponding to there being a band of energies associated with the outer electrons of the sodium ions, and another associated with the outer electrons of the chlorine ions. In covalence-bond crystals such as silicon and diamond only one broad band (around 14 eV wide for silicon[7] and 21 eV for diamond[8]) of electron-occupied energy levels

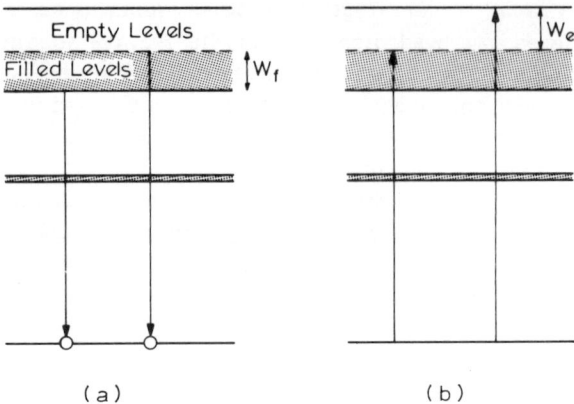

Figure 8.1 (a) Schematic representation of the process of X-ray emission by a metal after ionization of electrons in low lying energy states. The width of the emission spectrum is equal to the width W_f of the band of energy levels occupied by electrons. (b) The process of X-ray absorption. The width of the absorption spectrum defines the width W_e of the band of empty states

is observed, and it is reasonable to believe that the associated electrons are those in the valence bonds.

If a solid crystal is continuously irradiated with X-rays, it is found that there is a relatively uniform absorption of the X-ray quanta over certain energy ranges bounded by regions where the absorption falls off rapidly. This is understandable in terms of electrons gaining energy and being excited into a band of empty states as shown in Figure 8.1(b). By comparing the X-ray emission and absorption spectra for a metal it is found that there is a band of unoccupied states immediately next to the band of electron-occupied states, as for the case shown in Figure 8.1. In such a system the electrons can gain kinetic energy from interaction with an externally applied electric field since there are unoccupied energy states in their immediate vicinity. Electrical conduction can then take place. For valence-bond crystals and ionic crystals, however, it is found that the highest electron-occupied states are separated from the lowest unoccupied states by a large energy gap, which for some materials can exceed several electron volts in width. If an electric field is applied to such materials, the electrons in the occupied band of states are unable to gain kinetic energy and no electrical conduction can take place. If the valence electrons can be excited across the energy gap to the empty states by some other process, such as by heat agitation or from interaction with photons, then electrical conduction can occur.

By convention the band of unoccupied states is called the *conduction band*. For covalence-bond crystals the band of electron-occupied states is called the *valence band*, whereas in ionic crystals they are named after the ions whose associated electrons occupy the bands. The gap between the valence and conduction band is

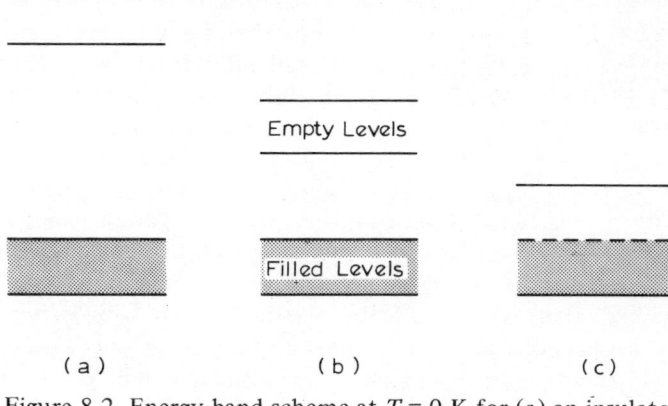

Empty Levels

Filled Levels

(a)   (b)   (c)

Figure 8.2 Energy band scheme at $T = 0$ K for (a) an insulator,
(b) a semiconductor, and (c) a metal

referred to as the energy band gap $\mathscr{E}_g$. If $\mathscr{E}_g$ exceeds a value of around 2 eV then at room temperature very few electrons are thermally excited into the conduction band and the material is classified as an insulator. For $\mathscr{E}_g$ less than 2 eV, some electrons occupy the conduction band at room temperature and the material is classified as a semiconductor. For metals, the conduction band is always partially occupied by electrons. The energy band schemes for insulators, semiconductors, and metals are shown in Figure 8.2, and correspond to the absolute zero of temperature.

Some of the electron properties that have been observed for biological materials such as proteins and DNA are described in Chapter 9. Many of the electronic properties have been discussed in terms of the transport of charge carriers in energy bands. In this chapter a basic outline will be given of the concepts of electron energy states and energy bands, together with a description of some of the dynamic properties of the charge carriers that are responsible for electronic conduction. Finally a description will be given of the energy band calculations that have been made for various protein and DNA structures. The treatment given here of the quantum mechanical and solid-state physical theories is only intended to provide an insight into the concepts required for an understanding of some of the electronic properties that have been observed for biological materials. More rigorous and detailed accounts can be found elsewhere (e.g. references 9–18) and these will be referred to.

THE ONE-ELECTRON APPROXIMATION

The motion of all the electrons in a solid is an extremely complicated process. Apart from interacting with the positively charged nuclei, the electrons also interact repulsively with each other so that the motion of any one electron is influenced by the motion of all the other electrons in the system. As yet no theory has been developed to describe the behaviour of all these electrons, and as a result a one-electron model has been developed as a first approximation of the more general many-electron systems. The one-electron model is based on the assumption that the effect on a given electron of all the nuclei and the rest of the electrons in the system can be described by an average effective electrical field. The potential energy of this one electron, the so-called 'effective one-electron potential'

$$V(r) = V(x, y, z) \tag{8.1}$$

depends only on the coordinates (x, y, z) of the electron. The problem is therefore reduced to the study of the properties of one electron in various electrical fields whose form depends on the system under investigation. Of particular interest will be the 'stationary' energy states of the electron, which correspond to stable electron orbits in the Bohr model of the atom. In quantum mechanics the wave nature of the electron is taken into account and the stable Bohr orbits are written in terms of the three-dimensional one-electron wave function

$$\psi(r) = \psi(x, y, z)$$

These wave functions take the general form

$$\psi(r) = S \cdot R(r)\Phi(x/r, y/r, z/r)$$

where S is a normalization constant referred to again later, $R(r)$ is the radial part and $\Phi(x/r, y/r, z/r)$ is the angular part of the wave function. The value of $|R(r)|^2$ gives the probability of finding the electron at distance r from the nucleus. The so-called Slater radial functions of the form

$$R(r) = A \cdot r^a \exp(-br) \tag{8.2}$$

are often used, where the parameters a and b can be determined using standard rules. The stable, stationary, states correspond to certain well-defined energy levels so that the sequence of one-electron orbitals

$$\psi_1, \psi_2, \psi_3, \ldots, \psi_n,$$

for the stationary electron states in the potential $V(r)$ correspond to the sequence of one-electron energy levels

$$\mathscr{E}_1, \mathscr{E}_2, \mathscr{E}_3, \ldots \mathscr{E}_n$$

The physical meaning of a wave function ψ is that the square of the wave function $|\psi|^2$ at some point in space is proportional to the probability of finding the electron at that point in space. In order to make the wave function correspond exactly to the probability of finding the electron at a point in space, then ψ must be 'normalized' by multiplying it by a numerical coefficient such that the equality

$$\int_v |\psi|^2 \, dv = 1$$

is satisfied.

A system of N electrons can then be described in terms of this one-electron model by assigning every electron to an orbital ψ_i. Each electron occupies the lowest possible energy level and in accordance with the Pauli exclusion principle, whereby only two electrons with opposing spins must be in each orbital. The values of the one-electron energy levels and orbitals are obtained from the one-electron Schrödinger wave equation

$$\tfrac{1}{2}\Delta\psi + (2m/\hbar^2)(\mathscr{E} - V)\psi = 0 \tag{8.3}$$

where $\Delta\psi = \partial^2\psi/\partial x^2 + \partial^2\psi/\partial y^2 + \partial^2\psi/\partial z^2$, and V is the effective one-electron potential of equation (8.1). In equation (8.3) m represents the effective mass of the electron and $\hbar = h/2\pi$, where h is Planck's constant. In principle, for any multi-atomic system, so long as the number, type, and interatomic spacings of the atoms are known, the form of $V(r)$ can be determined and equation (8.3) can be solved to give the stationary energy states. A more rigorous application of the one-electron model can be made using the self-consistent field (SCF) or Hartree–Fock method[19] which, apart from the form of the effective one-electron potential $V(r)$, uses an equation similar to the Schrödinger wave equation. Instead the effective potential, which is assumed to be the same for all the electrons, is written in the form

$$V = V_1 + V_2 + V_3$$

where V_1 is the field acting on the electron as a result of the coulombic attractions of all the nuclei, and V_2 is the field resulting from the coulombic repulsions of all the other electrons. The term V_3 is the so-called exchange potential and has no equivalent interpretation within the framework of classical physics. Since the coulombic and exchange potentials in the equations for each wave function ψ_i depend on the solutions for all the other wave functions in the system, then the Hartree–Fock method results in a set of interrelated equations that is solved by successive approximations to give self-consistency (hence the term SCF).

In a multiatomic system the relative motions of the nuclei, as well as those of the electrons, should be taken into account. However, the nuclei are of the order 10^4 times heavier than the electrons and as such they tend to move more slowly than the electrons. For many calculations the approximation is therefore made that the nuclei are considered to be stationary by comparison with the electron motions. This is called the Born–Oppenheimer approximation or the adiabatic approximation. It is also known that not all of the electrons in a multiatomic system are involved in the formation of chemical bonds or energy bands. The electrons in the atomic cores remain localized around their parent atoms and only the outermost valence electrons take part in interatomic interactions. This leads to the valence approximation where the effective one-electron potential acting on each valence electron is assumed to be the sum of the potentials of the atomic cores (nuclei plus core electrons) with the resulting potential of the remaining valence electrons in the system. The atomic wave functions decrease very rapidly in an exponential fashion with distance from the nucleus, as indicated by equation (8.2), so it is reasonable to assume that in the *immediate* vicinity of each atom the corresponding effective potential for the whole system will not differ too much from the potential of that atom alone. In this way each solution of the Schrödinger equation (8.3) for the whole multiatomic system in the immediate vicinity of a given atom can be expected to be close to that of that atom's own atomic orbitals. At any other point in the system we can expect the one-electron wave function to take the form of a linear combination of the atomic orbitals of all the atoms. For example, if we assume we have a multiatomic system, each atom of which can have any number of atomic orbitals, then we can arbitrarily number each orbital in a sequence

$$\chi_1, \chi_2, \chi_3, \cdots \chi_n$$

We can then suppose that any solution of the Schrödinger equation for the complete system can be written in the form

$$\psi = C_1\chi_1 + C_2\chi_2 + C_3\chi_3 + \cdots + C_n\chi_n \tag{8.4}$$

where specific values of the unknown coefficients C_1, C_2 etc. define a particular solution of the Schrödinger equation. This method is called the linear combination of atomic orbitals (LCAO) method. Where the atoms constitute a molecule the method is known as the molecular orbital LCAO (MO LCAO) method. A direct consequence of using the LCAO method is that an electron in a stationary state ψ cannot be thought of as being localized at any one atom, but should be considered instead as being delocalized over the entire system of atoms. To many chemists this may seem to be at variance with the well-established concept that chemical bonds are formed by localized valence electron pairs. In this respect it should be remembered that only the ground state σ-bonds of saturated chemical structures can be thought of as being localized. The well-known concept of the resonating bond structure of unsaturated conjugated chemical systems is an example of such a delocalized wave function, and the system of sp^3 hybrid orbitals in describing tetrahedral σ-bonds is an example of the LCAO method. The localization of certain chemical bonds can in fact be shown to be a direct consequence of the one-electron model, and this has been clearly described by the pioneers of quantum chemistry (e.g. references 20–22).

The Schrödinger wave equation (8.3) is usually written in the form

$$\hat{H}\psi = \mathscr{E}\psi \tag{8.5}$$

where $\hat{H}$ is referred to as the one-electron Hamiltonian operator of the Hamiltonian

$$\hat{H} = -\tfrac{1}{2}\Delta + V$$

The wave functions ψ_1, ψ_2 etc. of the stationary states are referred to as the 'eigenfunctions' of the Hamiltonian, and the corresponding energies $\mathscr{E}_1$, $\mathscr{E}_2$, etc. are the 'eigenvalues'. By multiplying each side of equation (8.5) by ψ^*, the complex conjugate of ψ, and integrating over all space, the eigenvalue of the Hamiltonian is obtained from its eigenfunction as

$$\mathscr{E} = \frac{\displaystyle\int_v \psi^*\hat{H}\psi \, dv}{\displaystyle\int_v \psi^*\psi \, dv} \tag{8.6}$$

The Hamiltonian possesses two important properties, namely linearity and self-consistency. The property of linearity corresponds to the situation where the equality

$$\hat{H}(C_1\chi_1 + C_2\chi_2 + \cdots) = C_1\hat{H}\chi_1 + C_2\hat{H}\chi_2 + \cdots$$

is satisfied for arbitrary functions $\chi_1(r)$, $\chi_2(r)$ etc., and arbitrary coefficients C_1, C_2, etc. This property is used extensively in the LCAO method, and it also follows that the product of the eigenfunction ψ by any number will also be an eigenfunction with the same eigenvalue. A solution can therefore always be chosen so as to be normalized, and in which case equation (8.6) takes the simpler form

$$\mathscr{E} = \int_v \psi^*\hat{H}\psi \, dv = \langle \psi|\hat{H}|\psi \rangle \tag{8.7}$$

The property of self-consistency corresponds to the equality

$$\int_v \chi_1^* H\chi_2 \, dv = \left\{ \int_v \chi_2^* H\chi_1 \, dv \right\}^* \tag{8.8}$$

being true for any two functions χ_1 and χ_2, where the asterisk again denotes the complex conjugate. Using an alternative notation, equation (8.8) is equivalent to the equality

$$\langle \chi_1|\hat{H}|\chi_2 \rangle = \langle \chi_2|\hat{H}|\chi_1 \rangle^*$$

where

$$\langle \chi_1|\hat{H}|\chi_2 \rangle = \int_v \chi_1^*\hat{H}\chi_2 \, dv$$

This property of self-consistency leads to a very important relationship for the eigenfunctions, namely that any two eigenfunctions ψ_1 and ψ_2 corresponding to the different eigenvalues $\mathscr{E}_1$ and $\mathscr{E}_2$ are orthogonal, such that

$$\int_v \psi_1^*\psi_2 \, dv = 0 \tag{8.9}$$

Returning to equation (8.4), we can now proceed to demonstrate the method of obtaining values for the unknown coefficients C_1, C_2, etc. Substituting the expansion of ψ in the form of equation (8.4) into equation (8.5) and using the linearity of the operator $\hat{H}$, then

$$\hat{H}\psi = C_1\hat{H}\chi_1 + C_2\hat{H}\chi_2 + \cdots C_n\hat{H}\chi_n$$

$$= \mathscr{E}(C_1\chi_1 + C_2\chi_2 + \cdots C_n\chi_n)$$

which on rearrangement of the various terms gives

$$C_1(\hat{H} - \mathscr{E})\chi_1 + C_2(\hat{H} - \mathscr{E})\chi_2 + \cdots C_n(\hat{H} - \mathscr{E})\chi_n = 0$$

This equation is then successively multiplied by the complex conjugate of the atomic orbitals $\chi_1, \chi_2, \ldots \chi_n$, and each resulting product is integrated over all space. For simplicity here we shall assume that the atomic orbitals of adjacent atoms do not overlap each other to any great extent, so that the (overlap) integrals of the type $\int_v \chi_i^* \chi_j \, dv$ are very small, and therefore χ_i and χ_j can be considered to be approximately orthogonal (see equation (8.9)). This process leads to the following system of linear homogeneous algebraic equations

$$
\left.
\begin{aligned}
(H_{11} - \mathcal{E})C_1 + H_{12}C_2 + \cdots H_{1n}C_s &= 0 \\
H_{21}C_1 + (H_{22} - \mathcal{E})C_2 + \cdots H_{2n}C_n &= 0 \\
\cdots \cdots \cdots \cdots \cdots \cdots \cdots \cdots \cdots \\
\cdots \cdots \cdots \cdots \cdots \cdots \cdots \cdots \cdots \\
H_{n1}C_1 + H_{n2}C_2 + \cdots (H_{nn} - \mathcal{E})C_n &= 0
\end{aligned}
\right\}
\tag{8.10}
$$

The parameters H_{ij} denote the matrix elements of the effective one-electron Hamiltonian $\hat{H}$ of the system

$$
H_{ij} = \langle \chi_i | \hat{H} | \chi_j \rangle = \int_v \chi_i^* \hat{H} \chi_j
$$

and so are expressed in terms of the known Hamiltonian $\hat{H}$ of the system, and also in terms of the known atomic orbitals $\chi_1, \chi_2, \ldots \chi_n$. The unknown coefficients C_1, C_2 etc. can then be determined if the unknown eigenvalues $\mathcal{E}$ are known. The values for $\mathcal{E}$ are readily obtained since for the system of equation (8.10) it follows that

$$
\text{Det}
\begin{vmatrix}
(H_{11} - \mathcal{E})H_{12} \cdots H_{1n} \\
H_{21}(H_{22} - \mathcal{E}) \cdots H_{2n} \\
\cdots \cdots \cdots \cdots \cdots \cdots \\
H_{n1}H_{ns} \cdots (H_{nn} - \mathcal{E})
\end{vmatrix}
= 0
\tag{8.11}
$$

This algebraic equation of nth degree has n roots $\mathcal{E}_1, \mathcal{E}_2, \ldots \mathcal{E}_n$, which give the n possible one-electron levels of the system. Substitution of each of these roots into equations (8.10) gives the values for $C_1, C_2, \ldots C_n$, and in this way the set of desired one-electron energy levels is obtained together with their associated one-electron eigenfunctions. These eigenfunctions are normalized by multiplying them by the factor S, where

$$
S = \{C_1^2 + C_2^2 + \cdots C_n^2\}^{1/2}
$$

The determinant of equation (8.11) is called the secular determinant, and the whole equation (8.11) is the corresponding secular equation. The integrals H_{ii} are termed the coulomb integrals α_i and represent the coulombic energy of attraction of an electron of wave function χ_i in the field of the atom i, and as such is a negative quantity. The integrals H_{ij} (where i $\neq$ j) are termed the resonance integrals β_{ij}, and represent the energy of an electron in the fields of atoms i and j involving the wave functions χ_i and χ_j. The values for β_{ij} are also negative.[10,11]

To demonstrate this procedure we can take the simple example of a homoatomic two-atom system with one valence electron per atom (e.g. the hydrogen molecule). Equation (8.4) then takes the form

$$\psi = C_1 \chi_A + C_2 \chi_B$$

where for convenience we have chosen to label each atom A and B, respectively. The system of equations (8.10) then has the form

$$(\alpha_A - \mathscr{E})C_1 + \beta_{AB}C_2 = 0$$
$$\beta_{BA}C_1 + (\alpha_B - \mathscr{E})C_2 = 0$$

(8.12)

and the secular equation (8.11) becomes

$$\mathrm{Det} \begin{vmatrix} \alpha_A - \mathscr{E} & \beta_{AB} \\ \beta_{BA} & \alpha_B - \mathscr{E} \end{vmatrix} = 0$$

(8.13)

Since atoms A and B are identical then $\alpha_A = \alpha_B = \alpha$, and $\beta_{AB} = \beta_{BA} = \beta$ so that equation (8.13) leads to the quadratic equation

$$(\alpha - \mathscr{E})^2 - \beta^2 = 0$$

which has the two roots

$$\mathscr{E}_1 = \alpha + \beta$$
$$\mathscr{E}_2 = \alpha - \beta$$

This gives the values of the one-electron energies for the two possible eigenfunctions that can be formed for χ_A and χ_B. Substituting $\mathscr{E}_1$ in the equation (8.12) gives $C_1 = C_2$ and on normalization we obtain for the first molecular orbital

$$\psi_1 = \frac{1}{\sqrt{2}}(\chi_A + \chi_B)$$

Since the resonance integral β has a negative value then $\mathscr{E}_1$ is below both the atomic energy levels α_A and α_B, so that an electron occupying this level will bind both atomic centres. The energy level $\mathscr{E}_1$ and the orbital ψ_1 therefore correspond to the bonding state. By contrast $\mathscr{E}_2$ is greater than the energy of each atomic level and $\mathscr{E}_2$ and ψ_2 refer to the anti-bonding or excited state. In the ground state the two valence electrons occupy the level $\mathscr{E}_1$ with opposing spins, and an energy equal to 2β is required to excite one of these electrons into the anti-bonding state $\mathscr{E}_2$, as shown in Figure 8.3.

If the molecule or system had contained N atoms, then each atomic orbital would have been split into N levels. Also, the number n of the basic atomic orbitals $\psi_1, \psi_2, \ldots \psi_n$ in the system cannot be less than one-half the total number of valence electrons. In fact the situation where the number of orbitals is equal to the number of valence electron pairs represents the trivial and chemically unimportant

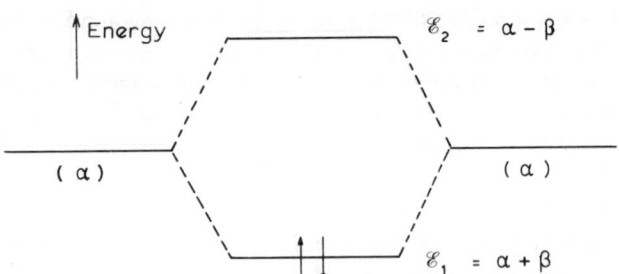

Figure 8.3 The energy level and electron occupancy scheme for a homoatomic two-atom system with one atomic orbital and one valence electron for each atom

case where the atoms are inert gases. The total number of levels obtained from the secular equations (8.11) therefore always exceeds the number of electron pairs that can occupy these levels. The lower lying, electron-occupied, levels represent the ground state, and the remaining empty levels correspond to the one-electron excited states of the system. For a crystal or large polymeric system a description of the distribution of the energy levels and the valence electrons can be obtained from a special application of the one-electron model known as the Band Theory.

BAND THEORY

When dealing with crystals or large polymers, the enormous number of atomic orbitals ($\sim 10^{23}$ per cm^3) leads to the appearance of a corresponding number of energy levels and eigenfunctions. To obtain the values of these energy levels would require solving a secular equation (8.11) of degree 10^{23}! As this would be an impossible task, another simplifying factor must be introduced into the theory. Such a simplification can be found and it relates to the periodicity of the atomic structure of the crystal or polymer, and in particular to translational symmetry involving single atoms or groups of atoms forming 'unit cells'. The effective potential of the one-electron Hamiltonian for the whole system will then have a periodic form matching the atomic periodicity. Since groups of atoms occupying unit cells can be identified, where similarly located atoms within these unit cells will have identical properties that repeat throughout the total system, then the secular equation can be considerably simplified. The degree of the secular equation will equal the number of atomic orbitals that make up the atomic structures of the atoms forming the unit cell. The one-electron eigenfunctions obtained are called 'Bloch functions'. Since they effectively represent a linear combination of all the atomic orbitals in the system, each Bloch function is *delocalized* over the entire crystal or periodic polymer describing the system.

Using Floquet's theorem, Bloch was able to prove that the solutions of the Schrödinger equation with a periodic potential are of the form

$$\psi_k = U_k(r)\exp(i k \cdot r)$$

where $\psi_k(r)$ is a function, having the periodicity of the potential, which depends on the wave vector k. The solutions ψ_k are known as Bloch functions and can be seen to be of the form of plane waves that are modulated with the periodicity of the unit cells of the system. We can demonstrate the essential features of this result using a simple model of N atoms equally spaced in a ring, as shown in Figure 8.4, where we can consider the effective one-electron potential to be periodic in 'a' and have the form

$$V(x) = V(x + ma)$$

with m being an integer. We could also have chosen a model in the form of a long line of equally spaced atoms, but in this situation the atoms at each end of the line would not have been representative of the periodic structure as a whole. The closed ring or 'Born–Kármán periodic boundary condition' is a commonly employed simplification. Because of the symmetry of this atomic system we are able to look for eigenfunctions of the form

$$\psi(x + a) = C\psi(x)$$

where C is a complex constant of unit magnitude. It then follows that

$$\psi(x + ma) = C^m \psi(x)$$

and for the eigenfunction to be single-valued, then

$$\psi(x + Na) = \psi(x) = C^N \psi(x)$$

The factor C therefore represents one of the N roots of unity and has the form

$$C = \exp(2\pi i m/N)$$

giving

$$\psi(x) = U_m(x)\exp(2\pi i x m/Na)$$

as a satisfactory solution, where $U_m(x)$ has periodicity 'a'. With the substitution

$$k = 2\pi m/Na$$

we have the Bloch result. In fact it is more usual to approach this problem in terms of group theory. Referring to Figure 8.4 it can be seen that atomic translations can

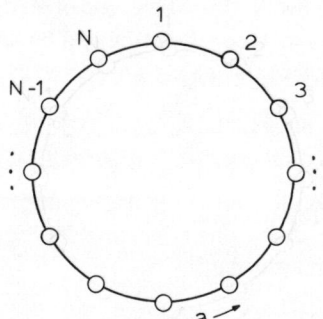

Figure 8.4 A periodic system of N atoms spaced distance 'a' apart in the form of a closed ring, or 'Born–Kármán' periodic boundary condition

occur for rotations of the ring of atoms through any angle that is a multiple of $2\pi/N$ radians. This corresponds to the $C_{\bar{N}}$ symmetry group. Our problem can then be defined as obtaining linear combinations of the atomic orbitals of the line of atoms, that are transformed into irreducible representations of this C_N symmetry group. It follows that there will be N such irreducible representations, and that upon rotation through an angle $2\pi/N$ each wave function is increased by the multiplying factor $\exp(2\pi in/N)$. Assuming that each of our similar atoms in Figure 8.4 have just one valence electron atomic orbital χ each, then the normalized Bloch functions take the form

$$\psi_1 = \frac{1}{\sqrt{N}}(\chi_1 + \chi_2 + \cdots \chi_N),$$

$$\psi_2 = \frac{1}{\sqrt{N}}\{\chi_1 + \cdots \chi_{N-1}\exp[2\pi i(N-2)/N] + \chi_N \exp[2\pi i(N-1)/N]\}$$

etc. or more generally

$$\psi_N = \frac{1}{\sqrt{N}}\sum_{m=1}^{N}\chi_m \exp[2\pi im(n-1)/N], \quad n = 1, 2, \ldots N.$$

To extend the situation to a three-dimensional model the atoms can be characterized by the radius vector

$$R = ma$$

where $|a| = a$. The mth atomic orbital then takes the form

$$\chi_m = \chi(r - ma) = \chi(r - R)$$

and using the identity

$$k = 2\pi(n-1)/Na$$

the Bloch functions can be given in terms of the radius vector r and the wave vector k as

$$\psi(k, r) = \frac{1}{\sqrt{N}}\sum_R \chi(r - R)\exp(ikR) \tag{8.14}$$

For a given periodicity 'a' and number of atoms N, the wave vector k depends solely on the Bloch function number, and so each Bloch function can be uniquely identified by the k value. The corresponding eigenvalues are found by substituting the Bloch function into equation (8.7) to give

$$\mathscr{E}(k) = \langle \psi(k, r)|\hat{H}|\psi(k, r)\rangle$$

This function is often called the 'dispersion law'. From the secular equation (8.11) and equation (8.14) the values for $\mathscr{E}(k)$ are determined by the coulomb and resonance integrals, α and β. The coulomb integral is given by

$$\alpha = \langle \chi_m|\hat{H}|\chi_m\rangle = \langle \chi(r - R)|\hat{H}|\chi(r - R)\rangle$$

and if only the interaction of adjacent atoms are taken into account, then the resonance integral is given by

$$\beta = \langle \chi_m |\hat{H}| \chi_{m+1} \rangle = \langle \chi(r-k)|\hat{H}|\chi(r-k-a) \rangle$$

For a chain of equally spaced sodium atoms, for example, where we only consider valence electrons in the 3s orbitals, the eigenvalue takes the form

$$\mathscr{E}(k) = \alpha + 2\beta \cos ka$$

The variation of the energy states $\mathscr{E}(k)$ as a function of the wave vector k is shown in Figure 8.5 for a chain of 25 sodium atoms. The reference energy level for Figure 8.5 is the 'vacuum level' of zero total energy, and represents the energy of an electron at rest an infinite distance away from the system of atoms. Since α and β have negative values, then the energy states in Figure 8.5 all have energy values below the vacuum level. It can be seen that only k values in the range

$$-\pi/a \leqslant k \leqslant \pi/a$$

need be considered in order to provide non-redundant information, and this range of values is known as the first Brillouin zone. Also shown in Figure 8.5 are the projections of the $\mathscr{E}(k)$ points onto the energy axis, which can be seen to result in a continuous band of width 4β in which all 25 energy levels of the line of 25 sodium atoms are located. This band is called the energy band, and in particular the valence band. It will also be appreciated that each energy level in this band can accommodate two of the 25 valence 3s electrons, and so the valence band is only half occupied by valence electrons. This agrees with the fact that sodium atoms form a metallic structure.

Using essentially the same formulation as described here, Andre (reference 18, pp. 1–21) has given the calculations for the energy band structures of an infinite

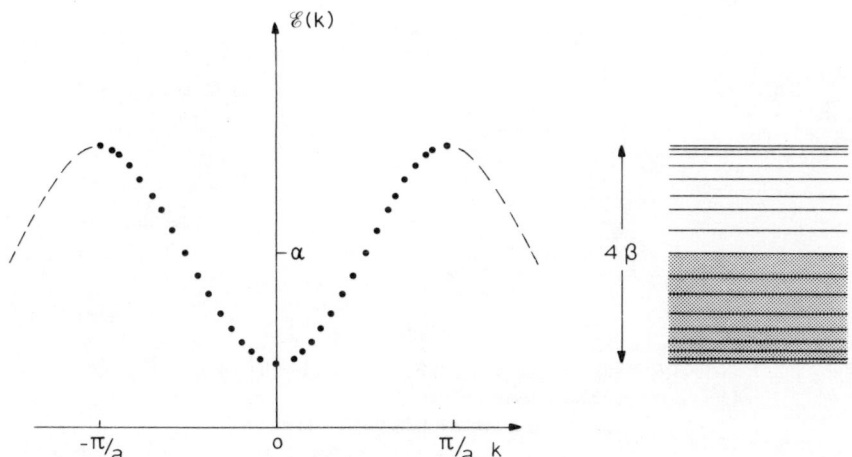

Figure 8.5 The variation of the energy states $\mathscr{E}(k)$ as a function of wave vector k for a chain of 33 sodium atoms spaced 'a' apart. The projections of $\mathscr{E}(k)$ on to the energy axis are seen to result in a band of energies of width 4β

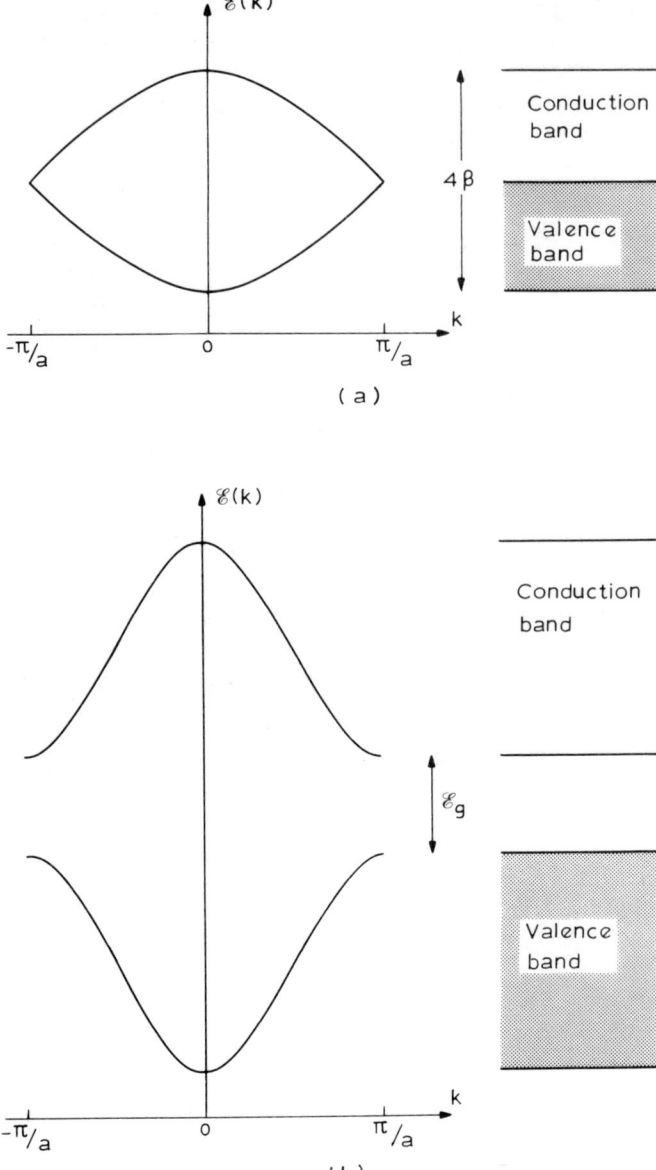

Figure 8.6 The energy band schemes and electron occupancies for (a) a linear π-electron conjugated carbon chain and (b) a linear σ-electron saturated carbon chain

conjugated π-electron system of the form

$$\cdots -C=C-C=C-C=C- \cdots$$

The unit cell was taken as containing two adjacent carbon atoms and the lattice vector was of length 'a'. The calculation yields two π-electron energy bands of the form

$$\mathscr{E}_1(k) = \alpha + 2\beta \cos(ka/2)$$

$$\mathscr{E}_2(k) = \alpha - 2\beta \cos(ka/2)$$

The π-electron energy bands are shown in Figure 8.6(a) from which it can be seen that for a large number of atoms in the chain there is an infinitesimal gap between the electron-occupied valence band $\mathscr{E}_1(k)$ and the empty conduction band $\mathscr{E}_2(k)$. This shows that the conjugated π-electron chain has metallic-like properties in the sense that the π-electrons are able to occupy empty states enabling them to 'conduct' along the chain. This is consistent with the chemist's concept of 'resonant' conjugated structures, and also with the diamagnetic properties of benzene-like chemical structures that result from circulating π-electron currents. The energy band structure for saturated σ-bond chains of the form

$$\cdots -C-C-C-C-C-C- \cdots$$

will be expected to have the form shown in Figure 8.6(b), which shows that it has the properties of an insulator with an energy gap $\mathscr{E}_g$ between the filled valence band and the empty conduction band. This is consistent with polyethylene being an insulator, for example. The energy gap $\mathscr{E}_g$ in Figure 8.6(b) will have the value

$$\mathscr{E}_g = (\alpha_2 + 2\beta_2) - (\alpha_1 - 2\beta_1)$$

where the suffices 1 and 2 refer to the relevant coulomb and resonance integrals for the ground and excited σ-orbital states, respectively. For aromatic compounds and polyenes the ground state π-electron resonance integrals β have values typically in the range -1.6 eV to -4.8 eV[10] and for σ-chains in paraffins $\beta \simeq -6.5$ eV.[23] The ground state coulomb integrals α, representing as they do the coulombic potential energies of the valence electrons in their orbits around the atomic cores, have the (negative) values of the orbital ionization potentials I. The value for the band gap $\mathscr{E}_g$ can also be given as

$$\mathscr{E}_g = I - A$$

where A is the electron affinity of the atoms. Values for I range from 7.5 eV to 9.5 eV for unsaturated hydrocarbons, and from 10 eV to 13 eV for saturated hydrocarbons.[10] Electron affinity values for organic molecules generally fall within the range 0.5–2.0 eV.

In considering the conduction properties of electrons in energy bands the concept of the density of states function $S(\mathscr{E})$ of the band is of importance. This density function $S(\mathscr{E})$ describes the number of allowed electron energy states per unit energy range in the band. We have seen that the N energy levels of the atomic chain are fully described within the length $2\pi/a$ of k-space, and since taking into account electron spin there are $2N$ energy states in this length, we have that

$$dN = \frac{Na}{\pi} dk$$

The number of states per unit energy range is then given as

$$S(\mathscr{E}) = \frac{dN}{d\mathscr{E}} = \frac{Na}{\pi} \frac{dk}{d\mathscr{E}}$$

We have seen that the valence band for an extended çonjugated π-electron system has an $\mathscr{E}(k)$ band of the form

$$\mathscr{E}(k) = \alpha + 2\beta \cos(ka/2) \qquad (8.15)$$

so that

$$\frac{d\mathscr{E}}{dk} = -a\beta \sin(ka/2)$$

$$= -a\beta \sin[\text{arc} \cos(\mathscr{E} - \alpha/2\beta)]$$

to give the density of states function as

$$S(\mathscr{E}) = -\frac{N}{\beta\pi} \{\sin[\text{arc} \cos(\mathscr{E} - \alpha/2\beta)]\}^{-1}$$

Since the resonance integral β has a negative value then $S(\mathscr{E})$ is a positive quantity.

The polymers of particular biological interest are those such as proteins and DNA. The basic repeating unit in the protein structures is the planar peptide residue of Figure 2.3. It can be seen from Table 2.1 that of the 20 common α-amino acids 18 of them have a carbon atom as the first atom in the side chemical chain R. In glycine the first and only side-chain atom is hydrogen, and in proline the side chain bends back onto the peptide unit. In the DNA double helix structure the basic repeat units are the base pairs, and one can proceed from one base pair to the next by the simultaneous translation of 0.34 nm along the DNA axis and by a rotation of $36°$ about this axis. Such considerations have been taken into account by Ladik and his associates in their calculations of the energy band structures for such biopolymers (reference 18, pp. 23 and 663). Also, the unit cell structures for the biopolymers are far more complicated than the simple examples of the linear carbon chains whose energy bands are depicted in Figure 8.6. The simplest polypeptide chain, that of polyglycine, has seven atoms comprised of four different types, and there are more than 60 atoms in each DNA dinucleotide. Such large atomic structures can cause problems for the quantum mathematician. The reason for this is that if all the interatomic interactions are to be included, then the number of integrals in the secular determinant of equation (8.11) increase in number roughly as the fourth power of the number of atomic orbitals included in the LCAO expansion. This means that even with the most modern computational facilities the number of calculations required are excessive and impracticable in terms of computer time. As a result much effort has been directed towards developing approximate LCAO—SCF schemes that are suitable for large molecules and polymeric systems.

There is a well-known theorem in quantum mechanics known as the 'Variation Principle' which effectively states that the more complex, in terms of the number of constituent orbitals, a wave function can be made then the lower will be its associated energy. To achieve such an ideal wave function ψ of lowest energy, then equation (8.4) would have to be in the form of an infinite expansion in order to give ψ its most complex or flexible form. However, the larger the number of atomic orbitals χ_i used for ψ the larger will be the number of integrals appearing in the secular determinant. The first approach in simplifying the calculations is therefore to limit the set of orbitals χ_i, and this is often done by using Slater-type atomic orbitals for each atomic orbital occupied by an electron in the ground state. For carbon atoms, for example, this means that only the 1s, 2s, and 2p atomic orbitals are used to describe ψ. Such a set of orbitals is known as the 'minimal basis set'. Other basis sets are possible, such as the 'double zeta', for example, which uses twice the number of orbitals used for the 'minimal basis set'. It should be mentioned that such simplifying procedures are often used in the so-called *ab initio* methods, and as such the final results may not be as perfect as their description may imply. Other simplifying procedures neglect atomic overlap integrals and introduce estimates for other integrals which are based on experimental data. Most of these computationally rapid approximations treat only the valence electrons and include the 1s electrons for example as part of the atomic core. They are given quite impressive code names such as the CNDO method, which (disappointingly perhaps) stands for 'complete neglect of differential overlap'. Developments of such procedures include the INDO (intermediate neglect), MINDO (modified INDO), and PNDO (partial neglect) methods, and have been described in the book by Pople and Beveridge.[24] An appreciation of the differences in the results obtained using some of these procedures can be gained from Table 8.1 which gives the net atomic charges calculated for the planar polyglycine structure using the INDO and MINDO methods,[25] together with the values obtained for Figure 2.6 using the simple Del Re method. It can be seen from this table that of the two 'modern' methods, the INDO procedure produces atomic charge distributions for the glycine structure in

Table 8.1   Net atomic charges for planar polyglycine obtained using the INDO and MINDO/2 procedures,[25] compared with the values obtained for Figure 2.6 using the Del Re method. The atomic notation is the same as that used for Figure 2.6

Atom	INDO	MINDO/2	Del Re (Figure 2.6(a))
N	−0.270	−0.676	−0.260
C_α	+0.223	+0.380	−0.013
C	+0.435	+0.812	+0.441
O	−0.448	−0.710	−0.461
H_1	+0.179	+0.276	+0.201
H_2, H_3	−0.060	−0.041	+0.046

262

Table 8.2 The resultant dipole moment μ_r and orientation θ for the peptide unit of polyglycine, using the data of Table 8.1 and the geometrical notation of Figure 2.7

	INDO	MINDO/2	Figure 2.7
μ_r	3.46 D	4.69 D	3.63 D
θ	45°	40.2°	46.7°

closer agreement with the Del Re method than does the MINDO/2 procedure. The INDO calculations lead to a realistic value of 2.78 Debye units for the carbonyl C=O group, whereas the MINDO/2 procedure gives the very large result of 4.41 Debye. From the charge distributions given in Table 8.1, the resultant dipole moment μ_r and its orientation θ can be calculated for the polyglycine peptide unit, and these values are given in Table 8.2, using the geometry of Figure 2.7. The results given in Table 8.2 again indicate that the INDO technique leads to the more realistic description of atomic charge distributions, and they also serve to show that the very simple Del Re method, outlined and used in Chapter 2, is capable of providing very useful and relatively accurate results and compares very favourably with the latest procedures available.

ELECTRONS AND HOLES – THEIR DYNAMICS AND ENERGY DISTRIBUTION

If we assume that the valence electrons experience a constant potential V throughout the system, then solutions to the Schrödinger equation will have the form

$$\psi_k = \exp(ikr)$$

which on substitution into equation (8.3) gives the resultant kinetic energy K.E. as

$$\text{K.E.} = \mathscr{E} - V = \hbar^2 k^2/2m$$

The kinetic energy is also given by

$$\text{K.E.} = mv^2/2$$

This shows that the wave vector

$$k = mv/\hbar$$

and so has a physical meaning in terms of the momentum of the electron.

The velocity of an electron in an energy band can be described even more generally in terms of its wave vector. From the general theory of wave mechanics the velocity v of a particle is equal to the group velocity of the waves representing the particle. Since the wave vector is inversely related to the wavelength associated with the electron, we have that

$$v = d\omega/dk$$

where ω is the angular frequency of the electron wave, and is related to its energy by

$$\mathscr{E} = \hbar\omega$$

The electron velocity can then be written in the form

$$V = (d\mathscr{E}/dk)/\hbar \qquad (8.16)$$

If an external electric field E is now applied to the system, each electron will experience a force

$$F = qE$$

and provided there are empty energy states in the near vicinity the electron will be accelerated according to the equation

$$\text{acceleration } a = F/m \qquad (8.17)$$

where m is the effective mass of the electron. We also have from equation (8.16) that the acceleration is given by

$$a = \frac{dv}{dt}$$

$$= \left(\frac{d^2 \mathscr{E}}{dk \cdot dt} \right) \Big/ \hbar$$

$$= \frac{1}{\hbar} \frac{d^2 \mathscr{E}}{dk^2} \cdot \frac{dk}{dt} \qquad (8.18)$$

When the field has acted on the electron for a small time dt, it will gain energy $d\mathscr{E}$ according to

$$d\mathscr{E} = Fv \, dt$$

$$= \frac{qE}{\hbar} \cdot \frac{d\mathscr{E}}{dk} \, dt$$

which leads to the relationship

$$\frac{dk}{dt} = \frac{qE}{\hbar}$$

and so equation (8.18) can be rewritten as

$$a = \frac{qE}{\hbar^2} \frac{d^2 \mathscr{E}}{dk^2}$$

Comparing this result with the form of equation (8.17) we see that the electron behaves as if it had an effective mass

$$m = \hbar^2 / (d^2 \mathscr{E}/dk^2) \qquad (8.19)$$

From the form of the $\mathscr{E}(k)$ curves of Figures 8.5 and 8.6 this gives the interesting result that the electrons in the lower half of a continuous energy band (where $d^2\mathscr{E}/dk^2$ is positive) have a positive effective mass, whereas at the upper half of a band they have a *negative* effective mass. This means that electrons occupying the upper half of a band interact with an electric field as if they had a positive rather than negative charge! This concept, which would not have been predicted from classical theories of electronic conduction, explains why certain metals such as beryllium, cadmium, cobalt, iron, and zinc exhibit positive Hall effect coefficients compatible with the free charge carriers having an effective positive charge. In these metals the valence energy bands are either more than half filled, or as a result of the overlapping of 4p and 4d electron energy bands, for example, the electrons occupying the most energetic states find themselves in a band structure where the factor $(d^2\mathscr{E}/dk^2)$ has a negative value.

So far we have considered the motions of electrons as if they only interacted with the external electric field. In actual fact electrons also interact with the thermal vibrations (phonons) of the atomic lattice in such a way that the electrons are scattered from the paths they would otherwise have followed. Such scattering effects lead to the concept of electrical resistivity and to allow for this we can introduce a relaxation time τ such that on average an electron is only free for τ seconds before it is scattered. If we assume that after each scattering action the velocity of the electron is lost or is so changed in direction as to have an effective value of zero, then after time τ the electron will on average acquire a drift velocity

$$v = qE\tau/m$$

and we can define an effective electron drift mobility μ_d as

$$\mu_d = \frac{v}{E}$$

$$= \frac{q\tau}{m}$$

$$= \frac{q\tau}{\hbar^2} \cdot \frac{d^2\mathscr{E}}{dk^2} \tag{8.20}$$

If there are n electrons per unit volume, each with a charge q, this electron drift motion will be equivalent to an electric current density

$$J = nqv$$

and the effective conductivity σ is given by

$$\sigma = \frac{J}{E}$$

$$= nq\mu_d$$

$$= \frac{nq^2\tau}{\hbar^2} \cdot \frac{d^2\mathscr{E}}{dk^2}$$

This shows that the conductivity of the system is directly proportional to the number of conducting charge carriers, on the relaxation time τ, and also on the form of the energy band. From equation (8.15) which can be taken to be a typical representation of the $\mathscr{E}(k)$ curve, we have that

$$\frac{d^2\mathscr{E}}{dk^2} = -\frac{a^2\beta}{2}\cos[\text{arc } \cos(\mathscr{E} - \alpha/2\beta)]$$

showing that $d^2\mathscr{E}/dk^2$ is directly proportional to β which in turn directly determines the width of the energy band. From equations (8.19) and (8.20) it follows that for narrow energy bands the electron's effective mass is large and its corresponding mobility is small. There is a limit to how small the charge carrier mobility can become before the general validity of the energy band theory breaks down. One way of expressing this is through the Heisenberg uncertainty principle

$$\Delta x\ \Delta(mv) \gtrsim h$$

which in terms of an energy band of width W can be written as

$$W\tau \gtrsim h$$

Substituting for τ, using the relationship $\mu_d = q\tau/m$, then for the band theory to be valid requires

$$\mu_d \gtrsim q\hbar/mW \tag{8.21}$$

which for $W \simeq 1.5$ eV and an effective electron mass equal to the normal rest mass of 9.1×10^{-31} kg gives

$$\mu_d \gtrsim 0.8 \times 10^{-4}\ \text{m}^2/\text{V s}$$

For free electrons in a nearly constant field V, then

$$\mathscr{E}(k) \simeq \hbar^2 k^2/2m$$

and the band width W is given by

$$W = \mathscr{E}(k = \pi/a) - \mathscr{E}(k = 0)$$

$$= h^2/8ma^2$$

so that equation (8.21) can be given as

$$\mu_d \gtrsim 4qa^2/\pi h$$

and for a typical lattice constant 'a' of the order 0.5 nm we have that for the band theory to be valid requires

$$\mu_d \gtrsim 0.8 \times 10^{-4}\ \text{m}^2/\text{V s}$$

The similarity of these two limiting drift mobility values of around 1×10^{-4} m^2/V s suggests a simple and approximate rule of thumb that for energy band widths greater than around 1.5 eV the effective mass of the charge carriers is less than that of the normal rest mass, whereas for smaller band widths the effective mass is greater. In silicon, for example, with a valence band width of around 14 eV, the effective charge carrier mass can be some six times less than that of the normal electron rest mass, and for germanium even smaller effective masses are applicable (reference 16, p. 337). We can also approach the question regarding the validity of the band theory by defining a mean free path length L between collisions as

$$L = v_{th} \tau$$

where v_{th} is the mean thermal velocity of the electron. The mean thermal electron kinetic energy $m v_{th}^2 / 2$ has the value $3kT/2$, so the mean free path length can be written as

$$L = \left(\frac{3kT}{m} \right)^{1/2} \tau$$

$$= (3mkT)^{1/2} \mu_d / q$$

The concept of coherent delocalized wave-like electron transport within an energy band becomes invalid if the mean free path length becomes so small as to be comparable with the interatomic dimensions, which is typically of the order 0.15 nm. In other words the validity of the band theory requires

$$\mu_d \gtrsim 1.5 \times 10^{-10} q / (3mkT)^{1/2}$$

which for the normal electron rest mass at 300 K requires

$$\mu_d \gtrsim 2.3 \times 10^{-4} \text{ m}^2 / \text{V s}$$

These simple calculations show that for mobility values less than about 10^{-4} m^2/V s, the concept of freely delocalized conduction electrons is not valid. For charge carrier mobility values less than this we have to envisage the conduction process in terms of a hopping mechanism, whereby the electrons jump or hop between the energy states.

The energy band scheme of Figure 8.6(b) in which the valence band is completely occupied by electrons represents the situation for insulators and semiconductors at very low temperatures. For such a filled valence band the resultant electron current is zero because for every electron with a wave vector k_i and velocity v_i, there is another electron with a wave vector $-k_i$ and velocity $-v_i$. For a band completely filled with N electrons the resultant current I is zero, and can be written as

$$I = -q \sum_{i=1}^{N} v_i$$

$$= 0$$

This can also be written as

$$I = -q \left(v_j + \sum_{i \neq j}^{N} v_i \right)$$

$$= 0$$

If we were to remove the jth electron from this band, then

$$I = -q \sum_{\substack{i \neq j}}^{N} v_i$$

$$= qv_j$$

and the missing electron, or electron 'hole', behaves as if it had a *positive* charge. Since all the electrons in the energy band strive to minimize their energies by occupying the lowest lying energy as possible, this indicates that the electron holes will occupy the states at the top of the valence band. We have seen that an electron occupying the top of the valence band will have a negative effective mass, so removing such an electron is equivalent to *adding* the mass of one electron. In this way an electron hole behaves as a particle with positive charge and positive effective mass.

At any temperature T above absolute zero there is a finite probability that electrons from the valence band will be excited across the band gap $\mathscr{E}_g$ into the conduction band. The number of electrons, in thermal equilibrium with the system, occupying states in the energy range $\mathscr{E} + d\mathscr{E}$ within the conduction band is given by

$$N(\mathscr{E}) \, d\mathscr{E} = S(\mathscr{E})F(\mathscr{E}) \, d\mathscr{E}$$

where as before $S(\mathscr{E})$ is the density of available states in the range $\mathscr{E} + d\mathscr{E}$, and $F(\mathscr{E})$ is the Fermi–Dirac distribution function

$$F(\mathscr{E}) = \frac{1}{1 + \exp(\mathscr{E} - \mathscr{E}_F)/kT} \tag{8.22}$$

in which $\mathscr{E}_F$ is the so-called Fermi level. The factor $\mathscr{E}_F$ acts as a normalizing parameter such that

$$\int_0^\infty N(\mathscr{E}) \, d\mathscr{E} = \int_0^\infty S(\mathscr{E})F(\mathscr{E}) \, d\mathscr{E} = N$$

where N is the total number of valence electrons in the system and the zero of energy is taken to be at the bottom of the valence band. From equation (8.22) it can also be seen that if an energy state exists at energy $\mathscr{E}_F$ then it has a probability of $\frac{1}{2}$ of being occupied by an electron. This scheme is illustrated in Figure 8.7, to show the form of the Fermi–Dirac distribution function and the associated distribution of electrons in the conduction band and holes in the valence band. The number of electrons n in the conduction band is given by

$$n = \int_{\mathscr{E}_c}^{\mathscr{E}_c + W_c} S(\mathscr{E})F(\mathscr{E}) \, d\mathscr{E} \tag{8.23}$$

where $\mathscr{E}_c$ is the bottom of the conduction band of width W_c. Similarly, since the probability of an energy state $\mathscr{E}$ being unoccupied is given by $[1 - F(\mathscr{E})]$ then the

268

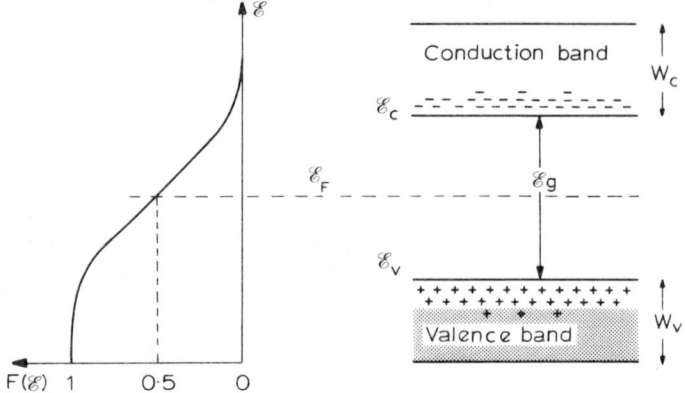

Figure 8.7 The form of the Fermi–Dirac distribution function $F(\mathscr{E})$ and the distribution of electrons and holes in the energy bands, where – represents electrons and + represents holes

number of electron holes p in the valence band is given by

$$p = \int_{\mathscr{E}_v - W_v}^{\mathscr{E}_v} S(\mathscr{E})[1 - F(\mathscr{E})]\, d\mathscr{E} \qquad (8.24)$$

To maintain electrical charge neutrality, and since every electron leaving the valence band is considered to enter the conduction band, then $n = p$. As a result of the exponential form of $F(\mathscr{E})$ then most of the holes occupy states near to the top of the valence band, and most of the electrons in the conduction band will be close to $\mathscr{E}_c$. This greatly simplifies the integrals of equations (8.23) and (8.24), and if it is further assumed that $\mathscr{E}_F$ lies more than $4kT$ away from the band edges $\mathscr{E}_v$ and $\mathscr{E}_c$, then it can be shown that

$$n = p = 2(2kT/h^2)^{3/2}(m_e m_h)^{3/4}\exp(-\mathscr{E}_g/2kT) \qquad (8.25)$$

where m_e and m_h are the effective masses of the electrons and holes, respectively.[16] If the electron and hole effective masses are assumed to be equal to the normal electron rest mass, then the pre-exponential term in equation (8.25) has the value 2.5×10^{25} m^{-3} at 300 K. For molecular solids such as anthracene, which give rise to very narrow energy bands, the pre-exponential factor has a value of the order 4×10^{27} m^{-3} and approximates to the molecular density.[26] The position of the Fermi level is given as

$$\mathscr{E}_F = \frac{\mathscr{E}_c + \mathscr{E}_v}{2} + \tfrac{3}{4}kT\log(m_h/m_e)$$

so that if $m_h = m_e$ then $\mathscr{E}_F$ is situated exactly at the midpoint of the energy band gap $\mathscr{E}_g$. The thermal excitation of electrons from the valence band to the conduction band results in the *intrinsic* generation of mobile electrons and holes, and the resulting intrinsic conductivity σ is given by

$$\sigma = |q|(n\mu_e + p\mu_h)$$

where μ_e and μ_h are the electron and hole mobility values, respectively. By comparison with the charge carrier concentrations, the mobility values do not in general vary significantly with temperature, so that from equation (8.25) the intrinsic conductivity can be written in the form of the standard semiconduction equation

$$\sigma = \sigma_0 \exp(-\mathscr{E}_g/2kT) \tag{8.26}$$

An Arrhenius plot of log σ versus $1/T$ produces a straight line of slope proportional to $\mathscr{E}_g$.

The approximations used in deriving equation (8.25) are equivalent to compacting the conduction and valence bands into two discrete effective density of states N_c and N_v located at the band edges $\mathscr{E}_c$ and $\mathscr{E}_v$, respectively, where

$$N_c = 2\left(\frac{2m_e kT}{h^2}\right)^{3/2} \quad \text{and} \quad N_v = 2\left(\frac{2m_h kT}{h^2}\right)^{3/2}$$

We shall see that for biopolymers the energy band gap $\mathscr{E}_g$ is at least of the order 4 eV, which from equation (8.25) corresponds to an intrinsic electron and hole concentration of the order 5×10^{-9} m^{-3} at 300 K. Such charge carrier concentrations would give rise to an insignificantly small conductivity. However, it is possible for electronic energy states to occur within the forbidden energy gap region. Such states could arise from the presence of impurity atoms or other chemical groups incorporated into the structure of the biopolymer, and in general

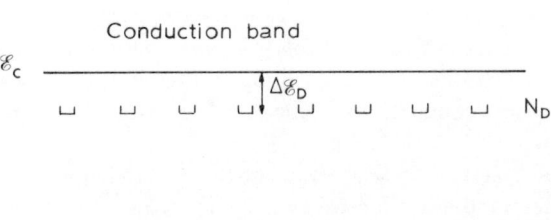

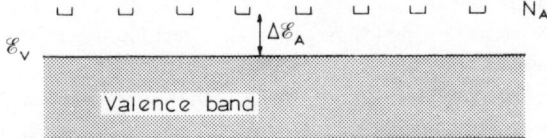

Figure 8.8 Energy level diagram showing a single set of N_A electron-acceptor states situated at an energy $\Delta\mathscr{E}_A$ above the valence band edge, and a single set of N_D electron-donor states situated $\Delta\mathscr{E}_D$ below the conduction band edge

such states can either act as acceptor levels for electrons leaving the valence band, or can act as donors to provide electrons in the conduction band. Such a situation leads to the *extrinsic* generation of conducting charge carriers and produces extrinsic conductivity. The corresponding energy level diagram is shown in Figure 8.8.

If the concentration N_A of electron-acceptor states is negligibly small so that they do not compensate for the donor states, then the number of electrons in the conduction band is controlled by the concentration N_D of the donors according to the equation

$$n = (N_c N_D)^{1/2} \exp(-\Delta \mathscr{E}_D / 2kT)$$

where again we have used the approximation that all the electrons in the conduction band are considered to occupy an effective density of states N_c at the conduction band edge $\mathscr{E}_c$. Similarly, if only electron acceptors exist within the energy band gap, then the concentration of holes in the valence band is given as

$$p = (N_v N_A)^{1/2} \exp(-\Delta \mathscr{E}_A / 2kT)$$

These equations correspond to the situation where the Fermi level lies midway between the donor or acceptor states level and the appropriate energy band edge. However it is often the case that the donor states are compensated by a smaller concentration of acceptor states, or *vice versa*, and as a result the Fermi level is situated nearer to the energy band gap centre. The extrinsic carrier concentrations then vary as $\exp(-\Delta \mathscr{E}/kT)$ and not as $\exp(-\Delta \mathscr{E}/2kT)$. The extrinsic conductivity will also vary with temperature according to these exponents. The modification of the conductivity of such semiconductors as germanium and silicon by the carefully controlled incorporation of donor and acceptor impurity atoms into their atomic structures forms an essential basis for modern solid-state electronics. For example, the addition of just 0.01 per cent of boron to pure silicon ($\mathscr{E}_g = 1.1$ eV) produces *p*-type silicon having a room temperature conductivity some 10^6 times greater than the intrinsic silicon. (At 150 K the *p*-type conductivity is greater by a factor of 10^{16}.) For a biopolymer with $\mathscr{E}_g = 4$ eV the incorporation of 0.01 per cent impurity, giving acceptor states 0.5 eV above the valence band edge, would result in a room temperature conductivity some 10^{24} greater than that of the pure biopolymer!

Electron energy states can also occur in the forbidden energy band gap $\mathscr{E}_g$ as a result of disorder in the atomic or molecular lattice.[17] We should also point out here that the formation of valence and conduction bands separated by a forbidden gap $\mathscr{E}_g$ does not require long-range order and periodicity of the atomic structure. This is evident from the fact that we can see through glass windows. Glass is a disordered amorphous material and the fact that it is transparent indicates that a continuum of randomly spaced electron states does not occur in the glass structure. Instead, a gap of at least 3 eV must exist between the band of electron-occupied states and the band of empty states so that the light in the optical range of frequencies cannot be used to excite the valence electrons into higher energy states. Metals absorb visible light because electrons in the partially filled valence band can be optically excited into vacant states within the valence band, and semiconductors like germanium and silicon have a metallic lustre because at room temperature there

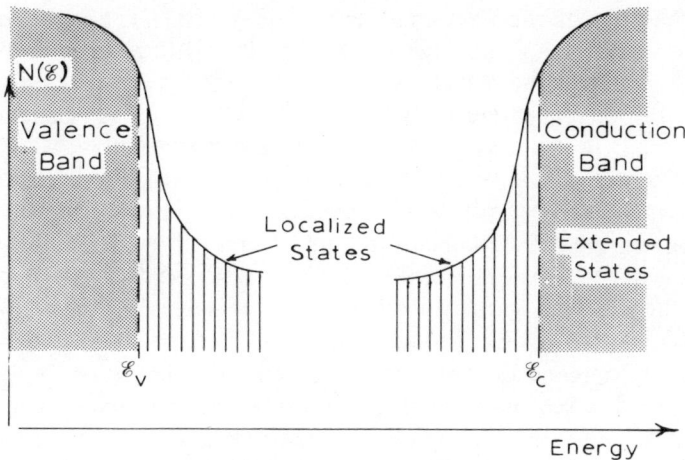

Figure 8.9 Tails of localized states extending from the valence and conduction band edges as a result of disorder in the atomic structure

are sufficient vacant states (holes) in the valence band, and electrons in the conduction band, for a large number of optically induced electronic transitions to occur. When a significant amount of disorder does exist in the atomic lattice, the major effect on the electron energy states is to produce tails of 'localized' states extending from the valence and conduction band edges, as shown in Figure 8.9. The localized states above the valence band edge act as traps in which the energy of a hole is quantized, and similarly the localized states beneath the conduction band edge act as electron traps. The holes and electrons can move from one localized state to another by the process of thermally activated hopping, rather like a type of Brownian motion, and not in the form of a coherent wave-like motion as for charge transport in the extended states within the valence and conduction bands. The transition between the localized and extended states occurs at the band edges $\mathscr{E}_v$ and $\mathscr{E}_c$, and these two levels have also been termed the 'mobility edges'[27] since in crossing from the extended to localized states the mobility becomes an activated process of the form

$$\mu = \mu_0 \exp{-(W/kT)} \qquad (8.27)$$

with a value much lower than that within the bands of extended states. The charge carrier mobility at levels just within the bands at $\mathscr{E}_v$ and $\mathscr{E}_c$ has been suggested by Mott[28] to have the values

$$\mu_e = \frac{\sigma_{min}}{qN_c}, \qquad \mu_h = \frac{\sigma_{min}}{qN_v}$$

where σ_{min} is a quantity referred to as the 'minimum metallic conductivity' given by

$$\sigma_{min} \simeq \frac{0.05q^2}{\hbar b}$$

and b is the scale of the fluctuation in atomic periodicity. For biopolymers we might expect $b \simeq 1.5$ nm, and in which case $\mu \simeq 20 \times 10^{-4}$ m^2/V s.

If the mobility of the charge carriers in the energy bands is limited solely by scattering arising from interactions with the lattice vibrations, then the mobility should be proportional to $m^{-5/2} T^{-3/2}$, where m is the effective mass of the charge carrier and T is the temperature. The distinction between band conduction and hopping charge transport can then sometimes, but not always, be made by determining the temperature dependence of the mobility through Hall effect or drift mobility measurements. From the form of equation (8.27) for hopping charge transport the mobility should increase with increasing temperature, whereas for band conduction the mobility should decrease as $T^{-3/2}$.

Apart from hopping over the potential barrier between two localized states, charge carriers can also tunnel through the barrier. Such a tunnelling phenomenon is considered to be of importance in some biological systems (e.g. references 29, 30). The concept of tunnelling has commonly been discussed in terms of the solution of the Schrödinger wave equation for an electron in a 'potential box'[9] of the form shown in Figure 8.10, where $\mathscr{E}$ is the total energy of an electron in a potential well of depth V. Superimposed on this potential well is shown a typical eigenfunction $\psi(x)$ of the one-electron Hamiltonian to show that this function extends in the form of exponential tails outside the potential well. The electron therefore has a finite probability of existing in the regions beyond the potential box and of penetrating (tunnelling) into a nearby unoccupied potential box (localized state). For a wide barrier of width w, the probability of electron transmission through the barrier is proportional[15] to $\exp -(Zw\gamma)$ where

$$\gamma = [2m(V - \mathscr{E})/\hbar^2]^{1/2}.$$

Modern concepts of tunnelling phenomena have been described by Duke,[14] and have been extended to take into account effects arising from vibrations of the atomic lattice which can modify the energy of the localized electron and hence its tunnelling behaviour.[31-33] The relevance of such concepts for large biological molecules has recently been discussed by Duke.[34]

When a charge carrier is travelling through the extended states of an energy band it is usually assumed that its velocity is so rapid that the vibrational motions of the atomic or molecular lattice can be ignored. However, if an electron or hole becomes trapped in a localized state it may remain there for a period greatly in excess of the

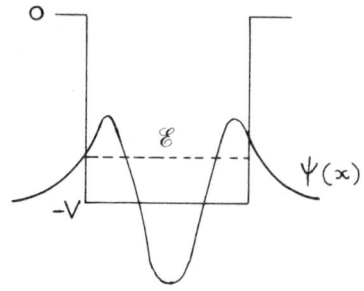

Figure 8.10 An electron of energy $\mathscr{E}$ in a potential well of depth V to show that its eigenfunction $\psi(x)$ extends beyond the boundaries of the well

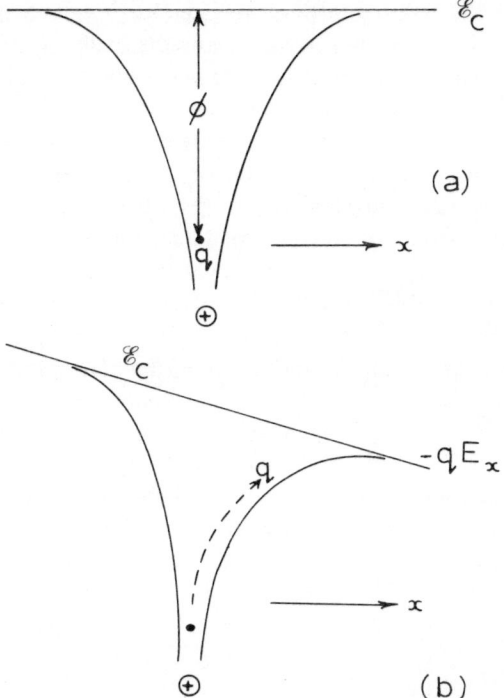

Figure 8.11 (a) An electron in a potential well below the conduction band edge $\mathscr{E}_c$. (b) The potential well is modified in the presence of an external field E in such a way that the potential barrier in the field direction is lowered

lattice vibration times in such a way that the lattice nuclei move to new equilibrium positions as a result of the electrostatic interaction with the stationary charge carrier. This can lead to the charge carrier becoming more deeply trapped as a result of 'digging itself' into a self-induced potential well. The entity which then migrates through the lattice is not just the charge carrier itself but also a large perturbation of the lattice vibrations (phonons). Such a combination is called a polaron.[35,36] The polaron moves through the lattice either by a hopping or activated tunnelling process. Each transition of the charge carrier is accompanied by a large number of phonon emissions and absorptions so that it is scattered many times during a single transition between neighbouring localized states.

Finally, we have noted that fields of the order 10^7 volts/metre and greater occur in many biological systems. Such large field strengths can greatly influence the way in which charge carriers behave in trapping sites. In Figure 8.11(a) an electron is shown localized in a potential well below the conduction band edge, and Figure 8.11(b) shows how this potential well is modified in the presence of an externally applied electric field E. In the absence of the field the electron has to surmount a potential barrier of height ϕ in order to enter the conduction band. With the field E

applied the barrier in the field direction is reduced by the potential qEx due to the field, so that the probability of the electron being thermally excited out of the trap into the conduction band is increased. This effect is known as the Poole–Frenkel effect.[37]

If the potential energy of the electron in the well results from a coulombic interaction between the electron and a positively charged lattice site (i.e. the trapping state is electrically neutral when occupied by an electron), then the total potential barrier in the presence of the field has the form

$$V = \frac{-q^2}{4\pi\epsilon_0\epsilon_r x} - qEx$$

in the field direction, and the maximum height of the barrier in the field direction is given by

$$\phi_m = -2(q^3 E/4\pi\epsilon_0\epsilon_r)^{1/2}$$

The resultant conductivity of the system, if it is dominated by the donor action of such electron-occupied localized states, is given by

$$\sigma = \sigma_0 \exp(-\phi/kT)\exp(\beta_{PF}E^{1/2}/kT)$$

where

$$\beta_{PF} = (q^3/\pi\epsilon_0\epsilon_r)^{1/2}$$

For fields of the order 10^7 volts/metre and $\epsilon_r \simeq 2.5$ the barrier ϕ is reduced by 0.15 eV in the field direction. Although this effect has been described in terms of electrons occupying electron traps, the same situation occurs for holes occupying hole traps at an energy ϕ above the valence band edge. Hill[38] has made a general analysis of Poole–Frenkel conduction and for the case where the trapping centres are far apart he derives a generalized equation for the conduction current J of the form

$$j \propto \alpha^2 \sinh \alpha \qquad (8.28)$$

or

$$j \propto \alpha^{-1}(\alpha \cosh \alpha - \sinh \alpha) \qquad (8.29)$$

where

$$\alpha = \beta_{PF}E^{1/2}/kT$$

and j is given by

$$j = JT^{-n} \exp(\phi/kT) \qquad (8.30)$$

Equation (8.28) corresponds to the trapped charge being released preferentially along the field direction, in which case $n = 4$ in equation (8.30), while equation (8.29) corresponds to spherically uniform charge release, in which case $n = 3$ in equation (8.30). The high field electrical conduction in the perylene–chloranil charge-transfer complex has been found to be well described in terms of equation (8.29) with $n = 3$ in equation (8.30).[39]

ENERGY BANDS IN BIOPOLYMERS

Proteins

As described at the beginning of this chapter, the first serious considerations of the possible biological significance of electronic conduction in energy bands of proteins were those expressed by Szent-Györgyi and Laki. In Laki's view the energy bands were possibly associated with the continuum of hydrogen bonds in the protein structure, in which the peptide groups formed the elementary cells. By 1949 the importance of hydrogen bonds as essential features of proteins had become well recognized, and another of Szent-Györgyi's students, Gergely, in collaboration with M. G. Evans at Manchester produced the first energy band calculation for a biopolymer.[40] The calculation was based on the hydrogen-bonded network

$$-C=O\cdots H-N-C=O\cdots H-N-$$

which runs perpendicular to the main polypeptide chains in the α-helix and β-pleated secondary structures of protein molecules. The β-pleated structure is shown in Figure 8.12 and it can be seen that the hydrogen bonds effectively act as links to form an extended π-electron conjugated pathway running perpendicular to the main polypeptide chains. The idea that such a π-electron system may give rise to electronic conduction pathways in protein molecules was first suggested by C. A. Coulson.[41] Each peptide unit

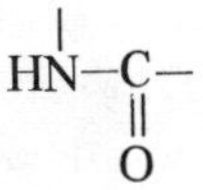

$$HN-C-$$
$$\|$$
$$O$$

can be considered to consist of a delocalized system of π-electrons associated with the π-orbitals of the C and O atoms together with the lone electron-pair orbital of the N atom. Evans and Gergely proposed that electron delocalization occurs across the hydrogen bonds, to give electronic coupling between all the peptide units in the hydrogen-bonded chains. This leads automatically to the result that for a hydrogen-bonded chain of n peptides the three discrete π-energy levels of each peptide unit will broaden into three energy bands of n closely spaced levels. Each peptide unit possesses four π-electrons: the two lone-pair nitrogen electrons together with two more electrons associated with the carbonyl π-bond. For a chain of n hydrogen-bonded peptides the four n π-electrons will fill the two lowest bands completely, leaving the uppermost band empty. In this way the hydrogen-bonded peptide chains cannot be thought of as metallic-like conduction pathways as is the

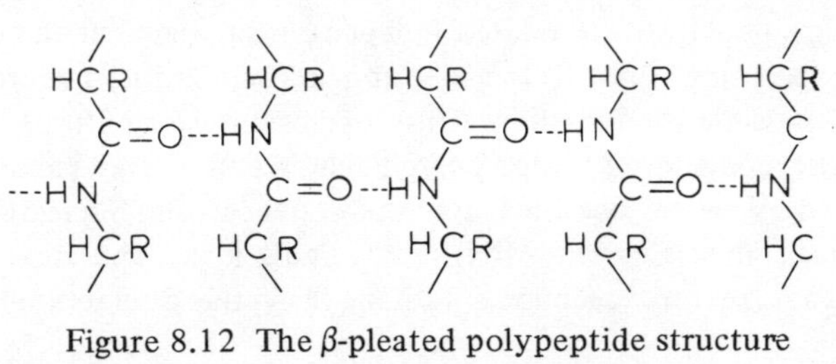

Figure 8.12 The β-pleated polypeptide structure

case for extended conjugated chains having partially filled energy bands of the form of Figure 8.6(a). For the polypeptide system of Figure 8.12 the intrinsic semi-conduction state can only be attained on excitation of electrons across the energy gap $\mathscr{E}_g$ between the uppermost electron-occupied band and the lowest empty band, as shown in Figure 8.13. On assuming a trigonal structure for the amino nitrogen atom, Evans and Gergely were able to derive a value of 3.05 eV for the energy gap $\mathscr{E}_g$, and values of 0.26 eV and 0.12 eV for the widths W_v, W_c of the valence and conduction bands, respectively. Calculations some twelve years later by Itoh,[42] who used Bloch orbital functions for the electrons in the hydrogen-bonded system, produced similar band widths and a value for $\mathscr{E}_g$ of 4.4 eV. More rigorous calculations soon followed[43-48] in which use was made of experimental atomic spectroscopic data for the peptides in assigning values for certain wave function integrals, and account was also taken of the lone-pair electrons of the oxygen atoms. The concept of π-electron delocalization across the hydrogen bonds was taken into account by including the empty 2p orbital on the H atom into the basis set of atomic orbitals. The final result of these calculations gives an energy band scheme, for an infinite hydrogen-bonded protein structure, of the form of Figure 8.13, which consists of two π-bands and one n-band, with band widths of around 1 eV and an energy band gap $\mathscr{E}_g$ of the order 5–7.5 eV. Such a wide band gap value definitely precludes the possibility of a significant intrinsic generation of mobile charge carriers in the α-helix and β-pleated protein structures.

The energy band calculations for protein structures so far described were based on the possibility of electronic delocalization through the hydrogen bonds between the peptide units. More recent calculations, however, show that the formation of energy bands of considerable width is possible as a result of atomic interactions along the main polypeptide chains. This approach follows from the original proposal by Brillouin[49] that the

$$-\overset{\displaystyle H}{\underset{\displaystyle H}{N}}-\overset{\displaystyle O}{C}-\overset{\displaystyle H}{\underset{\displaystyle R}{C}}-$$

groups of the polypeptide chain form the elementary cell for a periodic protein model (as for Laki's original idea[4]), with the side-chain groups (R) effectively acting as perturbations to produce impurity-type levels within the energy band gap. In their calculations of the electronic structure of polyglycine in the α-helix form, Fujita and Imamura[50] came to the conclusion that there was a considerable mixing of the π- and σ-orbitals in the energy bands. This indicated that the π- and σ-electron systems could not be treated independently, and that the earlier energy band calculations using only the π-electron protein structures required rein-vestigation and possible modification. This conclusion was confirmed by the later calculations of Beveridge *et al.*[25] for polyglycine, where energy band gap values of 5.7 eV and 6.14 eV were obtained for the α-helical and β-pleated. structures, respectively, using the so-called MINDO/2 calculations. The next obvious re-finement was to take into account simultaneously the interactions through the

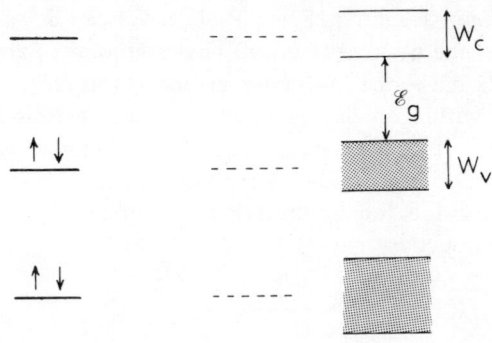

Peptide Unit   Polypeptide

Figure 8.13 Energy band structure formation
for the β-pleated polypeptide structure

hydrogen-bonded networks and those along the main polypeptide chains. The first
model studied in this way was of a two-dimensional polyformamide network[51]
which was followed by calculations for a polyglycine β-pleated conformation.[52]
The energy band widths depended on the molecular geometries used and whether
or not second neighbour interactions were taken into account. It has also been
found that the energy band widths are greater for atomic interactions along the
polypeptide backbone chain than those resulting from interactions through the
hydrogen-bonded networks. This can be seen in Table 8.3 which gives some
characteristic electronic parameters obtained by Suhai[53] and Ladik[54] for the
polyglycine parallel-chain β-pleated structure, using the MINDO/2 and *ab initio* SCF
LCAO CO methods, respectively.

The energy band widths calculated by Ladik for the main chain were corrected
for long-range correlation effects, and a value of 3.5 was used for the limiting high
frequency permittivity value ϵ_∞. As outlined in Chapter 2 a value for ϵ_∞ of around
2.0 may in fact be more appropriate when considering the interior regions of

Table 8.3   Electronic parameters for the polyglycine β-pleated structure

	Main polypeptide chain		H-bonded chain	
	Suhai[53]	Ladik[54]	Suhai[53]	Ladik[54]
Valence band width (eV)	1.23	1.91	0.31	0.29
Conduction band width (eV)	1.68	1.26	0.48	0.14
Electron drift mobility ($m^2\ V^{-1}\ s^{-1}$)	2.5×10^2	–	2.88×10^{-3}	–
Hole drift mobility ($m^2\ V^{-1}\ s^{-1}$)	1.4×10^2	–	1.16×10^{-3}	–

polypeptide and protein structures. From Table 8.3 it can be seen that the energy bands associated with the hydrogen-bonded chains are more narrow than those for the main polypeptide backbone. As shown earlier in this chapter, the mobility of free charge carriers within an energy band is inversely related to the carrier's effective mass, which in turn depends on the width of the band and on the strength of charge carrier–phonon interactions. The charge carrier–phonon interactions can be calculated in terms of the deformation potentials $V_{d,e}$ and $V_{d,h}$ for the electrons and holes, respectively, as

$$V_{d,e} = \frac{\delta \mathscr{E}_{c,l}}{\Delta}, \qquad V_{d,h} = \frac{\delta \mathscr{E}_{v,u}}{\Delta}$$

where $\delta \mathscr{E}_{c,1}$ represents the shift of the bottom of the conduction band edge as a result of longitudinal dilation Δ of the lattice, and $\delta \mathscr{E}_{v,u}$ is the corresponding shift of the top of the valence band edge. The dilation Δ is given as $\Delta = \delta\xi/\xi$, where $\delta\xi$ is the change of the lattice constant ξ along the direction of propagation of the longitudinal wave. In order to determine these deformation potentials, Suhai repeated the whole band structure calculations two times by distorting the lattice first along the main polypeptide chain and then perpendicularly to it in the direction of the hydrogen-bonded networks. The following expression[55] was used by Suhai for the drift mobility value calculations:

$$\mu = \frac{2\sqrt{2\pi}}{3} \cdot \frac{qC_l\hbar^4}{V_{d,e}^2 m^{5/2}(kT)^{3/2}}$$

where C_l is the elastic constant for longitudinal acoustic waves, and m is the effective charge carrier mass given by equation (8.19). By modifying the elastic constant values for diamond, and taking into account the differences in crystal structure and force constants, Suhai assumed values for the elastic constant C of 1.05×10^{11} N m^{-2} and 1.97×10^{10} N m^{-2} along the main chains and hydrogen-bonded networks, respectively. The drift mobility values so derived by Suhai are given in Table 8.3.

Although the absolute magnitude of the charge carrier drift mobilities may be in question in that they seem rather high, the results given in Table 8.3 indicate that the most favourable mechanism of electronic transport in proteins involves conduction along the direction of the main polypeptide chains rather than through the hydrogen-bonded pseudo-conjugated network. To chemists not accustomed to the concept of energy bands this result will not appear to be correct. General experience with the chemistry of organic compounds leads to the expectation that a conjugated bond system can readily conduct electronic charge from one site to another, sometimes over considerable atomic distances, as for example in the scheme

$$>C^{\delta+} = \overset{\curvearrowright}{C} - C = C - \overset{\curvearrowright}{C} = O^{\delta-}$$

where the effect of the electron-attracting carbonyl group causes electronic rearrangement within a linear conjugated chain. Also, from studies of the oxidation-reduction reactions that take place between metal ions that are separated by bridging ligands, it has been shown that whole electrons can 'conduct' over many bond lengths in conjugated hydrocarbon systems (e.g. references 56, 57). However, polypeptide chains, with their basic structure of the form

$$
\begin{array}{ccccccc}
 & H & & R & & O & H \\
 & | & & | & & \| & | \\
-N & - & C & - & C & - & N & - & C & - & C - \\
 & \| & & | & & | & & | \\
 & O & & H & & H & & R
\end{array}
$$

do not possess delocalized π-electrons associated with conjugated bonds, and in organic chemistry there does not appear to be a basis for the possibility of electronic transfer through such a chemical system. Evidence has in fact been found for the occurrence of electron transfer through peptide structures from studies of the reduction of polypeptide—cobalt complexes by chromous ions in solution,[58] but here the electron transfer from the chromium to the cobalt ion was considered to occur through the internal hydrogen-bonded amide groups. Such studies should now be repeated and extended, although it may be that the bridging polypeptide ligands may not be able to be made large enough for the features of a true valence energy band to be investigated. It is the presence of such an energy band, rather than whether or not we have a conjugated or saturated chemical structure that is of importance for us. We have seen that the hydrogen-bonded, pseudo-conjugated, network of the α-helix and β-pleated protein structures has an energy band scheme where an energy of up to 7.5 eV is required to excite an electron into the energy state where it becomes mobile along the whole network. For the main polypeptide chain the activation energy required to attain the conducting state may be as much as 11.6 eV.[54] Therefore at normal physiological temperature no significant possibility exists for free electronic conduction to occur in either of these two pure protein structures. But if an 'impurity' molecule were present which could either donate an electron into an empty conduction band, or extract an electron from an otherwise filled valence band, then free electronic conduction would be possible (Szent-Györgyi neatly demonstrates this in public lectures by removing a marble from an otherwise fully marble-packed tin box whereupon the remaining marbles can be heard to become mobile and able to 'rattle about'). The mobility of an excess electron in the unfilled conduction band, or of an electron 'hole' in the valence band, depends on the width of these bands and on their structure in 'reciprocal wave vector space'. Suhai's calculations[53] show that very significant electronic mobility can occur along the main polypeptide chains of protein molecules, and for this to occur requires a charge-transfer interaction with another molecule. A simple analogy arises for the polyethylene polymer, which has a completely saturated hydrocarbon structure. Many energy band calculations have been made for polyethylene[59] and there can now be no doubt that well-defined valence and conduction bands exist for this polymer. Incorporating neutral iodine into the polymer greatly increases its electrical conductivity, and this is considered to result from electron transfer from the polyethylene molecules to vacant acceptor levels in the iodine to produce mobile 'holes' in the polymer valence band.[60,61]

The fact that the widest energy bands are associated with atomic interactions along the main polypeptide backbone would suggest that the basic and underlying electronic conduction properties of protein molecules are mainly associated with their primary structures. The disruption of α-helix, β-sheet, and other structural features, as a result of dehydration for example, should not therefore destroy the essential electronic properties of proteins. This is of great advantage for the experimentalist since it is often more convenient to perform electronic measurements on compressed protein powders rather than on proteins in solution. By studying the electronic changes that occur as a result of increasing the water content of the test protein samples, it should be possible to make a meaningful extrapolation to the more relevant biological situation. It should also be remembered that many proteins are embedded in the hydrophobic environment of lipid membranes, so that measurements on dry protein powders may not be so biologically irrelevant as might at first have been thought the case. The precise details of the energy band structures will certainly depend on the exact conformation of the protein molecule when located in its correct biological environment, but for the purposes of demonstrating the existence and basic properties of electronic conduction pathways it is sufficient in the first instance to investigate extracted and purified protein material.

Apart from the consequences of conformational changes, hydration can also modify the energy band structure as a result of its influence on local permittivity values. We have seen that the energy locations and the widths of the valence and conduction bands are directly related to the values of the coulombic energies parameters α_i and β_{ij}, which in turn are inversely proportional to the value of the relative permittivity ϵ_r. For the parameter α_i, representing the coulombic energy of attraction of an electron in the field of its parent atom, the corresponding value for ϵ_r will be $\epsilon_r = 1.0$. However, for the resonance energies β_{ij} which can involve atomic nuclei not in immediate contact with each other, we can expect that the pertinent value for ϵ_r will be influenced by the presence or otherwise of polarizable molecules such as water. Hydration will increase the local value of ϵ_r and hence cause a reduction in the value for β_{ij}. From consideration of Figure 8.3 this in turn will tend to lead to a narrowing of the energy band widths. Also as a result of long-range correlation effects the energy band levels will depend on the value of the permittivity and hence on the local water content. Consideration of long-range correlation effects leads to the result that the perturbation of the energy levels is proportional to the factor $(1 - 1/\epsilon_r)$ where ϵ_r is the limiting high frequency value of the relative permittivity.[54] The reference level we can take is that where $\epsilon_r = 1$ and corresponding to zero shift or perturbation of the energy levels. For the particular case of the polyglycine main chain and using a value of $\epsilon_r = 3.5$, then compared with such a reference level the valence band edge $\mathscr{E}_v$ became raised by 0.69 eV and the conduction band edge $\mathscr{E}_c$ fell by 0.72 eV, resulting in a 1.41 eV reduction of the energy band gap $\mathscr{E}_g$.[54] The widths of the valence and conduction bands also became smaller by 0.19 eV and 0.12 eV, respectively. For larger values of ϵ_r approaching the high frequency limiting value for water of $\epsilon_r \simeq 4.2$, then we will expect this trend to continue, as shown schematically in Figure 8.14(a). Since

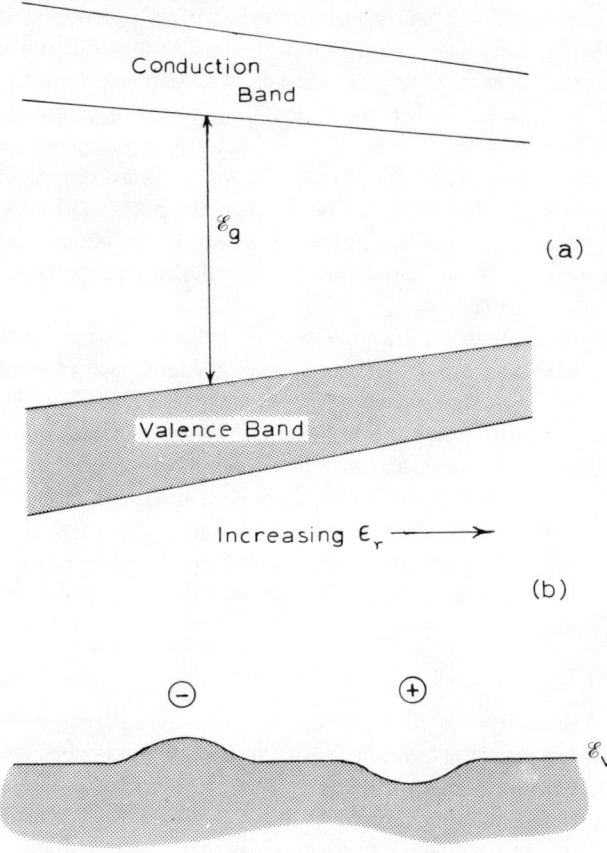

Figure 8.14 (a) The variation of the energy band
structures as a function of the relative permittivity.
(b) The influence of local charges on the structure
of the valence energy band. Similar effects will
occur for the conduction band

electrons always strive to minimize their energy by seeking the lowest possible
energy states, then the scheme represented by Figure 8.14(a) implies that in the
absence of other effects any mobile electrons and holes within the protein structure
will tend to migrate to the regions of highest local permittivity. In Chapter 2 it was
shown that the outer regions of protein molecules tend to have the largest
permittivity values, and especially those regions of greatest hydration. This may
have direct relevance to the fact that for many of the enzymes so far investigated in
terms of their molecular structure, the active site of the enzyme is located within a
'cleft' or 'crevice'. These crevices are large enough to allow access for water
molecules and as such they will represent regions of relatively high permittivity
towards which mobile free charges within the protein structure will be attracted.
The availability of such free charges may well be of importance in the overall
enzymic activity of many protein molecules.

Another factor that will be absent for dry protein powders is the presence of localized charged groups. When in solution, the basic and acidic ionizable side-chain groups distributed over the protein surface will become ionized as shown for cytochrome-c in Figure 3.3, for example. The presence of these localized charges will modify the energy band structure. In the vicinity of a negatively charged group, the resulting electronic coulombic repulsions will increase the potential energy of the electronic states in that region, whereas near the positive groups the coulombic attraction will tend to reduce the potential energy of the states. The result of such electrostatic interactions on the shape of the valence energy band is shown schematically in Figure 8.14(b).

Finally, before proceeding to the case of DNA it should be mentioned that Petrov et al.[62] have suggested that the approximations used in some of the energy band calculations for proteins have not been sufficient regarding the description of the excited states of the polypeptide chains. In particular they point out that most of the calculations use a minimal atomic basis set which includes only the Slater orbitals 1s, 2s, and 2p for the N, C, and O atoms, and the 1s orbital for the hydrogen atom, whereas it is known that the optical spectra of the second row elements of the periodic table involve the 3s and 3p states. For example, in the case of the carbon, nitrogen, and oxygen atoms in the polypeptide backbone optical transitions of the type:

$$(1s)^2(2s)^2(2p)^n \rightarrow (1s)^2(2s)^2(2p)^{n-1}3s$$

correspond to an energy of 7 eV, which is rather less than the forbidden energy band gap $\mathscr{E}_g$ value resulting from many of the energy band calculations given in the literature. Calculations[63] involving the 3s and 3p states for ethylene show that the lowest optical singlet excitation involves the $\pi \rightarrow 3s$ (7.11 eV) transition rather than the $\pi \rightarrow \pi^*$ transition of energy (7.66 eV). Also, for ethane the transitions $\sigma(2s, 2p) \rightarrow 3s$ at 9.16 eV are of lower energy than the $\sigma(2s, 2p) \rightarrow \sigma^*(2s, 2p)$ transitions. From such considerations, and the fact that the optical bands of proteins at 5.3 eV and 6.7 eV can be connected with the $n \rightarrow (3s, \pi^*)$ and $\pi \rightarrow (3s, \pi^*)$ transitions, respectively, Petrov et al. suggest that the energy bands of polypeptide chains have the form shown in Figure 8.15. Because of the mixing of the 3s and π^*-states in the conduction band, then an electron in the conduction band can migrate along the polypeptide chain even if an extended π-electron structure associated with hydrogen-bonded α-helices and β-pleated structures is absent. This is in agreement with the conclusions of Suhai[53] that the dominant conduction pathway is along the σ-bonded main polypeptide chain. Following the earlier concept of Brillouin,[49] Petrov et al. have attempted to estimate the positions of the electronic levels of the amino-acid side-chain groups. In particular they have concentrated on the possibility that the charged $-COO^-$ groups of the aspartic acid and glutamic acid side chains, as well as the terminal polypeptide chain carboxyl group, may be able to act as electron donors for the conduction band. With the water molecule having an ionization potential of around 12.7 eV, and the OH radical having an electron affinity of around 2.2 eV, Petrov et al. conclude that the removal of a proton from the $-$COOH group results in an increase in the

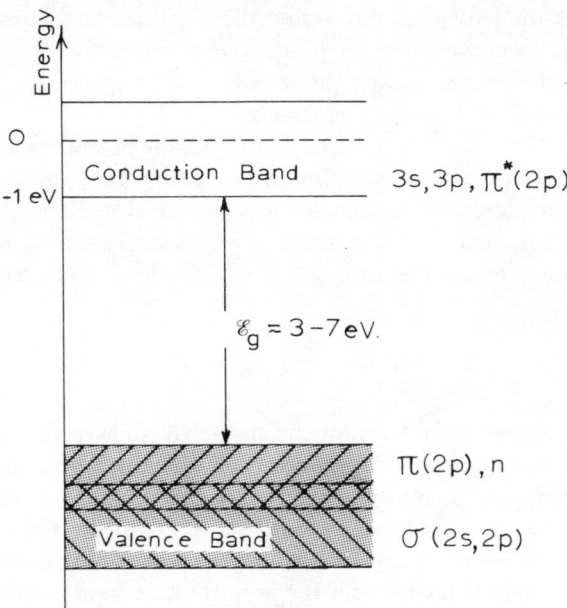

Figure 8.15 The energy band scheme for polypeptide chains as suggested by Petrov et al.[62]

energy of one of its electrons of about 11 eV. Also, from the electron affinity values of the order 3–4 eV given by Duke and Fabish[32] for the σ-bonded polymers, polystyrene and polymethylmethacrylate, it is concluded that the energy of an electron in a free —COO⁻ group is located close to the conduction band edge. As a result of an overlapping of its wave function with the 3s and π^*-orbitals of the polypeptide conduction band, this electron will be partly delocalized and can be transferred along the polypeptide chain to an electron-acceptor state situated elsewhere in the protein molecule. It is also noted that the —COO⁻ group can become an electron acceptor by forming bonds with other protein groups such as in the structure:

If the ionized carboxyl groups are involved in hydrogen bonds of the form

then they can no longer act as electron donors or acceptors. An analysis of the dynamics of such donor–acceptor electron transitions occurring through protein molecules in the presence of an electric field gives a good description of the gating current time evolution in membrane sodium channels.[64] Furthermore, comparing the kinetic data on gating currents and the data on the quantity of charge transferred during membrane depolarizations, then within the energy band scheme it can be shown that a single gating current event is compatible with the transport of an electron pair through the protein structure.

By contrast to this emphasis by Petrov *et al.* of the possibility that the $-COO^-$ groups can donate electrons into the protein conduction band, other studies[65] have indicated that keto-aldehydes, such as methylglyoxal, when incorporated into the structure of protein molecules can act as electron acceptors to produce mobile holes in the valence band. This concept is described in more detail in the next chapter.

DNA

When comparing the situation for proteins and DNA we have an intriguing state of affairs. Although the sequence of amino acids is known accurately for a great number of proteins, the complete conformational structure is known for only about 20 of the smaller protein molecules. For DNA, on the other hand, its periodic structure in terms of the Watson—Crick stacked base-pair double helix is fairly well known, but the sequence of the nucleotide base pairs in the different DNA molecules is still largely unknown. This poses a problem for theoretical calculations of the energy band structures for DNA.

Two different pathways for electronic delocalization have been considered. The first is due to Ladik[66] who calculated the overlap integrals between the π-type atomic orbitals of the different stacked base pairs, and found them to be significant in comparison with the intramolecular integrals for each base pair. As a result, Ladik concluded that the electronic system of DNA could effectively be represented by a delocalized Bloch-type wave function extending over the whole macromolecule. With this concept as a basis, band structure calculations were made for various DNA models, a review of which has been given by Ladik.[67] It was found that for the simple periodic DNA models such as the homopolynucleotides or those with repeated base pairs, the valence and conduction band widths were of the order 0.2–0.3 eV. For the more complicated periodic DNA models, such as single-stranded poly(GA) or double-stranded poly(AT, GC), the band widths were some ten times smaller. This result would indicate that the simple periodic DNA models can be expected to exhibit coherent electron transport, whereas a hopping-type electronic mechanism characteristic of transport in narrow energy bands will be appropriate for the more complicated structures. This conclusion was confirmed by Suhai[68] who, starting from the band structures and taking explicit account of the electron—phonon interactions, derived room temperature electron mean-free-path lengths of 1–5 nm for the simple models, and 0.35–0.6 nm for the more complicated ones. Recently, the first *ab initio* LCAO SCF CO calculations[69] have been performed for the homopolynucleotide polycytosine. The valence and conduction band widths of around 0.5 and 1.2 eV, respectively, obtained from these calculations were considerably broader than the values 0.2–0.3 eV obtained from the earlier semiempirical methods.[67] The value obtained for the energy band gap

$\mathscr{E}_g$ was 10.65 eV, which must represent a slight overestimate since the experimental ionization potential for cytosine is 8.9 eV.[70] To account for this in the calculations, Ladik has shown[54] that by using a corrected Fock operator and including long-range correlation effects, the band gap value is reduced towards a more realistic value (9.79 eV), and the valence and conduction band widths are decreased by about 10 per cent to values of 0.52 eV and 1.16 eV respectively. The electronic properties of native aperiodic DNA molecules are more likely to correspond to that for the complicated models, although there does appear to be a preference to have the same kind of base repeated several times in both native DNA and RNA.[71,72] Apart from the trend towards increasing band widths as a result of improved calculations for the DNA structures, another promising feature for the consideration of electronic delocalization in DNA is that the presence of charged ions such as Mg^{2+} associated with the base pairs has a pronounced effect on the energy band structures, considerably reducing the energy gap $\mathscr{E}_g$ and increasing the band widths.[73,74] A band broadening effect is also expected to occur as a result of the inhomogeneous electric field due to the $[PO_4^- \cdot \cdot K^+]$ charged double layers at the outer part of the DNA double helix.[75] The effect of hydrogen-bonded water molecules has been calculated to produce an extra π-band between the lowest filled bands leaving the other bands practically unchanged.[73] We will also expect the presence of water to influence local permittivity values and lead to the effect described by Figure 8.14(a).

As for the protein molecules, the second mechanism for electronic delocalization in DNA followed from an original suggestion by Brillouin.[49] In this theory the four different nucleotide bases act as 'impurities', either donating electrons into the conduction band or accepting electrons from the valence band associated with the sugar-phosphate backbone chain of the DNA molecule. This has great advantages for theoretical calculations, since the sugar-phosphate backbone is strictly periodic, whereas when considering the energy band structure of the aperiodic system of superimposed nucleotide bases, approximations to a more periodic structure have to be made. The first calculations of this problem indicated that conduction along the sugar-phosphate backbone was at least as probable as through the aperiodic system of the bases.[76] Moreover, on comparing the relative energies of the various bands, there is a possibility that in DNA a charge-transfer interaction may occur between the poly (base pairs) and the sugar-phosphate backbone, with the result that electronic delocalization occurs down the backbone and hole transport occurs through the base pairs.[74,75]

In other calculations, Altieri and Krizan[77] have used a self-consistent band model for DNA, with the band gap $\mathscr{E}_g$ fixed at a value of 2.4 eV so as to be consistent with experimental semiconduction data, and the assumption was made that the order of the bases did not affect the band structure. This semiempirical approach gave realistic energy band structures and free electron mobility values of the order 10^{-3} to 10^{-2} m^2 V^{-1} s^{-1} in reasonable agreement with the mobility value that can be estimated from Suhai's calculations.[76] It should be mentioned, however, that an $\mathscr{E}_g$ value of 2.4 eV is most likely to be incorrect. The experiment leading to this result may almost certainly have been influenced by extrinsic conduction effects. The charge carrier motion envisaged by Altieri and Krizan was one of tunnelling between the bases, or by conduction along the sugar-phosphate backbone.

By contrast with these theoretical approaches that lead to the conclusion that polynucleotides such as DNA are semiconductors with well-defined valence and conduction bands, there is another approach which considers that in these polymers the nucleotide bases should be treated as a set of loosely coupled monomers and not as a set of interacting units. This approach, due mainly to Rhodes,[78] leads to the situation where only excitons and not free electrons can pass between the bases. This concept of delocalized excitons rather than electrons gives a good model for describing the hypochromism of such macromolecules as DNA.[79] The result of a semiempirical treatment for poly(AT) by Rosen[80] gives very narrow band widths consistent with such an exciton model. In re-examining Rosen's calculations, Altieri and Krizan[77] reached the conclusion that the narrow energy band result was not the correct one, and instead they favoured results giving much broader energy bands, leading to the concept of electronic delocalization for these biopolymers.

The theoretical calculations for the proteins and DNA give strong support to the original suggestion by Szent-Györgyi that electronic conduction phenomena could play an essential rôle in biological activity. The long-term aim of the theoretician will be to find correlations between the electronic structure of the proteins and DNA and their biological functions. Many problems remain, especially in treating adequately the difficulties arising because of the aperiodic structure of naturally occurring biopolymers. For most of the calculations for protein structures it has been considered that the regularly repeating peptide units provide the primary feature of the energy band structures, and that the different side chains do not seriously influence the electronic properties. This particular point of view has not been proven one way or the other. In Table 8.3 charge carrier mobility values of the order $100 \ m^2 \ V^{-1} \ s^{-1}$ are given for conduction along the main polypeptide chains, whereas experimental evidence suggests a value no greater than around 5×10^{-3} $m^2 \ V^{-1} \ s^{-1}$.[65] This difference between experiment and theory may reflect the fact that the side chains do in fact create a significant disruption of the periodicity of the electronic structure of the protein molecules. The first attempts to investigate such a disorder effect have been those reported by Suhai and Laki[81] for the poly (glycine—alanine—serine) polymer, where it was found that the valence and conduction band systems each consisted of three closely spaced narrow bands. The occurrence of such narrow bands will lead to much smaller mobility values than those suggested by Table 8.3. This serves to show that the rôle of the experimentalist cannot be undervalued, since the results obtained by careful measurements of the electronic properties of biopolymers will assist the theoretician. Such progress is required before Szent-Györgyi's bold and exciting proposals[6] regarding the relationship between biological semiconduction and cellular regulation and cancer can be fully investigated. One of Szent-Györgyi's viewpoints, namely that induced protein semiconductivity can lead to the building of higher structures, has received strong theoretical support. Laki and Ladik[82] have shown that the creation of holes in the valence bands of protein molecules will allow for intra-band electronic excitations to occur which will greatly increase the electronic polarization and dispersion forces of attraction between protein molecules. Such increased attractive forces between biomacromolecules may help cells maintain their complicated structures and keep cell division under control.

Ladik has presented[54] an interesting description of some possible mechanisms of chemical carcinogenesis which involve the electronic properties of proteins and DNA. He points out that tumour development has statistically been shown to result from a purely chemical (as opposed to viral) origin for at least 90 per cent of the cases. According to present concepts cancerous information is contained in the DNA molecules of normal cells, but it cannot be expressed because the parts of DNA containing this information are suppressed by bound suppressor proteins. If a carcinogen binds to the suppressor protein then according to Bush[83] the protein is released from the DNA and the cancerous information becomes free to transcribe its 'message' to RNA which then translates it to new protein material. Recent findings also show that practically all chemical carcinogens bind directly to DNA.[84] Ladik, therefore, proposes that in order to begin to understand the molecular basis for carcinogenesis we should try first to understand how the binding of carcinogens causes the release of suppressor proteins from the DNA molecules. Whether or not the effects are local or long range, the understanding of the underlying mechanisms must require knowledge of the electronic and vibrational structure of DNA and proteins, including their charge transport properties. The possible charge-transfer interactions between the carcinogens and DNA or protein, as well as between DNA and proteins themselves, should also be investigated. It could be possible, for example, that the electron-donating action of a carcinogen results in the filling of an otherwise partially empty valence band in either the DNA molecule or its bound suppressor protein. This could in turn result in a reduction of the cohesive polarization forces holding together the DNA and protein molecules,[82] leading to the release of the suppressor protein. Progress in this area will require the development of refined calculations, and some of the possible approaches are described by Ladik.[54]

References

1. F. Möglich and M. Schön, *Naturwiss*, **26**, 199 (1938).
2. P. Jordan, *Naturwiss*, **26**, 693 (1938).
3. A. Szent-Györgyi, *Nature*, **148**, 157 (1941); *Science*, **93**, 609 (1941).
4. K. Laki, *Studies Inst. Med. Chem. Univ. Szeged*, **2**, 43 (1942); and private communication.
5. A. Szent-Györgyi, *Bioenergetics*, Academic Press, New York (1957); *Introduction to a Submolecular Biology*, Academic Press, New York (1960); *Bioelectronics*, Academic Press, New York (1968).
6. A. Szent-Györgyi, *Electronic Biology and Cancer*, Marcel Dekker Inc., New York (1976).
7. G. Wiech, *Z. Phys.*, **207**, 428 (1967).
8. T. Gora, R. Staley, J. D. Rimstidt, and J. Sharma, *Phys. Rev. B*, **5**, 2309 (1972).
9. N. F. Mott and I. N. Sneddon, *Wave Mechanics and its Applications*, Clarendon Press, Oxford (1948).

10. A. Streitwieser, *Molecular Orbital Theory for Organic Chemists*, J. Wiley & Sons, Inc., New York (1961).
11. J. D. Roberts, *Molecular Orbital Calculations*, W. A. Benjamin, New York (1962).
12. J. M. Ziman, *Principles of the Theory of Solids*, Cambridge University Press, Cambridge (1964).
13. J. C. Slater, *Quantum Theory of Molecules and Solids*, McGraw-Hill, New York (1965).
14. C. B. Duke, *Tunnelling in Solids*, Academic Press, New York (1969).
15. E. Merzbacher, *Quantum Mechanics*, J. Wiley & Sons, Inc., New York (1970).
16. A. J. Dekker, *Solid State Physics*, Macmillan, London (1971).
17. N. F. Mott and E. A. Davis, *Electronic Processes in Non-Crystalline Materials*, Oxford Press, Oxford (1971).
18. *Electronic Structure of Polymers and Molecular Crystals*, Eds. J. M. Andre and J. Ladik, Plenum Press, New York (1974).
19. C. C. J. Roothaan, *Rev. Mod. Phys.*, **23**, 69 (1951).
20. C. A. Coulson, *Trans. Faraday Soc.*, **38**, 433 (1942).
21. G. G. Hall and J. E. Lennard-Jones, *Proc. Roy. Soc. (London)*, **A205**, 357 (1951).
22. J. E. Lennard-Jones and J. A. Pople, *Proc. Roy. Soc. (London)*, **A210**, 190 (1951).
23. J. A. Pople and D. P. Santry, *Molec. Phys.*, **7**, 269 (1963).
24. J. A. Pople and D. L. Beveridge, *Approximate Molecular Orbital Theory*, McGraw-Hill, New York (1970).
25. D. L. Beveridge, I. Jano, and J. Ladik, *J. Chem. Phys.*, **56**, 4744 (1972).
26. W. Helfrich and P. Mark, *Z. Phys.*, **171**, 527 (1963).
27. M. H. Cohen, H. Fritzche, and S. R. Ovshinsky, *Phys. Rev. Lett.*, **22**, 1065 (1969).
28. N. F. Mott, *Phil. Mag.*, **26**, 1015 (1972).
29. J. J. Hopfield, *Proc. Nat. Acad. Sci. USA*, **71**, 3640 (1974).
30. J. Jortner, *J. Chem. Phys.*, **64**, 4860 (1976).
31. T. F. Soules and C. B. Duke, *Phys. Rev. B*, **3**, 262 (1971).
32. C. B. Duke and T. J. Fabish, *Phys. Rev. Lett.*, **37**, 1075 (1976).
33. C. B. Duke, T. J. Fabish, and A. Paton, *Chem. Phys. Lett.*, **49**, 133 (1977).
34. C. B. Duke, in *Proc. Conf. on Synthesis and Properties of Low-Dimensional Materials*, Ed. J. S. Miller, New York Acad. Sci., New York (1977).
35. H. Fröhlich and G. L. Sewell, *Proc. Phys. Soc. (London)*, **74**, 643 (1959).
36. W. Siebrand, *J. Chem. Phys.*, **41**, 3574 (1964).
37. J. Frenkel, *Phys. Rev.*, **54**, 647 (1938).
38. R. M. Hill, *Phil. Mag.*, **23**, 59 (1971).
39. R. Pethig and V. Soni, *J. C. S. Faraday I*, **71**, 1534 (1975).
40. M. G. Evans and J. Gergely, *Biochim. Biophys. Acta*, **3**, 188 (1949).
41. J. Ladik, private communication.
42. R. Itoh, *Ann. Report Res. Gp. Biophys. Japan*, **1**, 11 (1961).
43. M. Suard, G. Berthier, and B. Pullman, *Biochim. Biophys. Acta*, **52**, 254 (1961).
44. M. Suard, *Biochim. Biophys. Acta*, **64**, 400 (1962).
45. M. Suard-Sender, *J. Chim. Phys.*, **62**, 78 (1965); *ibid.*, p. 89.
46. S. Yomosa, *J. Phys. Soc. Japan*, **18**, 1494 (1963).
47. A. Pullman, *C. R. Acad. Sci., Paris*, **256** 5435 (1963).
48. J. Ladik, *Nature*, **202**, 1208 (1964).
49. L. Brillouin, in *Horizons in Biochemistry*, Eds. M. Kasha and B. Pullman, p. 295. Academic Press, New York (1962).
50. H. Fujita and A. Imamura, *J. Chem. Phys.*, **53**, 4555 (1970).

51. S. Suhai and J. Ladik, *Theor. Chim. Acta*, **28**, 27 (1972).
52. S. Suhai, *Theor. Chim. Acta*, **34**, 157 (1974).
53. S. Suhai, *Biopolymers*, **13**, 1731 (1974).
54. J. Ladik, in *Quantum Theory of Polymers*, Eds. J. M. Andre, J. Delhalle and J. Ladik, pp. 257–278. D. Reidel, Dordrecht-Holland (1978).
55. W. Shockley, in *Electrons and Holes in Semiconductors*, p. 287. Van Nostrand, New York (1950).
56. D. K. Sebera and H. Taube, *J. Am. Chem. Soc.*, **83**, 1785 (1961).
57. F. Nordmeyer and H. Taube, *J. Am. Chem. Soc.*, **88**, 4295 (1966); **90**, 1162 (1968).
58. K. D. Kopple and G. F. Svatos, *J. Am. Chem. Soc.*, **82**, 3227 (1960).
59. J. M. Andre, reference 18, pp. 1–21.
60. T. J. Lewis and D. M. Taylor, *J. Phys. D.*, **5**, 1664 (1972).
61. G. T. Jones and T. J. Lewis, *Faraday Symposia Chem. Soc.*, **9**, 192 (1974).
62. E. G. Petrov, I. I. Ukrainskii, and V. N. Kharkyanen, *Preprint ITP-77-38E*, Academy of Sciences of the Ukrainian SSR, Kiev (March, 1977).
63. R. J. Buenker and S. D. Peyerimhoff, *Chem. Phys.*, **8**, 56 (1975); **9**, 75 (1976).
64. E. G. Petrov, V. I. Teslenko, and V. N. Kharkyanen, *Preprints ITP-77-114P, 115P*, Academy of Sciences of the Ukrainian SSR, Kiev (1977).
65. R. Pethig, *Int. J. Quantum Chem., Quantum Biol. Symp.* **5**, (1978) (In print).
66. J. Ladik, *Acta Phys. Acad. Sci. Hung.*, **11**, 239 (1960).
67. J. Ladik, *Advan. Quantum Chem.*, **7**, 397 (1973).
68. S. Suhai, *J. Chem. Phys.*, **57**, 5599 (1972).
69. S. Suhai, Ch. Merkel, and J. Ladik, *Phys. Lett.*, **61A**, 487 (1977).
70. C. Lifschitz, E. D. Bergmann, and B. Pullman, *Tetrahedron Lett.*, **46**, 4583 (1967).
71. J. Josse, A. O. Kaiser, and A. Kornberg, *J. Biol. Chem.*, **236**, 864 (1961).
72. M. O. Dayhoff, *Atlas of Protein Sequence and Structure*, Vol. 5. National Biomedical Research Foundation, Silver Spring, Md. USA (1972).
73. B. F. Rozsnyai and J. Ladik, *J. Chem. Phys.*, **52**, 5711 (1970); **53**, 4325 (1970).
74. J. Ladik, *Int. J. Quantum Chem., Quantum Biol. Symp.*, **1**, 65 (1974).
75. J. Ladik, reference 18, p. 663, and *Int. J. Quantum Chem., Quantum Biol. Symp.*, **2**, 133 (1975).
76. S. Suhai, *Biopolymers*, **13**, 1739 (1974).
77. J. Altieri and J. E. Krizan, *J. Biol. Phys.*, **3**, 103 (1975).
78. W. Rhodes, *J. Chem. Soc.*, **83**, 3609 (1961).
79. M. Weissbluth, *Quart. Rev. Biophys.*, **4**, 1 (1971).
80. P. Rosen, *J. Biol. Phys.*, **1**, 244 (1973).
81. S. Suhai and K. Laki, private communication.
82. K. Laki and J. Ladik, *Int. J. Quantum Chem., Quantum Biol. Symp.*, **3**, 51 (1976).
83. H. Bush, *Biochemistry of the Cancer Cell*, p. 292. Academic Press, New York (1962).
84. R. Daudel in *Mutagenesis and Chemical Carcinogenesis*, Eds. P. and R. Daudel, Y. Moule, and F. Zajadela, C.N.R.S., Paris (1977).

Chapter 9

Electronic Properties of Biomacromolecules

SEMICONDUCTIVITY

Proteins

At about the same time as Szent-Györgyi was proposing that the semiconduction properties of proteins were of biological relevance, Baxter and Cassie[1,2] were demonstrating that moist specimens of wool behave as electrical conductors, with their conductivity σ varying with temperature according to the Arrhenius-type equation

$$\sigma = \sigma_0 \exp(- \mathscr{E}/kT) \tag{9.1}$$

in the manner commonly observed for semiconductors. As described in the previous chapter the parameter $\mathscr{E}$ in equation (9.1) is the activation energy of semiconduction, and σ_0 is a constant whose value depends on the test material. It should be noted that in the past equation (9.1) has often been written with the exponent of value $(- \mathscr{E}/2kT)$ since it has commonly been assumed that biopolymers are intrinsic semiconductors with the generation of charge carriers occurring by direct excitation of electrons across the band gap $\mathscr{E}_g$ ($\mathscr{E}_g = \mathscr{E}$ in this case). In using the more general equation (9.1), where $\mathscr{E}$ is simply the semiconduction activation energy, the derived values for $\mathscr{E}$ will be one-half the values that are often quoted in the previous literature on biological semiconductors. Equation (9.1) is adopted here to reinforce the opinion that very few, if indeed any, conduction studies on biopolymers have been related to the true intrinsic generation of electrons across the gap $\mathscr{E}_g$ between the valence and conduction bands. For wool the activation energy $\mathscr{E}$ was found to have a value of 1.1 eV, and the semiconductivity was considered to arise from an electron tunnelling mechanism between adsorbed water molecules rather than from energy band conduction associated with the protein structure. However, in 1946 Szent-Györgyi reported photoconductive effects for protein films containing dye molecules,[3] and the concept of biological semiconduction was given new emphasis. This led to many investigations of the electronic conduction proper-

290

Table 9.1   Values of $\mathscr{E}, \sigma_0$ and ρ (300 K) in equation (9.1) for various biopolymers, as derived from reference 5

Material	$\mathscr{E}$ (eV)	$\mathscr{E}$ (kJ mol^{-1})	σ_0 (mho m^{-1})	ρ (300 K) (ohm m)
Bovine plasma albumin	1.39	134	4×10^7	5.6×10^{15}
Cytochrome-c	1.30	125	6.3×10^6	1.1×10^{15}
Collagen	1.37	132	5×10^6	2.1×10^{16}
Elastin	1.46	140	1×10^6	3.4×10^{18}
Haemoglobin	1.33	128	1×10^7	2.2×10^{15}
Haemoglobin (denatured)	1.45	139	1.3×10^8	1.8×10^{16}
Lysozyme	1.31	126	7.9×10^6	1.3×10^{15}
Polyglycine	1.56	150	2×10^8	8.1×10^{17}
Poly-L-tyrosine	1.50	144	4×10^8	4×10^{16}
Tobacco mosaic virus	1.46	141	2×10^7	1.7×10^{17}

ties of various components and extracts of biological systems, such as proteins, DNA, and various cell fragments. The samples mainly took the form of dried compressed pellets or thin films, and in general the conductivity was found to obey the form of equation (9.1) with values for the activation energy ranging from *ca.* 0.8 to 1.6 eV. Work in the laboratory of D. D. Eley in particular, demonstrated that a wide variety of dry proteins and polypeptides exhibited roughly similar semi-conduction properties with activation energies $\mathscr{E}$ of the order 1.3–1.5 eV, σ_0 values in the range from around 10^6 to 10^8 mho m^{-1}, and room temperature resistivities of the order $10^{15} \sim 10^{18}$ ohm m (see Table 9.1). These experimental activation energies led to the compelling conclusion that they represented intrinsic generation of charge carriers across a band gap $\mathscr{E}_g$ of around 3 eV, in agreement with the theoretical value for $\mathscr{E}_g$ calculated by Evans and Gergely in 1949. Based on such an intrinsic semiconductivity model for the proteins, the charge carrier mobility values could be estimated to have the rather large value of around $1-10$ m^2 V^{-1} s^{-1}. With the advantage of our knowledge of the latest theoretical calculations of the energy band gap $\mathscr{E}_g$, we can now consider that the conclusion regarding the intrinsic semiconductivity was almost certainly not correct. The term 'intrinsic semiconductivity' used here refers to the concept of electron excitation across the band gap $\mathscr{E}_g$. It may well be the case that charge carriers are naturally produced in protein molecules as a result of internal charge-transfer interactions, for example. Such a mechanism could also be termed 'intrinsic' in that it would represent an inherent property of the protein molecule. For the purposes of adhering to the conventional terminology of semiconductor physics, this alternative will not be used. Reviews of the pioneering work in the 1950s and 1960s on the semiconductivity of biopolymers have been given in the literature,[4-7] and only some of the more significant aspects of this earlier work will be mentioned here before proceeding to the more recent developments.

One aspect of the earlier work on protein semiconductivity, which still remains to be clearly resolved, is the identification of the nature of the charge carriers

292

involved and in particular their dependence on the state of hydration of the protein. The studies on wool and keratin[2,8] produced the conclusion that the conductivity σ varied with the amount m of adsorbed water according to the relationship

$$\sigma(m) = \sigma_D \exp(\alpha m) \tag{9.2}$$

where σ_D is the dry-state conductivity and α is a constant. This relationship was later found by Spivey to apply for haemoglobin, as reported by Taylor,[9] and in fact describes the hydration-dependent conductivity for many insulators, including that of paper.[10] Equation (9.2) holds over a range of at least eight to ten orders of magnitude for σ, until at hydration contents of the order 25 wt per cent the conductivity begins to approach a limiting conductivity as shown in Figure 9.1.[6,11] King and Medley[8] considered that this effect arose from ionic conduction effects associated with salts absorbed in the proteins. They invoked Bjerrum's statistical theory of ion pairing to explain their results and considered the saturation value for σ to correspond to complete ionization of the salts. They also found for keratin samples, with at least 15 wt per cent adsorbed water, that the effect of passing an electric current through the samples was to cause the evolution of nearly an electrochemical equivalent of hydrogen gas. However, the expected amount of evolved oxygen was not found indicating that ionic conductivity was being observed rather than electrolysis of the sorbed water. On passing a steady-state current through haemoglobin samples with 7.5 wt per cent adsorbed water Rosenberg found that their conductivity and water content did not decrease with time, as should have been expected if hydrolysis and the creation of mobile protons

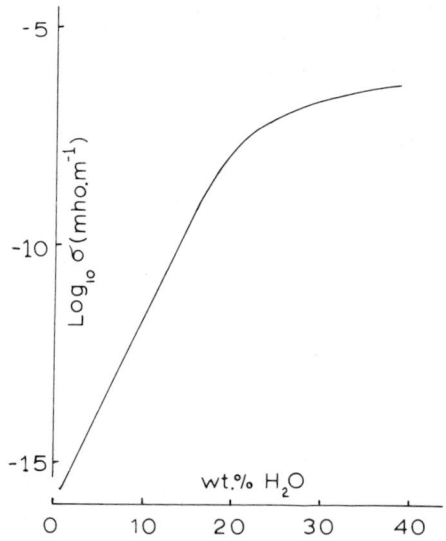

Figure 9.1 The variation of the d.c. conductivity with hydration typically observed for proteins and other biomacromolecules

had been occurring.[12] Similar measurements for haemoglobin samples containing 9.2 wt per cent water produced no evolution of gas, whereas at 18 and 40 wt per cent water, hydrogen was evolved.[13] From hydrolysis measurements, Murphy concluded that cellulose and fibrous proteins were ionic conductors.[14,15] Nylon, in certain aspects a polymer having similar characteristics to many biopolymers, was found in its dry state to be an electronic conductor below 363 K and to exhibit an increasingly protonic conductivity as the temperature was increased.[16] In a series of solid-state electrolysis experiments, Powell and Rosenberg[11,17] were able to determine the ratio of protonic to electronic conduction in various biomacromolecules (DNA, collagen, cytochrome-c, haemoglobin, lecithin, and melanin) as a function of water content. For hydration values from 6 to 50 wt per cent all the materials tested appeared to be mixed semiconductors in that both electronic and protonic charge carriers made significant contributions to the electrical conductivity. For each material the conductivity increased exponentially with water content as in the form of equation (9.2), and for the globular proteins the ratio of protonic to electronic conductivity increased linearly with hydration. For collagen, a fibrous protein, the electronic component of the conductivity appeared to increase linearly with hydration and in the dry state there was the possibility that it may have been a protonic semiconductor. Melanin was found to be a mixed semiconductor, with a protonic to electronic ratio of 65:35, which remained unchanged over the hydration range from 10 to 35 wt per cent. For DNA the conduction was electronic up to water contents of the order 35 per cent, whereafter there was a sharp transition to a dominantly protonic conductivity. Powell and Rosenberg also investigated the haemoglobin—methanol system, and as for DNA it exhibited an electronic conductivity below a threshold solvation state before showing a sharp transition to protonic conductivity. There were also indications that for haemoglobin the ratio of protonic to electronic conductivity was a function of the applied d.c. voltage, being mainly electronic below a threshold field of around 5×10^4 V m^{-1}. These measurements from the various laboratories serve to show that the nature of the conductivity of biopolymers in their dry or hydrated state is not yet clearly known. Powell and Rosenberg suggested that the protonic mode of conduction may be intrinsic to these biopolymers, in the same way that electronic conduction is intrinsic. However, it should be stated that in the measurements reported by Powell and Rosenberg no real distinction could be made between protonic and ionic conduction. A reasonable approach to take is that in the dry state most biopolymers can be considered to exhibit an essentially electronic conductivity involving the conduction of either electrons or holes in the appropriate energy bands, or via localized (trapping) states. With increasing water content, especially after the completion of one or two monolayers of sorbed water molecules, we may expect that protons and other ionic species are able to migrate around the surface of the hydrated proteins.

Although it is now well established that the conductivity of biopolymers increases markedly with increasing hydration, the underlying physical explanation for this remains to be definitely established. Rosenberg[18] was able to show for deionized bovine haemoglobin that the effective value of the semiconduction activation energy varied with the wt per cent of water m according to the equation

$$\mathscr{E} = \mathscr{E}_0 - \beta m \qquad (9.3)$$

where $\mathscr{E}_0$ was the activation energy in the dry state, and β was a constant. The value for σ_0 in equation (9.1) remained constant at around 7×10^5 mho m^{-1}, and at 7.5 wt per cent adsorbed water the conductivity increased by about nine orders of magnitude compared with the dry-state value, to give $\alpha = 2.65$ (reciprocal per cent water adsorbed) in equation (9.2). The value for the activation energy in equation (9.1) dropped from the dry-state value of 1.2 eV to 0.75 eV at the highest hydration level investigated. To account for this effect Rosenberg proposed that the addition of water, or any other solvent of high permittivity, reduces the energy required to generate free charge carriers by lowering the work required to separate counter-charges. This was in line with the previous suggestion by Vieth[10] that the conductivity activation energy was associated with an ionization energy being inversely proportional to the hydration-dependent permittivity of the material. The energy W required to remove an electron from one protein molecule to a neighbouring one is given by

$$W = (I - P^+) - (A + P^+) \qquad (9.4)$$

where I and A are the ionization energy and electron affinity values for an isolated protein molecule, respectively. The parameters P^+ and P^- represent the energies of polarization associated with the interaction on the surrounding molecules of the positive charge remaining on the donor molecule, and of the negative charge on the molecule that has received the 'roving' electron. The simplest approach is to consider the protein ions as spheres of radius R, in which case their electrostatic energy in vacuum is

$$\tfrac{1}{2} q^2 / 4\pi\epsilon_0 R$$

If these protein ions are now brought into the protein matrix of relative permittivity ϵ_r, their electrostatic energy becomes

$$\tfrac{1}{2} q^2 / 4\pi\epsilon_0 \epsilon_r R$$

The parameters P^+ and P^- will represent the difference of these two energies so that

$$P^+ = P^- = q^2 (1 - 1/\epsilon_r)/8\pi\epsilon_0 R \qquad (9.5)$$

Only a small error results in this approximation from ignoring induced dipole effects,[19] and this was the approach adopted by Rosenberg, apart from his oversight in ignoring the term P^+ in equation (9.4). With increasing water content the relative permittivity ϵ_r of the protein matrix increases so that the parameters P^+, P^- increase and the energy W of equation (9.4) decreases in value. From such considerations, and the fact that the parameter σ_0 was found to be invariant with water content, Rosenberg derived the following equation for the semiconductivity of hydrated proteins

$$\sigma = \sigma_0 \exp\left(-\frac{\mathscr{E}_0}{kT}\right) \exp\left[\frac{q^2}{kTR}\left(\frac{1}{\epsilon_r} - \frac{1}{\epsilon_r'}\right)\right] \qquad (9.6)$$

where ϵ_r and ϵ_r' are the relative permittivity of the dry and wet protein material, respectively. This equation is in qualitative agreement with equations (9.1) and (9.3) with

$$\frac{1}{\epsilon_r} - \frac{1}{\epsilon_r'} = \beta mR/q^2 \qquad (9.7)$$

and

$$\beta = kT\alpha.$$

Equation (9.6) predicts that with increasing water content (where the value for ϵ_r' will increase) then the conductivity σ should tend towards a maximum limiting value at any fixed temperature, in agreement with the experimental findings. Based on the data of King and Medley[8] that saturation of the conductivity occurs when the dry-state conductivity has been exceeded by a factor of the order 10^{10}, and using a dry-state relative permittivity value of 4, Rosenberg was able to estimate the value for the protein ion radius R in equations (9.5) and (9.6) to be $R = 0.28$ nanometres, which he considered to be a very reasonable value. This value for R leads to the result that $I - A \simeq 10$ eV for proteins, which is certainly of the correct order of magnitude. If the factor P^+ had been taken into account in Rosenberg's derivation of equation (9.6), then a factor 2 would appear in the numerator of the second exponent, and if the value $\epsilon_r = 3$ had been chosen for the dry-state permittivity as suggested in Chapter 3, then the following values would have been obtained; $R = 0.75$ nm, and $I - A = 4.9$ eV.

Apart from the electron-donor model of Eley et al., to be mentioned later, this work of Rosenberg represents the only in-depth assessment to have been attempted to account for hydration effects. For this reason it is appropriate to make a few comments on certain aspects of this work. Firstly, it should be noted that the value derived by Rosenberg for α (2.65) in equation (9.2) is far too high by comparison with more recent work, including his own,[11,20] where for most biomacromolecules α has the value typically in the range 0.9—1.5 (reciprocal per cent wt absorbed). A 7.5 wt per cent hydration content for many proteins results in an increase in the dry-state conductivity by a factor of around $10^3 \sim 10^4$ and not by as much as 10^9. However, as for the omission of the P^+ term in equation (9.4), this does not affect the main philosophy outlined by Rosenberg. Other problems do exist however. For example, King and Medley[8] found that formic acid ($\epsilon_r = 58$) was more effective, on a per mole basis, in increasing the conductivity of nylon and keratin than water ($\epsilon_r = 79$), and Eley and Leslie[21] found that adsorbed methanol ($\epsilon_r = 32$) caused a greater increase in the conductivity of bovine serum albumin than did water. Similar results were observed for the action of methanol and benzene ($\epsilon_r = 2.3$) on PBLG and DNA.[22] According to Rosenberg's model adsorbed materials of lower permittivity should have had a smaller effect on the protein conductivity. Also, as pointed out by Rosenberg himself,[18] the σ_0 factor in equation (9.1) may include a term due to an exponential dependence on the temperature of the mobility of the dominant charge carrier, as given by equation

(8.27) for example. Equation (9.1) could then also be written in the form

$$\sigma = C \exp - \left(\frac{\mathscr{E}_n + \mathscr{E}_\mu}{kT} \right)$$

where $\mathscr{E}_n$, $\mathscr{E}_\mu$ denote the carrier generation and mobility activation energies, respectively. It will then not be clear whether the experimentally derived equation (9.3) applies solely to either $\mathscr{E}_n$ or $\mathscr{E}_\mu$, or to both of them. Microwave Hall effect measurements on hydrated bovine plasma albumin[23] indicate that the charge carrier mobility is dependent on the water content, possibly being related to the extent of hydrogen-bond structures. This suggests that the precise interaction of sorbed molecules onto protein structures is an important factor determining conduction mechanisms, and may be relevant to understanding the action of other substances such as formic acid, methanol, and benzene. The values derived for the ionic radius R are of the order 0.28–0.75 nm, and represent a distance some 10–30 times smaller than the radius of many protein molecules whose conductivity varies with hydration in the same manner as equation (9.2). It will be of value to determine the magnitude of R for protein molecules other than haemoglobin using Rosenberg's theory, since it should have been expected perhaps that the value for R should approximate to the radius of a sphere having a volume equal to the molecular volume. Finally, we should also note that according to equation (9.7) the relative permittivity ϵ'_r of the hydrated proteins is given by

$$\epsilon'_r = \epsilon_r / (1 - m\epsilon_r \alpha k TR / q^2)$$

where ϵ_r is the dry-state permittivity. This predicts that ϵ'_r tends to infinity as m tends to the value

$$m = \frac{q^2}{\epsilon_r \alpha k TR} \text{ wt per cent}$$

and ϵ'_r assumes a negative value thereafter. This represents a physically unrealistic aspect of Rosenberg's model since this condition can be attained for values of m much less than 100 wt per cent.

An alternative approach has been adopted by Eley to explain the enhancement of the conductivity of proteins by adsorbed water and other materials. According to this scheme the conductivity increase arises from the injection of electrons into the conduction bands of the protein molecules.[21,24-26] With methanol having a stronger electron-donor quality than water, then adsorbed methanol should result in a larger conductivity increase as was observed for the case of bovine serum albumin.[21] The electron-donating action of water was considered to have been observed directly for bovine plasma albumin that had been complexed with p-chloranil, since the sorption of water to give approximately 1 BET monolayer coverage resulted in a 12-fold conductivity decrease.[24] The action of the chloranil was considered to be that of an electron acceptor, introducing holes into the valence band of the protein and leading to the dry-state conductivity increasing by a factor of 3×10^5. Electrons donated by the sorbed water molecules acted to neutralize these holes in the valence band. Ammonia gas exhibited a similar electron-donating action as the water, as expected from standard chemical considerations. Similar effects were later observed for the bovine plasma albumin–chlorophyll complex, where the removal of water caused the conductivity to increase.[27] The energy required to remove an electron from water and inject it into the protein energy band structure was considered[25] to decrease with increasing water content, as a result of interaction between the sorbed water molecules. This effect was required in order to explain why the conductivity increased exponentially in the form of equation (9.2), rather than linearly as a function of the number of injected electrons. It was suggested[28] that the first water molecules to be bound to proteins are held in special sites and donate electrons into the conductivity band of the protein molecule. The binding sites were suggested to be CONH groups adjacent to polar side chains. As the hydration content approaches the equivalent of 4 BET monolayers, it was expected that protonic conduction would set in within the hydration shell along the lines suggested earlier by Riehl.[29] It was also noted that the protonic conduction observed for keratin by King and Medley[8] corresponded to a hydration level of 2.5 BET monolayers and that the adsorbed water could also have a plasticizing effect to increase protein side-chain movements and facilitate proton transfer between adjacent NH and CO groups. Recent microwave dielectric and Hall effect measurements give some support to these concepts. For example, it has been observed[30] that at a frequency of 9.9 GHz the dielectric loss factor ϵ'' for proteins is relatively insensitive to the water content at low hydration values, as shown for bovine serum albumin in Figure 9.2. A similar trend and ϵ'' values was also observed earlier for this protein by Eley and Pethig.[31] As shown by Figure 4.14 normal water at 10 GHz exhibits a large dielectric loss associated with its molecular motions. The effect shown in Figure 9.2 can therefore be taken to indicate that the initial coverage of water involves the water molecules being very

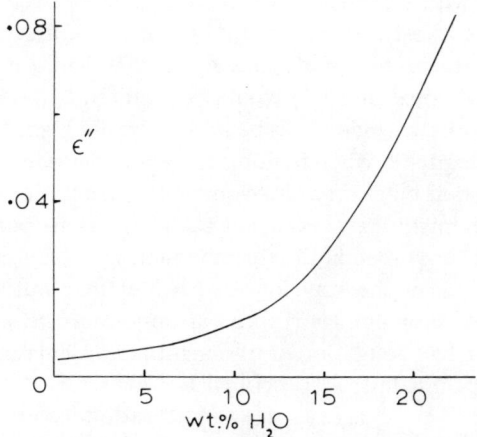

Figure 9.2 The variation of the dielectric loss factor ϵ'' with hydration typically observed for proteins (e.g. bovine serum albumin) at microwave frequencies

strongly and irrotationally bound to the protein molecule. We can imagine that such special binding sites are required before electron-donating charge-transfer interactions are possible. Evidence for such an electron transfer into the protein can be taken from microwave Hall effect measurements. Trukhan[32] observed for haemoglobin that the sign of the Hall effect changed from p-type to n-type as the samples were hydrated from the dry state, and a similar effect was observed for DNA by Eley and Pethig.[33] This is precisely the effect that would be observed if water were to act as an electron donor. The effect of protonic conductivity may have explained why the n-type Hall effects for cytochrome-c[32] and bovine plasma albumin[33] decreased with increasing hydration, where the expected p-type Hall effect for protons began to mask the electronic conductivity effect.

As for Rosenberg's dielectric-effect model, there are certain features of this electron-donor model which require clarification. The observation[18] that with increasing hydration the conductivity parameter σ_0 remained unchanged would suggest that the number of conducting charge carriers remained fixed, even though the conductivity increased dramatically. This in turn would imply that an electron injection mechanism is not occurring, and as mentioned in the discussion of Rosenberg's model, the decreasing conductivity activation energy could have resulted solely from a hydration-assisted reduction in the mobility activation energy. Both the p-type microwave Hall effect result of Trukhan,[32] and the earlier observation of a p-type thermoelectric power coefficient[34] would indicate that dry haemoglobin is a p-type semiconductor. The question then arises as to why the addition of electron-donating water molecules immediately increases the conductivity of haemoglobin, when its action on the p-type chloranil—albumin complex was first to decrease the conductivity.[24] Another problem arises from a consideration of the charge-transfer process that might occur between water and proteins. As described in Chapter 8, the ionization potential of a water molecule is around 12.7 eV and with a protein molecule having an electron affinity of no more than 3–4 eV at the most, it is difficult to imagine that any charge-transfer interaction between these two molecules would be significant. The polarization energies P^+ and P^- of equation (9.4) would be required to have values of the order 4 eV each, which can be considered to be unrealistically high. The model of Petrov *et al.* described in Chapter 8, whereby ionized carboxyl groups inject electrons into the protein's conduction band may have some relevance to this model of Eley *et al.* but in this case we might have expected such effects to occur at much higher hydration levels than those used in the semiconduction studies described here.

As an understanding of the way in which water may influence the solid-state physical properties of biopolymers is of great importance, it might be of value at this stage to make a few rough, order-of-magnitude, calculations for some of the protein conduction parameters. As described in Chapter 8, the conductivity σ can be described in terms of the charge carrier concentration n and mobility μ as

$$\sigma = ne\mu \tag{9.8}$$

From Table 9.1 we can take $\sigma = 2 \times 10^{-17}$ mho m^{-1} as a typical conductivity value for dry proteins at room temperature, and if we take an estimate for the mobility of 10×10^{-4} m^2 V^{-1} s^{-1} then the corresponding charge carrier concentration is of the order 10^5 m^{-3}. The microwave Hall effect measurements would imply that for dry haemoglobin and DNA these charge carriers are in fact holes in the valence band, and they will almost certainly have resulted from the existence of 'extrinsic' electron acceptors. According to the solid-state semiconductivity concepts outlined in Chapter 8, the concentration N_A of such (compensated) acceptors will be given according to the expression

$$p = (N_v N_A)^{1/2} \exp(-\Delta \mathcal{E}_A / kT) \qquad\qquad (9.9)$$

and will correspond to the scheme presented in Figure 8.8, with $\Delta \mathcal{E}_A = \mathcal{E}$ of Table 9.1 and $N_A > N_D$. If we assume that the effective density of states in the valence and conduction bands corresponds to the number of peptide units, then we have for a typical protein that $N_v = 6 \times 10^{27}$ m^{-3}. Then from equation (9.9), with $p = 10^5$ m^{-3}, $T = 300$ K and $\Delta \mathcal{E}_A = 1.4$ eV, we have that $N_A \simeq 2 \times 10^{29}$ m^{-3}. Such an acceptor concentration, representing as it does around 30 or more centres per peptide unit, is far too large for practical consideration. If the case for uncompensated acceptors had been considered, the calculations would have led to the same result. This strongly suggests that the exponent of equation (9.1) does in fact contain a mobility activation energy term. This in turn implies that the d.c. conduction studies determine the effects of an activated electron transfer between protein molecules and that the actual conduction properties of the protein molecules are being masked by the intermolecular barriers. An increasing water content could act to decrease the effective energy-barrier-limiting intermolecular charge transfer and lead to the result of equation (9.3) as observed by Rosenberg. In this respect it is of interest to note that microwave measurements[35] for bovine plasma albumin, which should not have been influenced by such intermolecular barriers, lead to molecular conductivity and charge carrier mobility values ($\sim 10^{-6}$ mho m^{-1} and 40×10^{-4} m^2 V^{-1} s^{-1}) which fit in well with the concept of a simple charge-transfer mechanism between the sorbed water molecule and the protein, and are in line with the d.c. conductivity activation energies determined for hydrated proteins.[18,36] Such considerations lead to the possibility that the appropriate activation energy value to be used in equation (9.8) should be of the order 0.85 eV rather than 1.4 eV, to correspond to a dry-state mobility activation energy of the order 0.55 eV. With increasing hydration, then according to this viewpoint, the mobility activation energy would decrease. To support this, both pulsed electron beam measurements[37] on wet protein films and microwave Hall effect measurements on hydrated proteins[23] demonstrated that the presence of water can greatly facilitate electronic mobility. The corresponding value for N_A in equation (9.8) would then be $N_A \simeq 6 \times 10^{10}$ m^{-3}. An extremely low hydration content of just 0.01 wt per cent water would correspond to the concentration of adsorbed water molecules being greater than 10^{24} m^{-3}, and even a very weak electron-donor action on the part of the water molecule would result in the electron concentration greatly exceeding the hole concentration of 10^5 m^{-3} calculated above. This would satisfactorily explain why dry proteins exhibit an immediate increase in conductivity on increasing their hydration content rather than first showing a decrease as observed for the p-type chloranil–albumin complex.[24] To complete this picture we need to check on the dry-state value of σ_0 appearing in equation (9.1) and given in Table 9.1. If we assume that the macroscopic mobility follows an activated law of the form

$$\mu = \mu_0 \exp-(\mathscr{E}_\mu/kT) \tag{9.10}$$

then from equations (9.1), (9.8), and (9.9) we can write

$$\sigma = (N_v N_A)^{1/2}\mu \tag{9.11}$$

From Table 9.1 we can take the typical value for σ_0 to be 10^7 mho m^{-1} which from equation (9.11) for $N_v = 6 \times 10^{27}$ m^{-3} and $N_A = 6 \times 10^{10}$ m^{-3} requires a mobility value

$$\mu = 5 \times 10^{-13} \text{ m}^2 \text{ V}^{-1} \text{ s}^{-1}$$

The value for N_A was determined on the basis of the mobility activation energy being of the order 0.55 eV, and from equation (9.10) this in turn would correspond to the microscopic (band conduction) mobility μ_0 having the value 9×10^{-4} m^2 V^{-1} s^{-1} at 300 K. This mobility value is very acceptable in terms of an energy band conduction mechanism. Returning to the work of Rosenberg,[18] it is perhaps significant that for dry haemoglobin he obtained a semiconduction activation energy of 1.15 eV, which is a little lower than the value of around 1.4 eV obtained by Eley and Spivey.[38] This might suggest that Rosenberg's samples did in fact contain some water. On increasing the hydration to his final stated value of 7.5 wt per cent, then according to equation (9.11) a total change in σ_0 by a factor of no more than 2.5 would be expected. This in fact is well within the error margin estimated by Eley and Spivey for their determination of σ_0 for a whole range of proteins in the dry state. The determination of σ_0 requires quite an extended extrapolation on the semi-log plot of conductivity against reciprocal temperature, and as such it is within the bounds of possibility that Rosenberg may have been mistaken in his conclusion regarding the invariance of σ_0 with increasing hydration. However, the basic premise of Rosenberg that on increasing the hydration content of a protein, and hence its permittivity, less energy will be required to separate counter-charges must be relevant to the underlying physical process of electrical conduction in proteins. As Figure 8.14(a) demonstrates, the 'dielectric effect' can certainly influence the energy band structures of proteins. Water is known to act as an electron donor in the photosynthetic process of higher plants, and in the form of an 'impurity' in the protein structure a water molecule can also be expected to introduce new electron donor and acceptor states. It is most likely that both the dielectric model and the electron-donor model, together with hydration-dependent mobility and ionic conduction effects, will be relevant to the total understanding of protein semiconduction. Many more careful experiments are required before such an understanding can be approached. Careful distinction between surface and bulk conduction effects will also have to be made for the electrical conduction measurements.

Apart from the microwave Hall effect measurements to be described later, other more recent work has been reported on the semiconduction properties of proteins. Eley and Thomas[39] examined the dry-state conductivity of bovine plasma albumin using four different electrode metals (Cu, Pt, Al, and In). Very similar results were obtained for each electrode material, with the parameters of equation (9.1) having the values $\mathscr{E} = 1.45$ eV and $\sigma_0 = 2 \times 10^8$ mho m^{-1}. This would suggest that all the previous conduction studies, summarized by Table 9.1, represented a bulk effect which was not influenced by effects associated with the electrode injection of charge carriers. Although Rosenberg[18] has provided evidence which would indicate that the conductivities investigated represented a bulk rather than surface leakage effect, the possible influence of sample surface effects remains to be fully investigated. Apart from d.c. steady-state measurements, the conductivity of proteins has been investigated over a wide frequency range extending from 10^{-5} Hz to 10 GHz. It has been found that the a.c. conductivity $\sigma(\omega)$ has a frequency dependence of the form

$$\sigma(\omega) = \sigma_{\mathrm{d.c.}} + A\omega^s$$

where $\sigma_{\mathrm{d.c.}}$ is the steady-state (d.c.) contribution, A is a constant and s has a value close to unity. It is well known[40,41] that a large number of poorly conducting materials exhibit this behaviour, and Pollak and Geballe[42] were the first to propose an interpretation in terms of charge carrier hopping between localized sites distributed randomly in space and energy. This behaviour is demonstrated for desiccated bovine serum albumin and cytochrome-c,[43] and for cytochrome-oxidase,[44] in Figure 9.3. The frequency-dependent a.c. conductivity component can be formally related to the dielectric loss factor ϵ'' by the relationship $\sigma(\omega) = 2\pi f \epsilon_0 \epsilon''$. Using the analogy that a solid containing a system of non-interacting electrons hopping over potential barriers will, at a constant temperature, be dielectrically indistinguishable from a solid containing a sparsely distributed set of dipolar entities, the background dielectric loss ϵ'' for the albumin and cytochrome-c samples was interpreted in terms of the existence of such hopping electrons.[43] This concept has been extended to show that the a.c. conductivity of such non-polar organic solids as anthracene and β-carotene can be interpreted in terms of hopping electron transport, rather than by classical dielectric loss mechanisms involving polar impurities, for example.[45,46] Low frequency dielectric dispersions in the region 10^{-2} Hz in the biologically relevant perylene–chloranil charge-transfer complex have also been ascribed to a hopping electron mechanism[47,48] and the same interpretation also seems likely for the dielectric properties of solid 1,4-butanediol.[49] A review of hopping electron concepts, and the development of a hopping electron model describing dielectric behaviour of poorly conducting solids, has recently been given by Lewis.[50]

Returning to the effects of water, Tomaselli and Shamos[51] have reported on the electrical properties of bovine Achille's tendon collagen as a function of temperature over the limited temperature range 269–324 K. At ambient temperature the conductivity was observed to change from 10^{-13} mho m^{-1} in the dry state to about 10^{-6} mho m^{-1} at a hydration content of 24 wt per cent. For all temperatures studied the conductivity increased exponentially with hydration according to

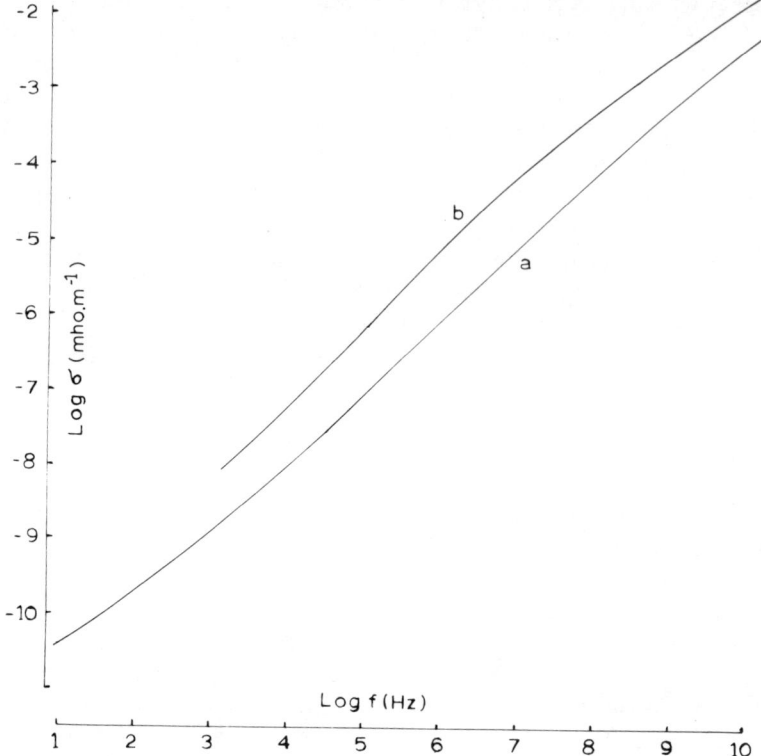

Figure 9.3 The frequency variation of the conductivity for (a) dry bovine serum albumin and cytochrome-c,[43] (b) desiccated cytochrome-c oxidase[44]

equation (9.2), and the pre-exponential factor (effectively the dry-state conductivity) was found to be independent of temperature, whilst the parameter α varied inversely with temperature. This extraordinary result means that the activation energy $\mathscr{E}$ of equation (9.1) was found to be zero for dry collagen (in fact close inspection of the results indicates a negative activation energy). Furthermore, the activation energy was found to increase linearly with hydration reaching a value of around 1.3 eV for a hydration of 12 wt per cent. These results are completely at variance with all previous measurements that have been reported of the hydration-dependent semiconductivity of proteins. The authors interpreted these results in terms of an impurity conduction mechanism, with the impurity (water) acting as an electron donor. In fact, for these results to be at all consistent with accepted semiconductor physics, at least two impurity types must have been present. A zero activation energy, even a small negative one, can be observed for extrinsic semiconductors in the so-called 'impurity-deionization'[52] or 'saturation'[53] range of temperature where all the electron acceptor and donor states are completely ionized and the temperature is too low for the intrinsic conduction to dominate. The small negative activation energy would then be consistent with band conduc-

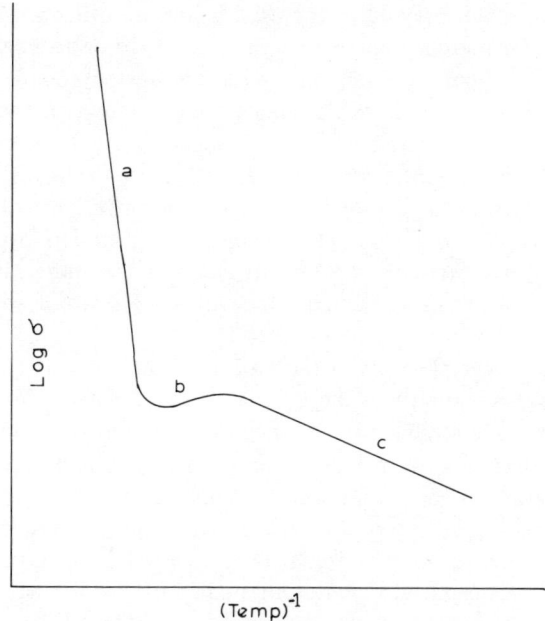

Figure 9.4 The conductivity–temperature characteristic commonly observed for semiconductors
to show the region of intrinsic conductivity (a),
the extrinsic conductivity region (c), and the
'saturation' region (b). The results of Tomaselli
and Shamos[51] for dry collagen suggest that they
were measuring region (b)

tion where the charge carrier mobility is limited by lattice vibrations to give
$\mu \propto T^{-3/2}$, for example. The variation of the conductivity with temperature that
results when saturation of the acceptor or donor properties of the impurities occurs
is shown in Figure 9.4. This behaviour is commonly observed for germanium and
silicon samples, for example. The room temperature conductivity of
10^{-13} mho m^{-1} observed for collagen would then be consistent with an impurity
concentration of 6×10^8 m^{-3} or less, with their associated energy states lying very
close to either the valence or conduction band of the collagen molecule so that at
room temperature they are practically all ionized, having either donated or
accepted their full complement of electrons.

The three-stranded collagen molecule, having a molecular weight of the order
3×10^5, is much larger than most of the other proteins investigated. This could
result in there being well-defined energy bands within the collagen molecules, with
the intermolecular barriers limiting long-range charge carrier transport being less
pronounced. Such factors would suggest that collagen may prove to be an important protein regarding further electrical conduction studies. Another protein of
possible potential interest is cytochrome-c. Nakahara et al.[54] have reported conductivity and activation energy results for carefully deionized cytochrome-c samples

that are much lower than previously reported values by other workers. At 328 K anhydrous films of ferricytochrome-c were observed to have the comparatively low resistivity of 6.5×10^8 ohm m, and the value for ferrocytochrome-c was only 6.5×10^6 ohm m. Both materials exhibited a semiconductivity activation energy of 0.6 eV. Since no effects attributable to the presence or otherwise of oxygen could be detected, it was concluded that the 100-fold difference in the resistivity between the two forms was attributable solely to whether haem groups of cytochrome-c were in the ferric (Fe^{3+}) or ferrous (Fe^{2+}) state. The conduction properties were considered to be consistent with the known electron transport rate through the mitochondrial electron transport chain, of which cytochrome-c is an integral component.

Following earlier suggestions by Szent-Györgyi that proteins can be converted into an electronically conductive state by the charge-transfer action of such compounds as dicarbonyls and unsaturated aldehydes, extensive investigations have been made of the electronic properties of proteins that have been complexed with methylglyoxal. It was found[55] that when casein interacts with methylglyoxal in methanol, a brown complex is produced which when isolated as a dry solid exhibits an increased electron spin resonance signal and a steady-state electrical conductivity some three orders of magnitude greater than that of the normal control casein. Electronic transference number measurements indicated that the increased conductivity was associated with an electronic rather than ionic conduction mechanism. This result was confirmed by later measurements on bovine serum albumin, casein, and lysozyme samples that had been complexed with methylglyoxal, where apart from a significant increase in free electron spin density and conductivity the appearance of a low frequency dielectric dispersion was observed.[56] It was considered that these effects arose from the action of methylglyoxal in accepting electrons from the valence bands of the proteins. The low frequency dispersion could be explained in terms of the polarizations arising from these holes being confined to conducting regions within the protein structures. According to this scheme the dielectric relaxation activation energy is associated with energy barriers between protein molecules, and these barriers also limit the steady-state conductivity of the protein–methylglyoxal samples. This model essentially represents another type of interfacial polarization whose essential features were described in Chapter 5. In the form described for the methylglyoxal–protein complexes, the electronic conduction properties of the protein molecules are more clearly emphasized. The range of experimental investigations on these protein complexes has been extended to include piezoelectric, hydration isotherm, microwave Hall effect, and dielectric studies,[57] all of which add to the evidence that when methylglyoxal is incorporated into the structure of a protein molecule, charge-transfer interactions occur resulting in the creation of mobile holes within the protein valence bands. The action on collagen is to increase the dry-state conductivity by more than a factor of 10^6, and the derived hole mobility is of the order 32×10^{-4} m^2 V^{-1} s^{-1}. The mobility for the lysozyme complex was derived to be of the order 100 times less than this value for the collagen complex, suggesting that the hole transport involved a hopping mechanism between localized energy states.

The large mobility value obtained for the collagen-complex may be related to the comparatively large size of the collagen molecule leading to the development of well-defined energy band structures. This idea was also expressed in connection with the studies of Tomaselli and Shamos described earlier. The methylglyoxal-complexed proteins remain brown even after vigorous washing procedures, indicating that some form of chemical bond is involved in attaching the methylglyoxal to the protein molecules. Also, if the lysine side-chains of the protein molecules are dimethylated, then incubation with methylglyoxal does not result in the proteins assuming a brown colour and neither do their electronic properties change significantly by comparison with the untreated proteins. Such considerations as these have led to the tentative conclusion that the primary interaction involves the methylglyoxal being covalently bound to the lysine and terminal amino groups. It is further thought that the binding involves a Schiff base (—C=N—linkage) as shown in Figure 9.5(a), and that this planar Schiff base becomes involved in a charge-transfer interaction with a neighbouring planar peptide unit as shown in Figure 9.5(b).

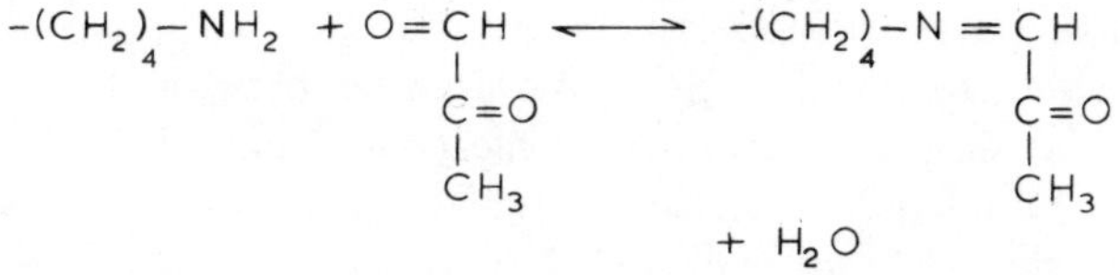

(a)

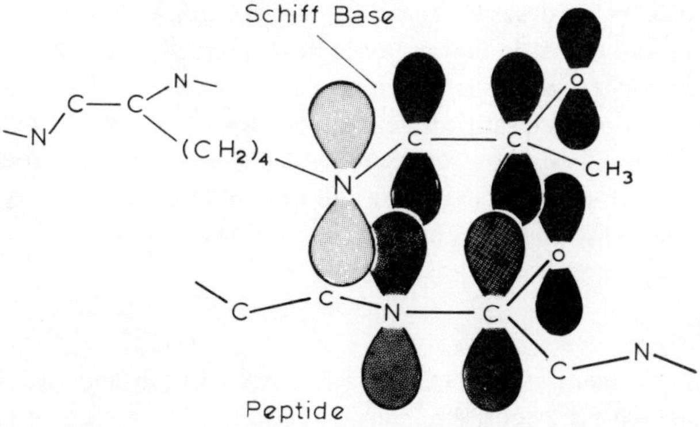

(b)

Figure 9.5 (a) The interaction of methylglyoxal with a lysine side chain to form a Schiff base plus water.
(b) A possible π-electron charge-transfer interaction that could occur between the Schiff-base-linked methylglyoxal molecule and a neighbouring peptide unit

From the construction of atomic models in the author's laboratory and else-where[277] it seems possible that the lysine side chain can in fact bend back so as to bring the Schiff base alongside the 'next-door' peptide unit. From an analysis of the various conduction measurements[57] it appears that in such a charge-transfer process only 0.005 electronic charge is required to be transferred from the peptide to the Schiff base. It is of interest to note that this is of the same order as the value (0.007) calculated, using a minimal basis set, for the relevant interaction between formamide and glyoxal molecules placed parallel to each other at a distance 0.3 nm apart.[58]

Finally, before proceeding to a discussion of the semiconduction and photo-conduction studies that have been made for the nucleic acids, it should be mentioned that so far great emphasis has been placed on interpreting the electrical conduction studies for biopolymers such as proteins in terms of what one may call 'classical semiconductor physics'. In other words, a clear demonstration of the existence of broad bands of delocalized electronic states, and electronic mobility within such bands, has appeared to be the main purpose behind the electrical conduction studies. In fact a much wider viewpoint should be taken which recognizes the fact that some biopolymers may be more accurately described as amorphous semiconductors where electronic transfer between localized energy states is often more significant than electronic conduction within broad energy bands. This was considered when analysing the properties of the protein—methylglyoxal complexes,[56] and will be referred to again when considering such biopolymers as melanin in a later section of this chapter. It has recently been demonstrated[59] that amorphous semiconductors can be doped to give n- or p-type electronic conduction, and so there will be no problems in using the amorphous—solid-state concepts to describe some of the electronic properties that have been observed for biopolymers. In certain biological systems containing complexed metal ions, a charge-transfer mechanism involving the hopping of electrons between metal ions may be the only relevant model to consider.[60] In general however, since we can consider biopolymers to represent a class of condensed matter where the solid-state concepts of both localized and extended electron energy states will be relevant, as broad a viewpoint as possible should be adopted.

Nucleic Acids

Extensive measurements for DNA and RNA have established that in the dry state these materials exhibit a reproducible d.c. conductivity with a temperature depen-dence of the form of equation (9.1). Values that have been obtained for the parameters σ_0 and $\mathscr{E}$ of equation (9.1), as well as the room temperature resistivity value, for samples of dry NaDNA are presented in Table 9.2.

For one of the studies included in Table 9.2 only the temperature-dependent resistance values were given,[67] so that only the activation energy value can be given. Although there appears to be a majority of the investigators reporting an activation value of around 1.2 eV, it can be seen from Table 9.2 that by comparison with the results obtained for the proteins (Table 9.1), there is a larger variation in

Table 9.2   Values of the semiconduction activation energy $\mathscr{E}$ and of the parameter σ_0 of equation (9.1), together with the room temperature resistivity, that have been obtained for dry NaDNA

Reference	$\mathscr{E}$ (eV)	σ_0 (mho m^{-1})	ρ (300 K) (ohm m)
61	0.9	1×10^{10}	1.3×10^5
62	1.21	3×10^5	7×10^{14}
63	1.22	6×10^4	5.1×10^{15}
64	1.2	$\sim 10^9$	1.4×10^{11}
65	0.95	$\sim 10^8$	9×10^7
66	1.18	$\sim 10^4$	6.6×10^{15}
67	1.22	—	—
68	1.1	6×10^6	5×10^{11}
22	1.0	2×10^4	3×10^{12}

the resistivity and σ_0 values for just NaDNA than is the case for a range of different proteins. It is almost certain that the lower values obtained for the activation energy and σ_0 resulted from the presence of water in the test samples. Also, accurate and reproducible measurements are difficult to obtain owing to the pyrimidic bases, for example, easily subliming. Sublimation and thermal decomposition effects were certainly observed by Eley and Leslie.[69] Another cause for differences in the results could also easily have resulted from the presence of different impurities in the various samples investigated. The most common nucleic acid to have been studied is calf thymus DNA, although others have used herring sperm and salmon sperm DNA. Duchesne et al.[61] essentially found no difference in the electrical properties of calf thymus DNA fibres and herring sperm DNA powder.

An interpretation of the observed electrical properties in terms of an intrinsic semiconduction mechanism would require the energy band gap $\mathscr{E}_g$ to be of the order 2.4 eV. This is much less than the theoretically derived values obtained for various DNA models, as described in Chapter 8. As for the situation for proteins we must conclude that the observed electrical properties reflect an extrinsic rather than intrinsic generation of mobile charge carriers. The semiconduction activation energy for dry RNA has been found to be similar in value to that for DNA,[66] whereas Snart[63] obtained the slightly higher value of 1.5 eV for RNA. The dry homopolyribonucleotides K-polyadenine, K-polycytosine, K-polyguanine and NH$_4$-polyuracine, were found to have activation energies in the range 0.9−1.1 eV with resistivity values at 400 K being of the same order as dry DNA and RNA.[66] An interesting feature emerged from the measurements on the nucleosides (adenosine, cytidine, thymidine, uridine, and guanosine) and nucleotides (the monophosphates of adenosine, cytidine, uridine, and guanosine), where it was found that the nucleotides exhibited much lower semiconduction activation energies ($\mathscr{E} = 0.7 \sim 1.1$ eV) than the nucleosides ($\mathscr{E} = 2.0 \sim 2.8$ eV, apart from guanosine with $\mathscr{E} = 1$ eV).[66,69] The phosphate group therefore appears to have a considerable effect on the semiconductivity of these materials, and it was suggested that the phosphate groups may form electron bridges between the bases in DNA,[69] or as a result of an intramolecular charge-transfer process inject holes into the valence band structure formed by the base pairs.[66] This second suggestion has received support from the theoretical calculations of Ladik[70] where on comparing the relative energies of the energy band structures for DNA it was found that the possibility exists for a charge-transfer interaction to occur between the poly(base pairs) and the sugar-phosphate backbone, with the result that electronic delocalization occurs down the backbone and 'hole' transport occurs through the base pairs. The p-type microwave Hall mobility value obtained for dry DNA[33] is consistent with these theoretical calculations which indicate that the hole mobility would be greater than that of the delocalized electrons. The semiconductivity of the nucleic acid bases (adenine, cytosine, guanine, thymine, and uracil) have also been studied, but little agreement arises from a comparison of the results obtained, with the activation energy values being reported to range from 1.8 to 2.5 eV,[66] 1.0 to 1.4 eV,[71] and 0.9 to 1.4 eV,[72] where for the last two reports[71,72] quite large differences are reported for the individual bases studied.

As was found for the proteins, the semiconduction activation energy and the resistivity of DNA has been found to decrease markedly with increasing hydration. Liang and Scalco[67] observed that on hydrating NaDNA the conductivity increased by a factor of 10^6 compared with the dry-state value, and the activation energy fell from 1.2 to 0.8 eV. Burnel et al.[73] found that the conductivity $\sigma(m)$ varied with the weight per cent water content m as

$$\sigma(m) = \sigma_D \exp(\alpha m^{1/2})$$

with the dry conductivity value σ_D having a value of about 10^{-14} mho m^{-1}. The conductivity $\sigma(m)$ approached a limiting saturation value at around 27 wt per cent. hydration, and the activation energy fell from 1.2 eV to around 0.6 eV. For hydration levels above 10 wt per cent, electrical polarization and non-ohmic conduction effects were observed, suggesting that protonic conduction was becoming dominant, in agreement with the findings of Powell and Rosenberg.[17] Solid-state electrolysis measurements on calf thymus NaDNA have demonstrated that the semiconduction activation energy approaches a value of zero for hydration contents exceeding 75 wt per cent.[74,75] The properties of various DNA salts have also been studied over the frequency range from 100 Hz to 1 GHz.[64,76] For NaDNA the semiconduction activation energy decreased with increasing frequency to the value 0.2 eV at 1 GHz, whereas for MgDNA a minimum value of the same order was observed in the region around 5 kHz. It is possible to understand this form of behaviour in terms of a model involving mobile charge carriers hopping over a set of potential energy barriers having a distribution of barrier heights.[46,50]

PHOTOCONDUCTION STUDIES

Apart from the dark semiconduction properties just described, various observations have also been reported of the photoconductivity of proteins and DNA, the first of which appearing to be those of Szent-Györgyi[3] who in 1946 reported that protein films stained with coloured dyes exhibited a large conductivity increase when illuminated. Photoconductivity was later observed in gelatin when irradiated with light of wavelength 366 nm[29,77] and on illumination with visible light films of serum albumin[78] and egg white[79] also exhibited an increase in conductivity. Photoconduction action spectra were studied by Nelson for haemoglobin and gelatin stained with dye.[80] For haemoglobin, similar wavelength dependencies for both the photocurrent and photoemission were obtained with the threshold wavelength being 310 nm, equivalent to an energy of 4 eV. Nelson concluded that this threshold energy represented a charge carrier generation as a result of energy absorption by the protein rather than by the haem group. The photocurrent was found to increase in proportion to the energy of the incident radiation, and was not related to the actual absorption spectrum. The threshold for gelatin occurred at 330 nm (3.76 eV) and this value was found to be independent of the dye content, although the actual magnitude of the photocurrent did increase as a result of increased staining with dye. Vladimirov and Timofeyev[81] studied thin films of trypsin and actomyosin and found the photoconduction threshold to occur at 320 nm (3.88 eV) in close agreement with the values obtained by Nelson, and also confirming his conclusion regarding the lack of involvement of the haem group in haemoglobin. The photocurrents were also found to be unrelated to the actual absorption spectra of the proteins, a result taken to indicate that the photocurrents were not due to the presence of aromatic amino-acid side chains. Photocurrents were also observed in solutions of tryptophan, tyrosine and glycyltryptophan when they were irradiated with energies below 340 nm (3.65 eV), and it was concluded that such effects arose from irreversible photochemical processes that possibly involved oxygen and produced free ions. Liang[82] investigated the d.c. dark conductivity and photoconductivity of two forms of polyglycine as a function of temperature. For polyglycine having the β-pleated structure the dark semiconduction activation energy was found to be 1.78 eV, and under white illumination the photoconduction activation energy value was 1.48 eV. For polyglycine having a possible α-helix structure, the corresponding activation energy values were found to be 1.42 eV and 0.77 eV. Eley and Metcalfe[83] observed non-transient photocurrents in protein and mitochondria samples not only in the visible range of optical excitation but also in the infrared. In the visible and ultraviolet, two types of photocurrent were observed. At energies above 5 eV for heavy beef heart mitochondria, above 4 eV for cytochrome-c, or above 3.5 eV for haemoglobin (with the illuminated electrode biased negative) the photocurrents rose steeply and approximately exponentially with increasing photon energy. This was not found for haemoglobin with the illuminated electrode biased positive, showing that electron and hole photoconduction effects are not similar for this material. Microwave photocurrents in the 4–5 eV region had been observed for DNA samples located in the microwave Hall effect apparatus[23] where sample electrodes were not used, and so it was considered that these high energy excited photocurrents were in fact representing a true bulk property of the test materials and not electrode injection effects. Furthermore the photoconduction action spectra appeared to correlate well with the optical absorption spectra. The infrared photocurrents exhibited peaks of intensity (0.83 eV for mitochondria, 1.38 eV for haemoglobin, and 1.71 eV for haemoglobin and cytochrome-c) and these were considered to arise from energy absorption at the haem groups rather than the protein structures. The photoconduction properties of dry cytochrome-c oxidase have also been studied.[84] The purest sample studied exhibited electron dark conductivity and photoconductivity, with the photocurrent exhibiting a temperature dependence of the form

$$i_{ph} = i_0 \exp(-\mathscr{E}_{ph}/kT)$$

The dark conductivity and the photocurrent exhibited a sharp change in slope on the reciprocal temperature plots at 343 K. For temperatures above 343 K the activation energy $\mathscr{E}$ of equation (9.1) and the photocurrent activation energy $\mathscr{E}_{ph}$ each had a value of 1.8 eV, whereas below this temperature $\mathscr{E}_{ph}$ had the value 1.0 eV and $\mathscr{E}$ was a little higher at 1.05 eV. The low temperature results were attributed to an impurity conduction mechanism involving the haem groups. Photocurrents were found for light irradiation through the ultraviolet, visible, and infrared energies, with the photocurrent increasing with the energy of the irradiation. Samples of bovine serum albumin were also studied and these exhibited a sharp photocurrent threshold at around 340 nm. For less pure samples of cytochrome-c oxidase larger d.c. dark conductivity and photocurrents were observed, but these were attributed to ionic impurity effects. For one sample the ionic conduction effects disappeared at applied electric field strengths greater than 3×10^4 V m^{-1}. No clear understanding could be made of the reasons why the impurities, which were mainly $(NH_4)_2SO_4$ and 'Tween 80', caused an increase in the photoconductivity.

Photoconduction and radiation effects have also been studied for DNA. Various photoconduction studies[67,85,86] have produced the result that the photoconduction activation energy $\mathscr{E}_{ph}$ and the semiconduction activation energy $\mathscr{E}$ (equation 9.1) are of similar value, with $\mathscr{E}_{ph}$ being possibly the smaller. This has been interpreted[67] in terms of the value for $\mathscr{E}_{ph}$ representing the energy required to hop from one DNA molecule to the next (i.e. the mobility activation energy) whilst $\mathscr{E}$ represents this value plus the energy required to generate charge carriers, which comes out at about 0.2 eV in agreement with the high frequency result reported by O'Konski.[76] An alternative suggestion[86] has been that the near equivalence of $\mathscr{E}$ and $\mathscr{E}_{ph}$ is associated with a mechanism involving the dissociation of singlet excitons to produce mobile charges at the sample interface. Gähwiler et al.[87] induced currents in thin DNA films using X-ray radiation. The induced currents were found to increase linearly with increased exposure rates up to 4×10^4 R min^{-1}. At low levels of applied voltage the conductivity was found to be ohmic, but the currents reached a limiting value of around 4×10^{-11} A as the applied voltage was increased to give fields of the order 2×10^6 V m^{-1}. On the basis of the saturation effect the product of the charge carrier mobility and the carrier lifetime ($\mu\tau$) was estimated to be of the order 10^{-11} m^2 V^{-1}.

HALL EFFECT MEASUREMENTS

Mention has been made several times of microwave Hall effect measurements. Before describing in more detail the results that have been obtained from such measurements on biological materials, a basic description of the Hall effect will now be given, together with a brief summary of the advantages and otherwise of the various techniques available for its detection. Particular emphasis will be given to the microwave technique since this appears to represent the most suitable for the investigation of charge carrier mobility effects in biological materials and as yet has not been fully exploited for that purpose.

We have seen that the total electrical conductivity of a biological material can result from contributions due to both ionic and electronic conduction so that in general the conductivity can be expressed in the form

$$\sigma = q(\Sigma_i z_i n_i \mu_i + \Sigma_e n_e \mu_e) \tag{9.12}$$

where i denotes the ion species with valency z_i and e denotes the electronic species, being either electrons or positive holes. The symbols n and μ denote the species number density and mobility respectively. On their own, electrical conductivity measurements can at best only provide information regarding the product of the number density and mobility of the most dominant charge carriers. It is only by combining conductivity measurements with Hall effect measurements that the polarity, number density, and mobility of the charge carriers can be determined, hence leading to a possible insight into the physical details of the electronic or ionic conduction mechanisms.

When an electrical conductor is placed in a magnetic field whose sense is perpendicular to the direction of the current flow, an electric field is developed across the conductor in the direction perpendicular to both the current and magnetic field. This electric field is the Hall field and the effect (Hall effect) is one of a family of related magnetoelectric and thermomagnetic effects. The inspiration leading to the original discovery of the Hall effect can be traced to paragraph 501 in Maxwell's *Treatise on electricity and magnetism*, which states that 'if the current be free to choose any path through a fixed solid conductor, then, where a constant magnetic force is made to act on the system, the path of the current through the conductor is not permanently altered'. In 1879, Hall published a paper[88] expressing his disbelief of Maxwell's statement and described the experimental phenomenon which now bears his name. He passed current along a thin gold leaf, 2 cm wide and 9 cm long, and found that the voltage between the opposite sides of the strip was permanently affected by a steady magnetic field applied normal to the

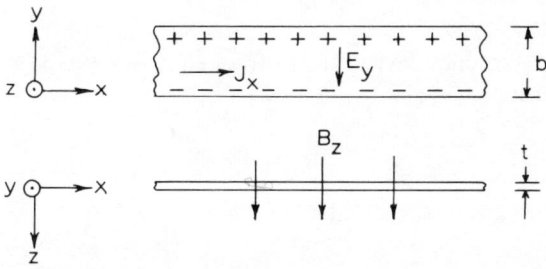

Figure 9.6 The Hall effect in a long strip, thickness 't' and breadth 'b'

direction of the current and the broad face of the gold leaf. In the third edition of Maxwell's treatise a footnote was introduced acknowledging Hall's discovery and the fact that Maxwell's earlier statement must be regarded as being only approximately true.

In Figure 9.6 a long conducting strip is shown carrying a steady current I_x and experiencing a steady uniform magnetic field B_z normal to its plane. If we assume that the current flow is due to electrons having an electronic charge q and a drift velocity v_x, then as a result of the magnetic field these conduction electrons will experience a deflecting force F_y (the Lorentz force) given by the vector product

$$F_y = -q(B_z \times v_x)$$

The deflected electrons accumulate on one side of the strip until the electric field E_y associated with this accumulated charge is large enough to cancel the force F_y. Thus, under equilibrium conditions

$$qE_y = -F_y$$
$$= q(B_z \times v_x)$$

or

$$E_y = (B_z \times v_x)$$

E_y is known as the Hall electric field.

The current density J_x is given by

$$J_x = I_x/(bt) = nqv_x \tag{9.13}$$

where n is the number of free electrons per unit volume. In this way

$$E_y = R(B_z \times J_x)$$

where $R = 1/nq$ is termed the Hall coefficient and is negative for the case where the electronic conduction is due to free electrons. A positive Hall coefficient corresponds to the situation where the current is carried by positive charge carriers (holes) as in the case for such metals as beryllium, zinc, and cadmium, and p-type semiconductors. In fact the Hall effect provides one of the clearest justifications for the use of the concept of the positive hole in electronic conduction theory.

Ohm's law can be written as

$$J = \sigma E \tag{9.14}$$

where σ is the sample conductivity. Hence from equations (9.13) and (9.14) we can derive the relationships

$$\sigma = nq\mu \tag{9.15}$$

and

$$\mu = R\sigma$$

where μ is the charge carrier mobility, defined as the ratio v_x/E_x, where v_x is the average acquired velocity of the charge carriers and E_x is the field producing the motion. The mobility value obtained from Hall effect measurements is termed as the Hall mobility, which we will designate as μ_H.

The measurement of the Hall effect and conductivity of a conductor therefore enables the charge carrier polarity, mobility, and number density to be determined. In fact, the measurement of these conduction parameters as a function of temperature has provided one of the most important techniques for the characterization and understanding of the electronic properties of modern semiconducting materials. A successful application of the technique for the study of electronic conduction in biological materials could therefore possibly prove of great significance.

In general, the interpretation of Hall effect measurements is not straightforward except for metals or degenerate semiconductors. In these materials the dominant charge carriers lie within narrow energy ranges and energy-dependent carrier scattering effects do not influence the results. This is not the case for non-degenerate semiconductors within which category most biological semiconducting materials will almost certainly fall. In the absence of an external electric field, the charge carriers are in random motion owing to their thermal energy. Most of the carriers will have energies near to the average kinetic energy value of $3(kT)/2$, but there will also be an appreciable number having energies over the whole range from zero to several times kT. The frequency of collisions and the amount of scattering per collision are functions of the energy of the carrier. The acceleration due to the application of an external electric field adds a component of velocity to the thermal velocity of the carriers and because of the energy dependence of the scattering this added velocity component is different for carriers in different energy ranges. Therefore, the current produced by the field is the result of an average of the different velocities of the carriers. With an applied magnetic field the force on the carriers, as well as the scattering, is a function of their velocities, so that the Hall field is a function of an average over the total velocity range of the carriers. This average is not, in general, the same as that for the current. The Hall coefficient must then be redefined as

$R = r/nq$

where r is known as the scattering parameter and is a function of the applied magnetic field, the scattering mechanisms effective in the sample, and the shape of the energy surfaces. The Hall mobility μ_H is then related to the conductivity mobility μ defined in equation (9.15) by the expression

$\mu_H = r\mu$

For small magnetic fields and spherical energy surfaces, r has the value $3\pi/8$ and $315\pi/512$ for lattice and ionized impurity scattering respectively. The value of r decreases and approaches unity as the magnetic field increases in magnitude. An anisotropy factor, which tends to reduce the value of r, must be taken into account for non-spherical energy surfaces. This anisotropy factor increases and tends to unity as the magnetic field increases. In this way, carrier densities and mobilities can be ascertained only to within the uncertainty of the value of r, which ranges from about 0.9 to about 1.8 for most cases observed so far. Matters are further complicated if there are two or more different types of dominant charge carriers. The derived Hall mobility then represents an average mobility, but not an obviously meaningful one.

So far the polarity of the dominant charge carrier has not appeared to be in doubt, but in fact this need not be the case. Re-entrant energy surfaces can produce anomalous Hall effect results for the sign and magnitude of the Hall mobility and in general the Hall mobility will be less than the conductivity mobility.[89] Also, if the energy bands are very narrow, the band states may be almost equally populated by charge carriers and the major electronic contributions are not necessarily from energy levels situated at the appropriate valence or conduction band edges.[90] For electrons, for example, the negative contributions to the effective mass may be more dominant than the positive contributions, resulting in a reversal of the conventional Lorentz force producing the Hall effect. In this way electrons could be mistaken as positive holes! For the situation where the charge transport is in the form of a hopping motion, the polarity of the Hall effect depends on the geometrical arrangement of the atomic orbitals between which the carrier jumps as well as on the nature and relative orientations of these orbitals.[91]

We can expect at least some of these complications to occur for Hall effect measurements on biological materials. This, together with the heterogeneous nature of biological materials, means that at best Hall effect measurements can provide only a rough indication of the number density and mobility of the dominant free charge carriers. Even so, such information will provide a valuable insight into the possible conduction mechanisms and help answer the question regarding the existence of broad energy bands in biological systems. Mobility values above about $1-10 \times 10^{-4}$ m^2 V^{-1} s^{-1} can be taken as an indication that the dominant charge carriers are moving in broad conduction bands. Values below this could result from protonic or ionic conduction, or else indicate an electronic mechanism involving localized electronic states rather than energy bands. The temperature variation of the measured mobility will also provide an insight into either the scattering processes involved in broad-band conduction or the activated mechanism involved in ionic conduction, or the hopping of electronic charge carriers between localized energy states. When used in conjunction with such experimental techniques as e.s.r. and n.m.r., Hall effect measurements should provide an additional important tool for the study and understanding of the electronic properties of biological materials.

The method of measuring the Hall effect depends primarily on the material under test. The measurement of the Hall voltage developed across the test sample has been by far the most popular method, but in general this only gives accurate results for single crystal specimens of relatively low resistivity and large mobility values. The sensitivity limit for Hall voltage measurements corresponds approximately to the ability of determining mobility values down to 1×10^{-4} m^2 V^{-1} s^{-1} for samples whose resistance is less than 10^{12} ohms. Many organic and biological materials have effective bulk resistances greater than this, and surface leakage effects coupled with space charge effects make Hall voltage measurements impracticable especially if single crystal specimens cannot be obtained. If large single crystals are available, then by using a suitable electrode geometry the Hall field can be shorted-out so that the Lorentz force produces a transverse Hall current. The main advantage of such a Hall current measurement technique is that a guard ring can be used for high resistance samples to separate the surface leakage effects from the true bulk properties of the sample. A main disadvantage for biological materials is that large single crystals of the test material are required. An ultimate limitation for all the d.c. Hall effect measurement techniques is imposed by the presence of thermal noise in high resistivity samples, and noise associated with the high impedance electronic circuitry. By using a.c. techniques with alternating magnetic and electric fields at frequencies high enough to avoid $1/f$ noise and with a band width of detection small enough to enhance the signal-to-noise ratio, the sensitivity of Hall detection can be greatly improved over the d.c. techniques. The main disadvantage is that interfacial polarization effects can occur at the electrode–sample interfaces. If polycrystalline samples are used, problems associated with intercrystalline resistance and interfacial polarization effects can also produce deleterious results.

As the frequency of the applied electric field to the test sample increases, the nature of the corresponding Hall effect changes. In 1845 Faraday observed that when a beam of light is passed through a transparent dielectric medium, the plane of polarization of the beam is rotated when a magnetic field is applied along the direction of propagation. A similar effect can be observed at microwave frequencies in both ferrites and semiconductors, and is referred to as the Faraday effect or Faraday rotation. The angle of rotation of the incident beam is generally given by the empirical law

$$\theta = VBd$$

where B is the magnetic intensity, d the specimen thickness and V is referred to as the Verdet coefficient. When the propagation direction differs from the magnetic field direction by an angle of ϕ, the angle of rotation is given by

$$\theta = VBd \cos \phi$$

An important characteristic of the Faraday effect, as observed by Faraday himself, is that contrary to natural rotation the sense of rotation depends on the direction of the magnetic field and not on the direction of propagation of the incident beam. Consequently, reflecting a beam back and forth through the specimen progressively increases the Faraday rotation.

The Faraday effect arises through the coupling of the incident beam radiation with electrons or bound nuclei, the resulting precession of the spin magnetic moments of which about the applied magnetic field gives rise to different velocities of propagation for the right and left circularly polarized components of the incident linearly polarized beam. The plane of polarization of the resulting beam emerging from the specimen will therefore be rotated compared with that of the incident beam. If there is absorption of the beam in the specimen then the absorption coefficient will be different for each sense of circular polarization, and the emerging beam will then be elliptically polarized. In semiconductors at microwave frequencies the largest contribution to the Faraday effect comes from the presence of conduction electrons (mobile free charge carriers), but an interpretation of the underlying mechanism depends on the mean free relaxation time of these charge carriers. If the charge carrier collision frequency is less than, or of the same order as, the incident microwave frequency, the charge carriers will be able to execute orbital motions associated with the cyclotron resonance effect, so providing the desired gyromagnetic action and the resulting Faraday rotation. At normal temperatures, however, charge carrier mean free paths are too small to permit such cyclotron orbital motions and the observed Faraday-type rotation arises from a different mechanism, namely the Hall effect.

Rau and Caspari[92] were the first to successfully measure the Hall effect at microwave frequencies using the Faraday rotation technique. Measurements were made at 9 GHz on n and p-type germanium single crystals at room temperature. The rotation of the plane of polarization of the microwave power passing through the samples was measured directly and Rau and Caspari derived an expression for this rotation as

$$\theta = \tfrac{1}{2}(\mu_0/\epsilon_0\epsilon_r)^{1/2}\sigma_0\mu_H Bd$$

where μ_H is the Hall mobility, σ_0 the d.c. conductivity, B the magnetic field strength, d the sample thickness, ϵ_r the sample relative permittivity at the measurement frequency, and μ_0 and ϵ_0 are the permeability and permittivity of free space respectively. Good agreement between experiment and theory was obtained. Donovan and Webster[93] later extended the theory of anisotropic semiconductors to take account of anisotropic polarization effects.

Such measurements of the Hall effect do not make use of the fact that reflecting the incident microwave power back through the sample results in the microwave being further rotated in the same direction. If one had an unlimited quantity of the sample material and the means of maintaining a constant magnetic field over an unlimited length, then there would be no advantage in using such reflections to increase the measurement sensitivity. To understand the full extent of the problem, it is useful to have some idea of the magnitude of the rotation expected for a standard semiconducting material such as germanium. A germanium sample 10^{-2} m thick, of resistivity 0.1 ohm m, when placed in a magnetic field of 1.0 tesla (10^4 gauss) will produce an electronic Hall effect rotation of about 1 degree. The electron Hall mobility value for germanium is around 0.35 $\text{m}^2 \text{ V}^{-1} \text{ s}^{-1}$. To obtain the same degree of rotation for a typical high resistivity sample (resistivity 10^2 ohm m, electron mobility $10 \times 10^{-4} \text{ m}^2 \text{ V}^{-1} \text{ s}^{-1}$) the sample length would need to be increased to a value exceeding 100 metres, a result requiring no comment regarding its practicability. In this case it would obviously be particularly advantageous to allow the microwave power to be reflected back and forth through a much thinner sample as many times as possible. The easiest way of achieving this situation is to place the sample in the electric field antinode of a resonating cavity. Such a resonant cavity may be characterized by a quality-factor (Q-factor) which can be interpreted as expressing the number of times incident microwave power is reflected back and forth within the cavity before being finally absorbed in the cavity walls. Carefully designed and constructed microwave cavities can have

Q-factors exceeding a value of 10^4, resulting in an increase of the measurement sensitivity by this factor over the direct rotation measurement method. Portis and Teaney[94] were the first to describe the essential features of a bimodal cavity suitable for Hall effect measurements, and its use for biological materials was first attempted by Trukhan,[95] and then later by Eley and Pethig,[23] and Bogomolni and Klein.[96] The technique has also been used to measure the Hall mobility in various organic solids exhibiting one-dimensional conduction properties.[97-99] Detailed descriptions of the experimental technique and theory have been given[23,94,98] and a more comprehensive review of the method has been given elsewhere.[100] Although the microwave technique can give rise to many problems regarding an accurate interpretation of the results, it has many positive advantages as far as the investigation of biological materials are concerned. These include the ability to investigate polycrystalline materials, since heterogeneities associated with inter-crystalline and intracrystalline defect effects are eliminated. Electrical contact effects are also eliminated since the method is an electrodeless one. Measurements are also possible over a wide temperature range and for different atmospheres, since the microwave resonant cavity can be designed to incorporate these facilities. A brief account of the microwave Hall effect measurements that have been reported in the literature for biological materials will now be given.

Very few laboratories have investigated the microwave Hall effect in biological materials and much more work remains to be accomplished before the technique can provide a significant contribution towards the understanding of the relevance of solid-state electronic concepts in biology. The Hall mobility values that have been obtained in two different laboratories for a few biopolymers are presented in Table 9.3. It can be seen that there is reasonable agreement regarding the order of magnitude of the Hall mobility. Some of the larger values[23] were considered to

Table 9.3   Hall mobility values obtained at 10 GHz[95] and 9 GHz[23] for various biopolymers at room temperature. The polarity of the mobility is indicated by n (electrons) or p (holes)

Material	Hall mobility (10^{-4} m^2 V^{-1} s^{-1}) at 10 GHz[95]	at 9 GHz[23]
Cytochrome-c		
'dry'	$1 \pm 0.8, n$	—
'damp'	0	$0.1 \sim 0.5, n$
DNA		
'dry'	0	$0.3, p$
'damp'	$0.5 \pm 0.2, p$	$\sim 9.5, n$
Haemoglobin		
'dry'	$2 \pm 1.2, p$	
'damp'	$2 \pm 1, n$	$0.1 \sim 0.5, n$
Bovine plasma albumin		
'dry'	—	$\sim 20, n$
'damp'	—	$0.8, n$

result from an underestimate of the sample resistivities calculated by Trukhan,[95] and these mobility values for damp DNA and dry bovine plasma albumin were taken as evidence for the existence of band conduction in these materials. The Hall mobility values for the biopolymers also appears to be dependent on the water content. For DNA and haemoglobin a reversal of the polarity of the Hall effect from p- to n-type occurs as the hydration content increases from the dry state, whereas for cytochrome-c and bovine plasma albumin the n-type Hall mobility decreases in magnitude. The effect for DNA and haemoglobin is consistent with the concept that adsorbed water acts as an electron donor. No explanation as yet has been advanced for the reducing mobility value observed for cytochrome-c and bovine plasma albumin as the water content increased, although it could signify an increased protonic conductivity.

Measurements have also been made on rat liver mitochondria and spinach chloroplasts,[31] and the effects of known respiratory inhibitors were investigated for the mitochondrial 'electron transport chain'. Potassium cyanide and rotenone were found to reduce the Hall mobility but no effect was observed for antimycin-A. The marked effect of potassium cyanide, and the low mobility value obtained for the lipid extract of the mitochondria, suggested that electronic conduction through the electron transport chain of cytochrome molecules was being observed. On estimating the fractional volume occupied by the transport chain in the mitochondria samples, volume-corrected Hall mobility values of around 65×10^{-4} m^2 V^{-1} s^{-1} were obtained. Such a large mobility value definitely corresponds to a conduction mechanism involving charge transport within energy bands. These mobility values were very much larger than that observed for isolated cytochrome-c molecules (see Table 9.3) and this may reflect the fact that the protein conformations of the various cytochrome molecules within the mitochondrial electron transport chain are particularly suitable for long-range electron transport and that this ideal conformation was lost in the extracted soluble cytochrome-c samples. The Hall mobility values for the chloroplast samples were found to be of the order $0.5-0.8 \times 10^{-4}$ m^2 V^{-1} s^{-1} and p-type. It was thought possible that these relatively low mobility values were due to a possible damaged state of the quantasomes, but in fact similar values were later reported by other workers.[96] The measurements for mitochondria were extended to an investigation of the four enzymic complexes that constitute the electron transport chain of heavy beef heart mitochondria.[101] Of the four complexes, cytochrome oxidase exhibited the largest Hall mobility, with a value of the order 20×10^{-4} m^2 V^{-1} s^{-1} based on a 100 per cent protein basis, and there appeared to be a direct relationship between the observed Hall effect signal and the capacity of the complex to oxidize ferrocytochrome-c. Photomicrowave Hall effects have also been investigated. Trukhan et al.[102] have reported a p-type Hall mobility value of 15×10^{-4} m^2 V^{-1} s^{-1} for the pigment epithelium of the eye. Bogomolni and Klein[96] investigated the photo-induced Hall effect in films of whole and broken chloroplasts from plants and algae, and of chromatophores from photosynthetic bacteria. The Hall mobility values obtained were p-type and of value around 1×10^{-4}

m^2 V^{-1} s^{-1}, and these values were interpreted in terms of a tunnelling or hopping process, rather than as charge transport in energy bands.

Several, as yet unresolved, problems are associated with the microwave Hall effect measurement technique, and these arise because both the theory and experimental procedures have not been fully developed. For example, it is not clear to what extent dielectric displacement currents arising from dipolar losses in the test samples influence the Hall effect signals. Chai and Vogelhut[102] observed the microwave Hall effect in haemoglobin as a function of hydration by measuring the Hall rotation directly. The observed angle of rotation increased linearly with the amount of adsorbed water until a hydration level of 23 wt per cent was attained, whereafter the rotation decreased with increasing hydration. This effect was attributed to the dielectric displacement currents associated with the microwave dipolar losses of the adsorbed water. Eley and Pethig[23] observed a similar effect for bovine plasma albumin, and having calculated that the displacement current Hall effect would be smaller than the electronic Hall effect, explained their results in terms of the influence of hydrogen-bonded structures. Clearly, this problem deserves further consideration since if the displacement currents do give rise to Hall effect signals larger than those arising from mobile charge carriers, then although the technique would have great value for hydration effect studies, it would be of little value for exploring the electronic solid-state properties of biopolymers.

A more serious problem arises from the calculations of Fletcher[103] which indicate that Hall effect signals from the metal end-walls of the microwave resonant cavity, in which the test samples are located, would exceed those from the test samples if the Hall mobility value is less than around 1×10^{-4} m^2 V^{-1} s^{-1}. As can be seen from Table 9.3 this would seriously affect the interpretation of many of the microwave Hall effect results so far reported in the literature for biopolymers. This conclusion by Fletcher is at variance with the minimum detectable mobility value of 7×10^{-6} m^2 V^{-1} s^{-1} reported by Sayed and Westgate,[98] and with the p-type value of 3.5×10^{-6} m^2 V^{-1} s^{-1} determined by Ong and Portis[99] for TTF–TCNQ samples. In an attempt to understand more fully the possible influences of dielectric displacement currents and cavity end-wall effects a microwave Hall effect measurement system has been constructed to operate at a frequency of 33 GHz.[104] It is hoped, and still expected, that microwave Hall effect measurements will prove to be a useful addition to the various techniques being employed in the study of the solid-state electronic properties of biopolymers.

CHARGE-TRANSFER INTERACTIONS

Mention has been made several times in this chapter of charge-transfer interactions in which electronic charge is transferred from a 'donor' molecule to an 'acceptor' molecule. Such interactions have important significance for many biological processes, and for this reason an outline of some of their basic features will be given here. When two molecules interact to produce a new compound as a result of the formation of new chemical bonds, electronic charge is often redistributed amongst the atoms of the original molecules that form the framework of the new compound. On performing a 'book-keeping' exercise of the total electronic charge of the system it may well be found that the atoms that can be identified with one of the original molecules have collectively given up charge to the atoms of the other molecule. This effect can be interpreted as representing a form of charge-transfer process. Another example occurs when two water molecules are linked by a hydrogen bond to form a water dimer. In this dimer formation a fraction of an electronic charge ($\sim 0.02q$) is transferred across the hydrogen bond from the hydrogen atom of one water molecule to the oxygen atom of the other.[105] Although such interactions are obviously of fundamental importance for the understanding of many biochemical processes, this is not the type of charge-transfer process that will be described here. Instead, we shall concentrate on those interactions where charge is transferred from one molecule to another in the *absence* of the formation of a chemical bond between them. Such an association of two molecules can be thought of as a 'complex' rather than as a new compound, and for this reason such an association is often referred to as a 'charge-transfer complex'. The formation of such complexes is often characterized by the appearance of a broad absorption spectrum that occurs in the optical frequency range.

The action of bringing together two colourless or nearly colourless compounds sometimes leading to the appearance of strong colours has been known for many decades. The early studies[106] of the interaction of silver ions with aromatic or unsaturated compounds, and the somewhat more spectacular studies by Benesi and Hildebrand[107] of the interactions of iodine with aromatic compounds, soon led to extensive studies of molecular complexes now recognized to be of the electron-donor—acceptor type. The idea that electron transfer from a donor molecule to an acceptor molecule is involved in these types of molecular complexes was suggested by Bennett and Willis[108] in 1929, although the final result was envisaged to be the formation of a donor—acceptor covalent bond. Briegleb[109,110] later proposed that the essential mechanism was that of electrostatic attraction between molecules with permanent dipole moments and between non-polar molecules which become polarized by induction. Although such dipole-induced dipole interactions are now considered to aid in the stability of such complexes they cannot by themselves be the cause of the marked changes in the electronic absorption spectrum frequently observed on their formation. Other workers[111,112] supported the concept of an electronic interchange between the components of the complex, and suggested that the colours which appear in a liquid medium arise from those collisions between the donor and acceptor molecules whose relative orientations favour intermolecular electronic transfer. By about this time the physical basis of light absorption was generally better understood, so that colour formation was no longer taken to imply solely that new compounds had been formed as a result of covalent bonding between the constituent molecules of the complex. Furthermore, X-ray studies especially those of Powell and Huse,[113,114] had shown quite clearly that the separation distances between the donor and acceptor molecules in the crystalline complexes were only slightly less than the van der Waals distance and hence at least two times greater than normal covalent bond lengths.

The development of a theoretical understanding of the formation of such molecular complexes was aided in particular by the ideas of Weiss[115] and Brackmann.[116] Weiss proposed that they were of an essentially ionic structure of the form B^+A^-, and concluded that a low ionization potential for the base B and a high electron affinity for the Lewis acid A should favour a stable complex formation. Brackmann attributed the complex formation to 'complex resonance' and emphasized that the light absorption process causing the colour should not be considered as being located in one of the molecular components but rather as a property of the complex as a whole. Mulliken[117,118] was able to incorporate both of these main ideas into what is now generally accepted to be the most successful theory describing the phenomena of what he termed to be charge-transfer complexes. Mulliken's theory is essentially that the complexes are formed between electron-donating and electron-accepting molecules, with these complexes existing in two states: a ground state and an excited state. The characteristic colours of the complexes arises from the transfer of an electron from the donor to acceptor molecule on the absorption of light. This description of the components of the complex in terms of an electron donor and electron acceptor is preferable to that of Lewis bases and acids, since the original concepts outlined by Lewis[119] envisaged the more restrictive scheme of the donation and acceptance of electron pairs rather than of single electrons.

It would be impossible, and of no particular value for our purpose, to attempt a comprehensive account here of the work that has been accomplished in the study of charge-transfer complexes. Many excellent books and reviews of the subject have been written,[120-127] and emphasis will be given here to only those theories and dielectric and electronic properties which can be considered to form background material for the more general concept of biological semiconduction described in this chapter. Mulliken suggested[117] that the study of charge-transfer complexes may lead to a better understanding of intermolecular interactions in biological systems, but by far the strongest and most consistent advocate of this concept has been Szent-Györgyi through his various books[128-131] and papers (e.g. references 132−135) on the subject.

Following the theory first described by Mulliken[117,118,123] the electronic ground state of a donor−acceptor complex having a 1:1 stoichiometry can be described in terms of the wave function ψ_g as

$$\psi_g = a\psi_0(D, A) + b\psi_1(D^+A^-) \tag{9.16}$$

For a weak or loosely bound molecular complex we have $a^2 \gg b^2$, so that ψ_g can be regarded as a resonance hybrid of two wave functions with the major contribution coming from the no-bond function $\psi_0(D, A)$. This no-bond function corresponds to the two molecules D and A being in close proximity to each other but having no charge transference between them, so that it consists essentially of van der Waals, electronic dispersion, and dipole−dipole interaction forces. The contribution $\psi_1(D^+A^-)$ is termed the ionic or dative form, and is the wave function of the two molecules bound together by the coulombic attractive forces arising from an electron being totally transferred from the donor molecule D to the acceptor molecule A. The transfer of the electronic charge is often considered to be from the highest filled orbital of the donor to the lowest unfilled orbital of the acceptor molecule. In the formulation of equation (9.16) such contributions as those arising from locally excited electronic states and higher energy charge-transfer states have been neglected. If all the wave functions are normalized so that $\int \psi_g^2 \, d\tau = 1$, $\int \psi_0^2 \, d\tau = 1$ and $\int \psi_1^2 \, d\tau = 1$, then the coefficients a and b are related as follows:

$$a^2 + 2abS_{01} + b^2 = 1$$

where $S_{01} = \int \psi_0 \psi_1 d$ represents the overlap integral of ψ_0 and ψ_1. The molecular complex in its excited electronic state ψ_e is described by

$$\psi_e = a^* \psi_1 (D^+A^-) - b^* \psi_0 (D, A) \tag{9.17}$$

where $a^* \simeq a$ and $b^* \simeq b$. This is the state which determines the characteristic colour of the complex. From normalization considerations we have that

$$a^{*2} - 2a^*b^*S_{01} + b^{*2} = 1$$

and for weak complexes $a^{*2} \gg b^{*2}$. An electronic transition from the ground to the excited state can be attained by the absorption of light of appropriate wavelength. The absorption spectrum will also consist of locally excited states of both the donor and acceptor molecules, so that the total spectrum is a characteristic of the complex as a whole, as postulated by Brackmann.[116] For weak complexes the ground state is essentially non-ionic in nature and on absorbing light of energy $h\nu_{ct}$ the excited ionic state is attained according to the scheme

$$DA \xrightarrow{h\nu_{ct}} D^+A^-$$

For complexes between strong electron donors and acceptors, however, we have in equations (9.16) and (9.17) that $a^2 \ll b^2$ and $a^{*2} \ll b^{*2}$ leading to the class of complexes where the ground state is now ionic in form. Excitation with light energy can now lead to 'back charge transfer' of an electron from the acceptor to the donor. Such systems are often termed ion-radical complexes, and are of great interest because they exhibit unusual electric, magnetic, and optical properties.

The energy levels of the molecular complex can be obtained by solving the Schrödinger wave equation

$$H\psi_N = W\psi_N$$

where H is the Hamiltonian operator for the entire set of nuclei and electrons of the complex, and W is the energy. The solution of such equations gives a lower energy W_g corresponding to the ground state, and an upper energy W_e for the excited state. Their energy difference corresponds to the light energy $h\nu_{ct}$ required for the charge-transfer transition, so that

$$h\nu_{ct} = W_e - W_g$$

We can gain some physical insight into the value required for $h\nu_{ct}$ by first considering just the electron transfer process

$$D + A \xrightarrow{\Delta E} D^+A^-$$

The energy ΔE required for this transfer of charge from the donor to acceptor molecule is given by

$$\Delta E = I_d - A_a - \Delta \tag{9.18}$$

where the quantity $(I_d - A_a)$ represents the ionization energy I_d required to remove an electron completely away from the highest filled orbital of the donor minus the energy gained in then placing this electron in the lowest unoccupied orbital of a distant acceptor molecule of electron affinity A_a. We are at present ignoring the polarization parameters P of equation (9.4). The factor Δ represents the coulombic energy released in bringing the two ionic species D^+ and A^- together to their equilibrium distance apart d. From equation (5.28) we have that

$$\Delta = \frac{q^2}{4\pi\epsilon_0 \epsilon d}$$

For solid π-electron donor–acceptor complexes, d is of the order 3.3 x 10^{-10} m,[136] so that taking the intermolecular permittivity ϵ to be equivalent to that of free space we have that $\Delta \simeq 420$ kJ/mol (4.4 eV). For n–σ^* complexes such as trimethylamine–iodine[137] and trimethylamine–iodine chloride[138] the value for d is much less at around 2.3 x 10^{-10} m, giving $\Delta \simeq 6.3$ eV. From these examples we can see that Δ is a very significant term in equation (9.18). This sequence of charge transfer can be shown schematically as a potential energy diagram of the form of Figure 9.7.

For the very weak donor–acceptor complexes we will then expect

$$h\nu_{ct} \simeq I_d - A_a - \Delta \tag{9.19}$$

But in actual fact, for any complex formation there will be both attractive and repulsive electrostatic interactions involved when the donor and acceptor molecules are forced together into the configuration of the complex. These classical electrostatic interactions will accompany the formation of the complex (D, A) from D and

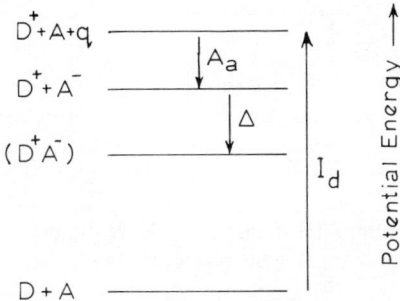

Figure 9.7 Schematic potential energy diagram to show the relationships between the various parameters of equation (9.18)

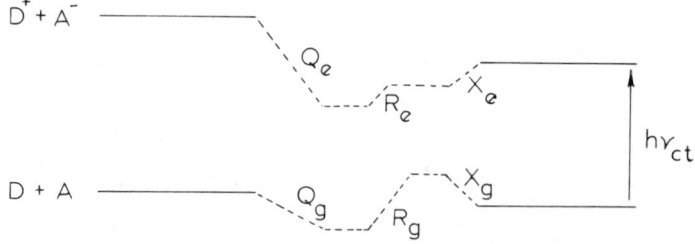

Figure 9.8 Schematic potential energy diagram to show the factors involved in the formation of a gas-phase charge-transfer complex between the electron donor D and the acceptor A. Q, R, and X are the attractive, repulsive, and resonance energies respectively, and g and e denote the ground and excited states, respectively

A in the ground state, and also the complex (D^+A^-) from D^+ and A^- in the excited state. Alterations in the energies of the ground and excited states will also arise due to the factors b and b^* of equations (9.16) and (9.17), and these are normally termed as the resonance interactions. These effects are shown schematically in Figure 9.8. The scheme of Figure 9.8 corresponds to the gas phase. If the complex is formed in solution, then the solvation energy of the complex must be considered, and this may be different for the ground and excited states. For complexes in the solid form, polarization energies must be taken into consideration to account for the interaction between the crystalline lattice and the complex in both its ground and excited states, and these take the form of the parameters P of equation (9.4). These two further considerations are shown in Figure 9.9, and demonstrate the fact that in general we can expect the spectral characteristics of a complex to change on passing from the gaseous state to being put into solution or prepared in the solid state.

For the weak complexes in particular, there will be loose coupling between the donor and acceptor molecules in the ground state which will tend to permit a

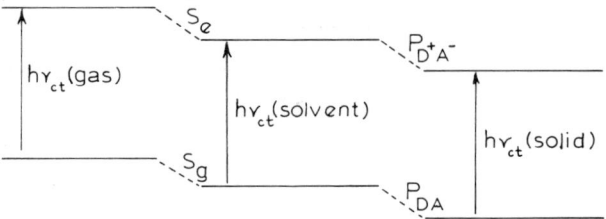

Figure 9.9 A schematic diagram to show the effects that occur in passing from the gas phase, through the solution phase, to the solid phase for the formation of a charge-transfer complex. S_g and S_e are the solvation energies of the complex in its ground and excited states respectively, and P_{DA} and $P_{D^+A^-}$ are the polarization interaction energies between (DA) and (D^+A^-), respectively, in the crystal lattice

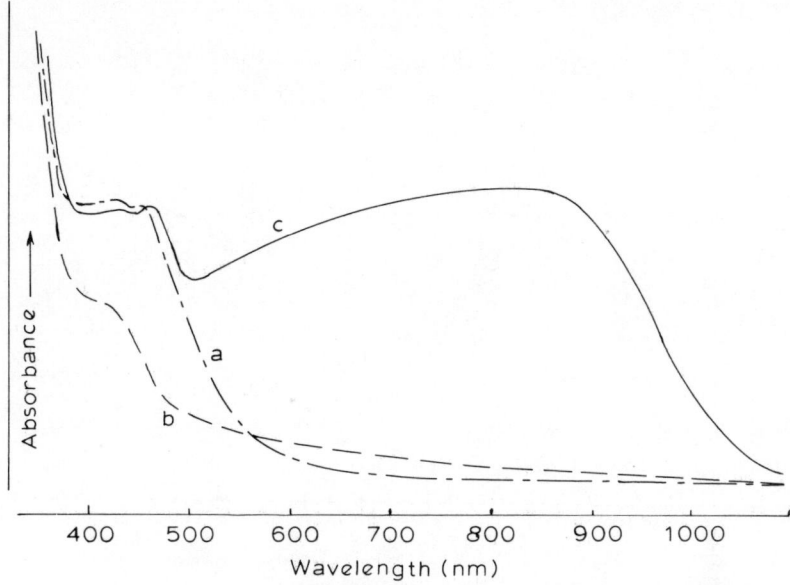

Figure 9.10 Optical absorption spectrum for (a) perylene, (b) chloranil, and (c) the perylene–chloranil complex, derived from diffuse reflectance spectra in the author's laboratory

continuous range of relative orientations of the molecules forming the complex. The molecular motions of charge-transfer complexes can be quite considerable,[139,140] leading in general to a continuous variation in the energy of the ground state. We can therefore expect a spread of absorption frequencies for the weaker complexes, leading to a broad and featureless charge-transfer absorption band with the maximum absorption corresponding to the transition for the most probable ground state alignment of the donor and acceptor molecules. A typical example of such a broad charge-transfer absorption band occurs for the perylene–chloranil complex, as shown in Figure 9.10.

It has been found for a large number of weak complexes containing a common acceptor molecule, that the charge-transfer absorption energy obeys a law of the form

$$h\nu_{ct} \simeq I_d - C_1 + \frac{C_2}{I_d - C_1} \qquad (9.20)$$

where C_1 and C_2 are characteristic parameters of the acceptor molecule.[120,123,141] For the hydrocarbon–iodine complexes, for example, it is found that the values $C_1 = 5.2$ eV and $C_2 = 1.5$ eV^2 give the best fit of the spectral results when applying equation (9.20).[123] Although equation (9.20) is quadratic in I_d, for many weak complexes and especially for a homologous series of donors, the variation in the last term of the equation is quite small. This leads to the charge-transfer absorption energy $h\nu_{ct}$ being essentially a linear function of ionization energy I_d of the donor, as suggested by equation (9.19). This behaviour is

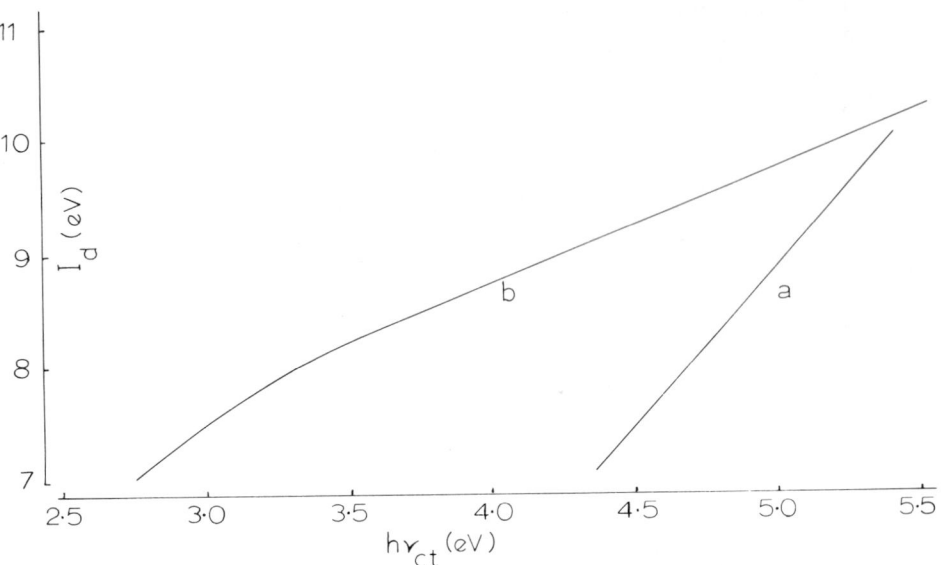

Figure 9.11 The variation of the charge-transfer absorption energy $h\nu_{ct}$ against the ionization energy I_d of the donor for (a) strong amine–iodine complexes, and (b) weak hydrocarbon–iodine complexes

typical of the weak hydrocarbon–iodine complexes and is shown in Figure 9.11. Similar linear relationships between I_d and $h\nu_{ct}$ will not be expected to occur for donors of markedly different type, since differences in molecular orientation and separation distance will certainly occur.[123,142] For the stronger complexes such as those between the amines and iodine, equation (9.20) does not appear to be applicable. An altogether different and more complicated relationship is required to describe such spectral data as that shown in Figure 9.11 for the amine-donor, iodine-acceptor complexes.[141] Equations (9.19) and (9.20) involve the vertical ionization potentials, and Collin[143] has suggested that the adiabatic values would be more appropriate for the strong charge-transfer complexes. This refinement does not appear to lead to a better description of the spectral data.[123] However, the large number of spectral data which can be described in the form of equation (9.20), taken together with the reasonableness of the values for C_1 and C_2, does lead to the conclusion that the charge-transfer absorption spectra are well described in terms of electronic transitions from an essentially no-bond ground state to an ionic charge-transfer excited state. In their summary of a rather detailed analysis of the charge-transfer theory, Hanna and Lippert[144] conclude that charge-transfer resonance is both a necessary and sufficient condition to explain the unique spectral features of donor–acceptor complexes that possess charge-transfer absorption bands. Furthermore, it is concluded that for both the weak and strong complexes, not only the charge-transfer resonance interactions but also both electrostatic and repulsive interactions must be included in considering the stability of the ground state.

The electronic conductivity of a charge-transfer complex is usually found to be markedly greater than the conductivity of either molecular component. This is to be expected, since for electronic conduction to take place the complete separation of electron-hole pairs must occur and in the formation of a charge-transfer complex such charge separations have already been initiated. This corresponds to a reduction of the activation energy required for the generation of free charge carriers. The more strongly interacting complexes, corresponding to those in which the ground state is largely ionic, will exhibit the largest reduction in charge generation activation energy. This is exemplified by the complex formed between chloranil and phenylenediamine where the room temperature conductivity (10^{-4} mho m^{-1}) of the complex is greater than that of either component by a factor of at least 10^8.[145] The strength of the charge-transfer interaction depends directly on the extent of overlap of the electronic wave functions of the two molecular components of the complex. On this basis Mulliken predicted[117] that the conductivity should increase with increasing pressure, and this has been verified in several laboratories where it has been found that for moderate pressure increases the conductivity increases approximately exponentially with increasing pressure. A particularly good example of this effect occurs for the pyrene—TCNE complex where the conductivity increases by seven orders of magnitude with a pressure change from atmospheric to about 30 kbar (3×10^9 Pa).[146] The electronic properties of a great number of complexes have been investigated, and the types studied typically include those formed by the interaction of aromatic hydrocarbons with halogens, amines with quinones, and hydrocarbons with tetracyanoethylenes. Most of these interactions involve π-electrons. Of particular interest are the various complexes that can be prepared using the tetracyanoquinodimethane (TCNQ) ion-radical. Polymeric charge-transfer complexes can also be prepared by interacting aromatic hydrocarbons or halogens with polymers such as polyacenaphthylene, polystyrene, and polyphenylene for example, where the polymer usually acts as an electron donor. A good example is poly-N-vinyl carbazole (PVCA)—iodine which exhibits a maximum conductivity of around 10^{-2} mho m^{-1}, a free electron spin concentration ranging from 10^{15} to 10^{18} spins g^{-1} depending on the iodine concentration, a Hall mobility of 0.5×10^{-4} m^2 V^{-1} s^{-1} and a Seebeck coefficient of 0.25 mV/$^\circ$C.[147] This polymeric complex was also found to exhibit interesting photovoltaic effects.[148] Reviews of the electronic studies that have been made on a wide variety of charge-transfer complexes have been given by Gutmann and Lyons,[149] Foster,[122] and Rembaum[146] and cover periods up to 1967, 1969, and 1970, respectively. The dielectric properties of charge-transfer complexes in solution have been reviewed by Price[150] and Kulevsky[151] and cover the periods up to 1973 and 1975, respectively.

Vincent and Wright[152] studied the steady-state semiconductivity and photoconductivity of a series of single crystals of 24 complexes in which the electron donor and acceptor molecules had been systematically varied. The results of the semiconduction and Seebeck coefficient measurements indicated that the semiconductivity was extrinsic and possibly due to metallic impurities. A strong relationship was found between the crystal structure and the photoconductivity of the various complexes, and it was considered that for the first time a basis had been developed for predicting the relative magnitudes of the photoconductivity of such complexes. The conclusion regarding the extrinsic nature of the semiconductivity was supported by measurements[153] of the high electrical field conductivity of the perylene–chloranil and perylene–DDQ complexes where the semiconductivity was investigated for the temperature range 150–345 K with applied fields up to 10^7 V m^{-1}. Evidence was found to indicate that the high field conduction mechanism was a bulk process involving the Poole-Frenkel effect described in Figure 8.11 and by equations (8.28)–(8.30), where electrons were considered to be hopping between localized trapping sites associated with the impurity. The trapping energy states, of density around 5×10^{20} m^{-3}, were calculated to be located at an energy of around 0.72 eV below the conduction band for the complex. This energy value for the trapping site is similar to the semiconduction activation energy that has been determined for a wide variety of molecular complexes. The effective charge carrier mobility was found to be of the order 10^{-8} m^2 V^{-1} s^{-1}, and is a value compatible with a hopping electron conduction mechanism. The dielectric properties of the perylene–chloranil complex have also been studied over the frequency range 10^{-4}–10^6 Hz, and for temperatures between 190 and 320 K.[47,48] A dipolar-type dispersion at sub-hertz frequencies was found, an effect not found for perylene or chloranil on their own, with a relaxation activation energy similar to the semiconduction activation energy. Of the various interpretations attempted to describe the observed dielectric dispersion, an activated hopping electron model was considered to be the most applicable.

Recent studies[55-57] of the brown complexes produced when proteins are interacted with methylglyoxal have given results with features directly comparable with those observed for the perylene–chloranil complex. The electronic properties of these protein–methylglyoxal complexes have been described earlier in this chapter. Evidence for the possibility that proteins can form charge-transfer complexes has also been obtained from the conduction studies of Eley *et al.* on the chloranil– and chlorophyll–bovine plasma albumin complexes,[24,27] and from spectroscopic and conduction measurements on the bovine serum albumin–chloranil complex.[154] Apart from chloranil and chlorophyll, the albumin protein molecule has been found to form complexes with β-carotene, β-naphthaquinone and various known carcinogenic molecules, and to result in an electrical conductivity larger than that of the pure protein.[27,155] Various other spectroscopic studies of amino-acid complexes in solution indicate that amino acids and proteins can form charge-transfer complexes as a result of the donor action of the lone-pair electrons on the terminal nitrogen in the unionized form of the amino acid or peptide.[124] This serves to indicate that, apart from the more usual π-electron interactions found in most molecular complexes, for biomolecules n-electron charge-transfer interactions may be significant. Nagchaudhuri and Suri[156] have studied the spectroscopic properties of the complexes formed between the aromatic amino acids and nicotinamide adenine dinucleotide (NAD), and molecular orbital calculations were used to determine the extent of charge transfer and to show the significance of the mutual orientation of the donor and acceptor in determining the stability of the complexes. It was hoped that such studies would enhance a better understanding of the way in which NAD interacts with enzymes. Apart from the books[120-127] cited which describe the various aspects of molecular charge-transfer interactions, the particular relevance of such studies for biological systems has been discussed in considerable detail in other volumes of collected works.[157,158]

PIEZOELECTRIC EFFECTS

One of the earliest investigations of the piezoelectric properties of biological materials were those conducted by Martin[159] on keratin in the form of Merino wool and human hair. He found that if a number of similar wool or hair fibres were held together and all pointing in the same direction, then a fibre pulled out of the bundle by the root-end carries a positive charge, whereas a fibre pulled out by the tip carries a negative charge. Martin was able to show that this tribo-electric effect originated from the scales or cuticle cells, rather than from the central cortical cells of the fibres. On applying ten pounds compressive force to a $\frac{1}{4}$-inch long bundle of wool fibres, a potential difference of one volt was observed to develop across the fibre lengths, with the root-ends assuming a negative polarity. If the test samples were arranged to have equal numbers of fibres tip first and root first, then no piezoelectric potentials were found. On applying a shear force to the fibres, a small piezoelectric effect was observed, indicating that the associated dipoles were oriented at a small angle to the fibre axis. More recent measurements of the thermally stimulated current in porcupine quills, where the current flows from the tip to the root in an external short-circuit, produce the same conclusion regarding the polarity of the structure dipoles in the keratin fibres at normal physiological temperatures.[160]

The term 'piezoelectricity' is derived from the Greek word meaning 'to press' and hence its literal meaning is 'pressure-electricity'. For true piezoelectric effects to be present in a material it must possess a structure having a non-random orientation of permanent dipole moments, or otherwise contain trapped electric charges. Neglecting the effect of trapped charge, then only crystalline and not amorphous materials should exhibit piezoelectricity. Crystals can be divided into 32 classes on the basis of the symmetry of their atomic structure. The criterion that determines the existence of piezoelectricity is a lack of a centre of symmetry, and of the 32 crystal classes 21 satisfy this condition, with 20 exhibiting piezoelectricity. (The non-symmetric cubic pentagonal icosi-tetrahedral system has other symmetry features which combine to preclude the effect.) For a crystal with a centre of symmetry, then no combination of uniform stresses can be found which will cause a separation of the centres of gravity of the positive and negative charges and produce the necessary induced dipole moment. A crystal with no symmetry at all would have 18 piezoelectric constants, whereas the most symmetrical type, a cubic crystal, has just one piezoelectric constant. A complete designation of piezoelectric constants, in terms of tensor notation has been given by Mason for the various crystal classes.[161]

There are four different piezoelectric coefficients which relate the electric polarization or field to the mechanical stress or strain, and they are:

$$d = \left(\frac{\partial D}{\partial F}\right)_{E=0} = \left(\frac{\partial P}{\partial F}\right)_{E=0} = \left(\frac{\partial S}{\partial E}\right)_{F=0}$$

$$e = \left(\frac{\partial D}{\partial S}\right)_{E=0} = \left(\frac{\partial P}{\partial S}\right)_{E=0} = -\left(\frac{\partial F}{\partial E}\right)_{S=0}$$

$$g = -\left(\frac{\partial E}{\partial F}\right)_{D=0} = \left(\frac{\partial S}{\partial D}\right)_{F=0}$$

$$h = \left(\frac{\partial E}{\partial S}\right)_{D=0} = \left(\frac{\partial F}{\partial D}\right)_{S=0}$$

where D is the electric displacement, P the polarization, E the electric field, S the mechanical strain, and F the mechanical stress. The coefficients d and g refer to the electric polarization and field, respectively, caused by unit stress under either short-circuit ($E = 0$) or open-circuit conditions. The coefficients e and h refer to similar effects as d and g, caused instead by unit strain. The most commonly determined parameters are the direct piezoelectric coefficient d, and the inverse piezoelectric coefficient e.

It can be shown[161] that the four piezoelectric coefficients are related to each other through two different elastic constants G^E and G^D, and through two different relative permittivities ϵ^F and ϵ^S. If the electrical connections to the crystal are open-circuited, then the elastic constant of compliance G^E corresponding to a zero electric displacement is obtained, while if the connections are short-circuited one obtains the constant G^D for zero field. Also, if the permittivity of the crystal is measured when it is clamped so firmly that it cannot change in dimension, then the clamped or constant strain permittivity ϵ^S is derived. If the crystal is allowed unrestricted compliance then additional energy can be stored in the crystal in mechanical form giving rise to an increased permittivity representing the constant stress or free permittivity ϵ^F. The difference between ϵ^S and ϵ^F is determined by the electromechanical coupling factor k_e, defined as the square root of the energy stored in mechanical form, for a given type of displacement, to the total input electrical energy. For a given mode of motion

$$k_e = d(\epsilon^F S^E)^{1/2}$$

and

$$\frac{G^D}{G^E} = \frac{\epsilon^S}{\epsilon^F} = 1 - k_e^2$$

For ferroelectric crystals k_e may have values up to 0.9 whereas for most polymers k_e is usually less than 0.1 and we have that $G^E \simeq G^D$ and $\epsilon^S \simeq \epsilon^F$. It also follows that the piezoelectric coefficients are related through

$$\frac{e}{d} \simeq \frac{h}{g} \simeq G; \quad \frac{d}{g} \simeq \frac{e}{h} \simeq \epsilon$$

Both the permittivity ϵ and elastic coefficient G are complex quantities, expressed as

$$G = G' + iG''$$

$$\epsilon = \epsilon' - i\epsilon''$$

and they exhibit relaxation phenomena as outlined in Chapter 1. It also follows that the various piezoelectric coefficients are complex quantities.

Since biological polymers are not readily obtained in the form of large single crystals, their piezoelectric properties are usually observed in the uniaxially oriented state in the form of thin films as shown in Figure 9.12 where the direction of polymer orientation is defined by the z-axis and the plane of the film surface is parallel to the x-axis. For biological materials such as bone and tendon where the sense of the polar axis of individual crystallites is in the same direction in an assembly of uniaxially oriented crystallites, then the class of symmetry is that of the C_∞ class and the corresponding piezoelectric matrix is of the form

$$d = \begin{pmatrix} 0 & 0 & 0 & d_{14} & d_{15} & 0 \\ 0 & 0 & 0 & d_{24} & d_{25} & 0 \\ d_{31} & d_{32} & d_{33} & 0 & 0 & 0 \end{pmatrix}$$

where as a result of the symmetry

$$d_{14} = -d_{25}; \quad d_{15} = d_{24}; \quad d_{31} = d_{32}$$

In this conventional notation use is made of the identities $x = 1$, $y = 2$, $z = 3$, $xx = 11$, $xy = 12$, etc., and the number of indices in the tensor is further reduced by replacing 11 by 1, 22 by 2, 33 by 3, 23 by 4, 13 by 5, and 12 by 6. For example, the coefficient $d_{14} = P_1/F_4$ defines the polarization P_1 produced in the x-direction by the shear stress F_4 applied in the yz plane, and d_{25} is the ratio of the polarization P_2 produced in the y-direction to the shear stress F_5 applied in the xz plane. The coefficient d_{25} is often the most practicable one to measure and determines the so-called face-shear piezoelectricity. For materials like bone and tendon, the coefficients d_{14} and d_{15} are typically 20 times larger or more than the coefficients d_{33} and d_{31}, with values around 10^{-12} C/N.[162]

Figure 9.12 The usual sample geometry for studying piezoelectric effects in biopolymer thin-film samples

For single crystals of cellulose of crystal class 'monoclinic C_2', the piezoelectric matrix takes the form

$$
d = \begin{pmatrix} 0 & 0 & 0 & d_{14} & d_{15} & 0 \\ 0 & 0 & 0 & d_{24} & d_{25} & 0 \\ d_{31} & d_{32} & d_{33} & 0 & 0 & d_{36} \end{pmatrix}
$$

For a large number of such crystals assembled so that their z-axes are all aligned but their polar directions are randomly distributed, and their x- and y-axes are distributed at random in a plane perpendicular to the z-axis of the sample, then

$$
d = \begin{pmatrix} 0 & 0 & 0 & d_{14} & 0 & 0 \\ 0 & 0 & 0 & 0 & d_{25} & 0 \\ 0 & 0 & 0 & 0 & 0 & 0 \end{pmatrix}
$$

and the face-shear piezoelectricity becomes the measurable quantity.

The origins of piezoelectricity in polymers are considered[163] to arise from essentially three sources as follows:

(1) The first, and perhaps most important, origin is that arising from internal strain effects for materials that are asymmetrical in their chemical structure and hence have 'built-in' dipoles. A typical example is that observed for uniaxially oriented samples of α-helical polypeptides such as collagen and keratin. On account of internal, microscopic, strains that result from the application of a shear force the orientations of the peptide dipole moments become altered, and an electric polarization results which appears as the piezoelectric effect.

(2) Another source arises from strain dependences of spontaneous polarization. Such polarizations need not represent a permanent thermodynamically stable state of the material, but could be due to the presence of trapped excess electrons or ions, for example. Such a situation will not give reproducible piezoelectric effects if the test material is heated, since this will tend to cause movement or complete removal from the sample of the trapped charge.

(3) If the test material is of heterogeneous density and contains a non-uniform distribution of trapped charges, then the resulting non-uniform stress dependence of strain and charge redistribution will also result in piezoelectric effects. A typical example of such a material is an electret, which contains an antisymmetrical charge distribution to give it electric polarity.

To these more generally accepted origins of piezoelectricity, the possibility should also be considered that the effect could also arise in protein structures as a result of non-uniform energy band widths resulting from either structural irregularities of the polypeptide chains or from local charge effects as shown in Figure 8.14(b). For conventional semiconductors such non-uniform energy bands can produce inhomogeneous field effects similar to those produced by non-uniform elastical strains.[164]

Piezoelectric effects of type (1) above have been observed in wood, bone, nerve membranes, DNA, and in proteins such as collagen, fibrin, myosin, actin, keratin, and synthetic polypeptides such as PMLG and PBLG. A valuable and well-referenced review of these measurements up to 1974, together with an account of experimental procedures, has been given by Fukada.[165]

A recent development in the theory has been the consideration of induced piezoelectric effects in isotropic materials as a result of the application of a constant electric field.[166,167] These induced effects will be superimposed on any inherent piezoelectricity. In an isotropic material in which there is no inherent piezoelectricity, nor polarization proportional to the square of the electric field strength E^2, then as outlined by Zimmerman[167] the electric polarization P and mechanical deformation S can be described in terms of the electric field E and mechanical stress F as

$$P = \chi E + 2\gamma EF \qquad (9.21)$$

$$S = sF + \gamma E^2 \qquad (9.22)$$

In these equations χ, s, and γ are the electric susceptibility at zero stress, the mechanical compliance at zero electric field, and the electrostriction constant, respectively. Defining the constants

$$P_0 = \chi E_0, \; S_0 = \gamma E_0^2$$

caused by the constant applied field E_0, then equations (9.21) and (9.22) can be written in the form

$$P = P_0 + d_i F$$

$$S = S_0 + d_i E$$

where d_i is the coefficient of electric field-induced piezoelectricity given by

$$d_i = 2\gamma E_0 \qquad (9.23)$$

and is linearly proportional to the bias field E_0. There is an alternative interpretation of this effect, approached in terms of changes in the material's permittivity ε as a result of strain-induced dimensional changes. The change in capacitance C of the test sample can be written in the form

$$\frac{\partial C}{\partial S} = C[(1/\epsilon)(\partial\epsilon/\partial S) + 1 - \delta_1 + \delta_2] \qquad (9.24)$$

The area of the sample increases by the fraction $(1 - \delta_1)S$ and the thickness decreases by $\delta_2 S$. For amorphous or polycrystalline materials, then $\delta_1 = \delta_2$ and

$$\frac{\partial C}{\partial S} = C[(1/\epsilon)(\partial\epsilon/\partial S) + 1]$$

When a constant potential V is applied across the sample of area A, the apparent piezoelectric strain constant 'e' is

$$e = \frac{1}{A} \frac{\partial}{\partial S} (CV)$$

$$= E_0 \left(\frac{\partial \epsilon}{\partial S} + \epsilon \right)$$

The induced piezoelectric stress constant $d_i = es$ is then given by

$$d_i = E_0 \left(\frac{\partial \epsilon}{\partial S} + \epsilon \right) s \qquad (9.25)$$

where s is the elastic compliance and E_0 is the constant applied field from the constant potential V. It can be seen from equation (9.25) that d_i varies linearly with the applied field with a proportionality factor that increases with the permittivity of the test material. Comparing equations (9.23) and (9.25), then

$$2\gamma = \left(\frac{\partial \epsilon}{\partial S} + \epsilon \right) s$$

Zimmerman proceeds to show that an additional effect can be expected if the test material conducts electrical current. This can be included into equation (9.24) in terms of the imaginary component ϵ'' of the permittivity, but a simpler approach is through consideration of the changes in sample resistivity ρ caused by the deformation changes. The change in electrical resistance R can be written

$$\frac{\delta R}{\delta S} = R \left(\frac{1}{\rho} \frac{\delta \rho}{\delta S} - 1 + \delta_1 - \delta_2 \right)$$

where for isotropic materials we have as for equation (9.24) that $\delta_1 = \delta_2$. When a constant potential V is applied across the sample, both the conduction current and the stored charge are modulated by the strain, and the apparent piezoelectric strain constant includes both the effect of changing capacitance and of changing resistance, so that for

$$e = e' - je''$$

we have

$$e' = \epsilon E_0 \left(\frac{1}{\epsilon} \frac{\delta \epsilon}{\delta S} + 1 \right)$$

$$e'' = J_0 / \omega \left(\frac{1}{\rho} \frac{\delta \rho}{\delta S} - 1 \right)$$

In these equations J_0 is the average current density in the sample given by $J_0 = E_0 / \rho$ and ω is the frequency of the sinusoidally modulated strain. The current density J_0 may also include displacement current densities proportional to dP/dt.

Zimmerman terms the factor e'' as the 'current-induced piezoelectric effect' and for $\rho\epsilon\omega < 1$ it will be of significant magnitude.

Although many investigations of the piezoelectric properties of bone and tendon have been reported in the literature, the underlying molecular mechanisms can only really be studied by measurements on purified and well-oriented collagen. One of the extensive studies of oriented collagen has been that of Fukada et al.[168] who determined the two piezoelectric constants d and e, the elastic constant c, and the permittivity, at a frequency of 10 Hz as a function of hydration and over the temperature range 123–323 K. With water contents greater than 25 wt per cent, the constants d, e, and c exhibited different temperature dependencies than those with lower water contents. This is consistent with X-ray diffraction studies[169] which indicate that at 26 wt per cent water content the sorption sites of the helical collagen structure become saturated, and water in excess of this quantity is accommodated within the intermolecular spaces.[170] This excess water may produce several effects in that its presence will tend to act as a plasticizer and also increase the effective permittivity of the amorphous regions of the sample. This in turn will influence the induced charge at the measuring electrodes due to the piezoelectricity of the crystalline phases of the sample. Also, with increasing temperature, ionic conduction effects in this excess water can be expected to neutralize the piezoelectric polarizations of the crystalline phases. The piezoelectric effects observed by Fukada et al.[168] could be explained in terms of such concepts, as well as in terms of reorientational motions of the molecular structure of collagen. For samples containing up to about 12 wt per cent water content, distinct dispersions were observed in the piezoelectric stress and strain constants, e and d, at 193 K and 273 K. The low temperature dispersion was considered to be associated with local mode relaxations of the main polypeptide chains, and supporting evidence for this was given by the fact that a maximum in the dielectric loss factor ϵ'' also occurred at this low temperature. The dispersion at 273 K was considered to arise from motions of the polypeptide side chains. As pointed out by Onaral,[171] Fukada et al. did not take electrode polarization effects into account in their determination of the complex permittivity of the hydrated collagen. Such electrode effects would have seriously influenced only the low frequency measurements and this will not influence the discussion of the work given here. Referring to Chapter 5 and the results of Chang and Chien described in Figure 5.15, it can be seen that there appears to be some confusion in the literature regarding the origin of the dielectric loss peak at around 200 K for collagen containing about 12 wt per cent water. Chang and Chien interpreted their results in terms of dipolar losses associated with the polypeptide side-chain motions and the disruption of hydrogen bonds. There is now certainly sufficient evidence from both elastic relaxation[172] and n.m.r. studies[173] to indicate that hydrated collagen exhibits two different relaxation phenomena at around 200 K and 273 K, and that they are both correlated with hydration effects. It is also clear that as yet no definite interpretation has been given of these relaxation effects, and in this respect further studies are to be encouraged. It is known[174] for PBLG that relaxations of the side chains occur at around 295 K, and it has clearly been shown that the piezoelectric elastic and dielectric parameters for PBLG exhibit dispersions closely associated with these side-chain molecular motions.[165]

The piezoelectric stress constant e_{25} of drawn films of poly-γ-methyl-D-glutamate (PMDG), cast from solutions in either α-helix promoting or non-helicogenic solvents, has been measured at 110 Hz for temperatures between 93 K and 473 K by Koga et al.[271] X-ray diffraction patterns and the dynamic mechanical properties of the films were also investigated. From an analysis of the various piezoelectric and mechanical relaxation processes it was concluded that the piezoelectric properties of PMDG arise from strain-induced distortions of the α-helical form of the backbone or main chain. The maximum value of $-e'_{25}$ and the orientation function of the α-helix were found to be linearly related, and an extrapolation of e'_{25} to unit orientation function corresponded to a value for the dipole moment of each peptide residue of 2.4 debye units. Taking account of the fact that a correction must be made to this value to correspond to there being complete specimen crystallinity, this dipole moment value is in good agreement with that expected for the peptide unit, as discussed in Chapter 2. In what appears to be the most detailed investigation of the piezoelectric properties of a biopolymer to date, Furukawa et al.[272] investigated the piezoelectric 'd' and 'e' constants, together with the elastic constant and permittivity, for oriented poly-γ-benzyl-glutamate (PBG) films with various elongation ratios as a function of frequency (10^{-4}–10^{4} Hz) and temperature (150–380 K). Both the L- and D-form of the polymer were investigated (see Figure 2.15). It was shown that the piezoelectricity of PBG films originates from the piezoelectric and optically active symmetry of PBG crystallites and their orientation distribution. This conclusion was based on the fact that the d_{14} component, which in fact is the only component (d_{14} = $-d_{25}$) of the piezoelectric matrix for a uniaxially oriented system, was found to be proportional to the degree of orientation of the α-helices, and that the sign of the d_{14} and e_{14} components reversed to be negative on changing from the PBDG to the PBLG polymer. The PBG films were essentially a two-phase system consisting of PBG crystallites dispersed randomly within the non-piezoelectric amorphous regions at the crystallite boundaries. The PBG crystallites, in turn, consisted of α-helices packed in an approximately hexagonal lattice array with each helix surrounded by its bulky side chains. Near room temperature these side chains become free to rotate independently of the helical main chains. It was found that the piezoelectric relaxation was influenced by such thermal motions of the side chains, with the effect being relaxational at the lower frequencies ($\sim 10^{-2}$ Hz) and retardational at higher frequencies (~ 100 Hz). On the assumption that the α-helical main chains surrounded by the side chains were responsible for the origin of the piezoelectricity, such relaxation phenomena were interpreted in terms of the relaxation of the local elastic field in the main chains. With respect to the influence of the side-chain motions, it should be added here that hydration effects will be significant. The presence of adsorbed water molecules will alter the relaxation frequency of the side chains, and the associated conductivity increase should also influence the piezoelectric effect, as in fact has been observed by Date.[273] At the atomic-scale, both the dipolar side-chain ester group and the main-chain peptide dipole could have produced the piezoelectric properties of PBG. However in order to exhibit piezoelectricity a uniaxially oriented system should have an asymmetric, optically active, structure, and by themselves the side chains do not possess this quality. Consequently, Furukawa *et al.* regarded the PBG crystallite as a two-phase system, with the rod-like non-relaxing main-chain piezoelectric phase being dispersed in a relaxing non-piezoelectric (side-chain) medium. Using an equivalent model which incorporated the amorphous intercrystallite phase, together with the two-phase PBG crystallites, a good description could be made of the frequency characteristics actually observed for the films. Although structural evidence to support such a model has yet to be established, it does represent the simplest equivalent structure to explain the piezoelectric properties of PBG films. There is evidence that PMLG exhibits a low frequency piezoelectric relaxation that can be interpreted in terms of a system consisting of a non-relaxing piezoelectric phase and a relaxing non-piezoelectric phase. Hayakawa *et al.*[274] have shown that such a composite system should be characterized by a piezoelectric strain constant where $d'' > d'$. Measurements on uniaxially drawn PMLG have shown that this inequality does hold for frequencies below 10^{-1} Hz and above 10 Hz.[275] It could well be that the three-phase model of Furukawa *et al.* could be used to successfully explain the complete frequency variation of d'' and d' observed for PMLG.

Fukada *et al.*[160] have reported the simultaneous measurements of the piezoelectric strain coefficients d'_{14} and d''_{14}, and of the thermal depolarization current of bovine horn keratin as the temperature was slowly increased from 293 K to 473 K. The coefficients d'_{14} and d''_{14} increased with rising temperature and reached a maximum at around 448 K, with there being evidence of a relaxation effect at around 373 K. An initial decrease of d' was considered to arise from the influence of water originally present in the horn sample, and the increase of d' thereafter was ascribed to the increasing flexibility of the keratin molecules. At 448 K a sharp decrease in d' occurred, and this was considered to arise from denaturation effects where the helical keratin molecules became uncoiled into a random configuration and in doing so losing the non-centrosymmetric structure required to give piezoelectricity. Repeated measurements of the denatured horn sample produced a zero piezoelectric effect for all temperatures. The measurement of the depolarization current, taken at the same time as the piezoelectric measurements, showed that it rapidly reversed in sign at about 403 K for directions both perpendicular and parallel to the bone growth axis. The sense of current flow below 403 K was axially toward the root and radially outward, whereas above 403 K both of these current components reversed in direction. Samples which had been thermally recycled below 403 K, and reheated, produced depolarization currents only above 403 K and in one direction only, showing a step-wise denaturation process. Every indication suggests that bovine horn keratin has two independently oriented dipole systems, which denature below and above 403 K, respectively. This effect could directly correspond to there being two differently hydrogen-bonded dipolar residues as was considered the case from earlier measurements for porcupine quill keratin.[175]

Maeda *et al.*[176] have determined the piezoelectric, dielectric, and elastic constants of bovine cortical bone at 10 Hz as a function of hydration from 123 K to 423 K. The piezoelectric constant of the dry bone was found to be practically independent of temperature. Bone samples with 7.4 wt per cent water content were found to have a larger piezoelectric constant than dry bone at temperatures below 273 K, but a value smaller than dry bone at higher temperatures. The low temperature effect was attributed to the increase of crystallinity of the collagen fibrils with hydration, and the higher temperature effect was considered to arise from ionic conduction associated with the mobility of the adsorbed water. The elastic constants of the hydrated bone were found to be greater than those for dry bone below 273 K, but smaller above 273 K. This was attributed to the adsorbed water forming weak hydrogen bonds within and between the collagen fibrils. At high temperatures, these bonds would be easily broken and the elastic constant would become smaller. At lower temperatures however, below around 263–223 K, the adsorbed water acts as a 'filler' which would tend to increase the elastic constant. The piezoelectric constant d''_{14} of hydrated bone was found to have three peaks on the temperature scale. The peak at the highest temperature (~373 K) was attributed to vaporization effects of the water and the peak at around room temperature was explained in terms of the increase of ionic conductivity accompanied by the increase of the mobility of the adsorbed water. The peak at about 153 K, which also occurred for the dielectric and elastic constants, was considered to be related to the motion of the collagen molecules coupled with the adsorbed water. The activation energy for this low temperature relaxation was determined to be about 0.35 eV, which is very close to the value of 0.34 eV that Chang and Chien obtained for the relaxation process in their tendon samples shown in Figure 5.15. By using a model which assumes a parallel combination of collagen fibres and hydroxyapatite crystals, Maeda et al. were able to derive an estimate for the piezoelectric constant of bone which was in close agreement with their observed value.

Apart from the relevance of piezoelectricity to the growth, development, and healing of bone structures[177] as described in Chapter 7, there are many other possible ways in which the piezoelectric properties of biopolymers could be related to biological phenomena, especially when one considers the large number of biological materials which exhibit piezoelectricity. When a finger is drawn lightly across the palm of one's hand a very distinct 'tingling' sensation is experienced in the palm. Taken at the molecular level, at the scale of a collagen molecule, the force per unit area produced by the 'barely touching' finger is of enormous magnitude, easily exceeding a value of 10^6 newtons. It is tempting to consider that the known piezoelectric property of collagen is directly associated with the 'tingling' sensation experienced. The conversion of mechanical vibrations into electrical signals that occurs in the sensation of hearing, and the sensory properties of hair, whiskers, and spines that adorn much of the animal kingdom, must also surely rely on piezoelectric effects? Such considerations as these, as well as the possibility that thermoreceptor, photoreceptor, mechanoreceptor, and certain membrane mechanisms are based on piezoelectric or pyroelectric effects, have been discussed by several authors.[178–181] Caserta and Cervigni[182] have proposed an interesting piezoelectric theory of enzymic catalysis which appears to relate directly to the electro-mechano-chemical principles of bioenergetic processes formulated by Green and Ji.[183,184] The basic premise of this theory is that enzyme molecules can be considered to represent composite piezoelectric semiconductors, which hopefully by now should not be considered a completely unrealistic proposal. As a result of these solid-state physical properties, enzymes (other than the transferases) are able to manipulate energy in such a way that they lower the energy barrier required in the substrate–product conversion process to be catalysed by the enzyme. The basic process involved is the amplification and channelling of the energy associated with the low frequency vibrational modes known to occur in enzyme structures.[185] As a result of the mechanical–electrical coupling arising from the piezoelectric effect, the acoustic mode energy W required to be channelled from the enzyme to the substrate for the substrate–product conversion is given by

$$W = (1 - K^2)W_s$$

where K is the piezoelectric coupling coefficient and W_s is the energy required for the substrate conversion in the absence of the enzyme. For values of K approaching unity then W will become small. The value for K depends on the local polarizability, and as mentioned in Chapter 8, the fact that water molecules have access to the active sites in crevice-like features of enzyme structures may be of relevance. The energy associated with the acoustic vibrations is at least 100 times lower than the energy required for enzymic reactions, so some form of amplification mechanism is required. Such an amplification mechanism is well known to occur in conventional piezoelectric semiconductors, where a small band of the acoustic thermal lattice vibrations can be selectively amplified by charge carriers whose drift velocity is greater than the sound velocity (see, e.g. reference 186). The band of frequencies that can be so amplified is centred around an optimal frequency f_m, given to a first approximation as

$$f_m = \frac{1}{2\pi}(\sigma q v_s^2 / \epsilon_0 \epsilon_r \mu kT)^{1/2}$$

where v_s is the sound velocity, which in biological materials is of the order 10^3 m s^{-1}.[187] At this frequency f_m the amplification is a maximum when the electric field stress E_m is given by

$$E_m = (v_s/\mu)\{1 + 2(\sigma \mu kT/\epsilon_0 \epsilon_r q v_s^2)^{1/2}\}$$

and we can see that the process is a highly selective one, and as such would not be out of place in the general scheme of selectivities and subtleties which characterize the animate from the inanimate. Using literature values[101] for the electronic conductivity σ, mobility μ, and relative permittivity ϵ_r, Caserta and Cervigni were able to show that the main conditions for the process of acoustic wave amplification to occur in enzymes can easily be fulfilled in realistic biological situations. Also, they were able to describe the way in which this amplified energy could be used for the enzyme-controlled substrate–product conversion process.

SUPERCONDUCTIVITY

The possibility that superconductive phenomena may play a biological rôle is at present a controversial subject in several laboratories. Unlike the situation for normal electronic conductors, electrons in a superconductor are not free to move independently of each other but exist as coupled electron pairs constrained to be in the same quantum state. As a result of this pairing-up of electrons, electron scattering effects are minimized with the result that the flow of electron current can occur without the generation of heat and hence with no electrical resistance. Such an effect could obviously have far-reaching consequences if it could be detected in biological systems at physiological temperatures. In conventional superconductors the electron pairing results from interactions between the electrons and lattice phonons. In 1964, Little proposed that suitably constructed organic polymeric systems would be capable of sustaining superconductivity as a result of an electron-pairing mechanism involving electron–exciton interactions.[188] Little estimated that such a polymer, consisting of a conducting conjugated hydrocarbon backbone and side chains in the form of highly polarizable dye molecules, would be superconducting up to temperatures of the order 2200 K. Such high temperatures would obviously not be realistic for organic systems for reasons of thermal stability, but this estimate of the critical temperature does serve to indicate that the concept of the existence of superconducting biopolymers at physiological temperatures lies well within the limit of the applicability of Little's theory. The existence of superconductivity in aromatic compounds was first speculated upon by London,[189] and Ladik et al.[190] have provided a theoretical basis for the superconductive behaviour of DNA. A theory has been developed to show that the critical temperature below which superconductivity can occur can be directly related to the ratio of the net internal energy of the superconducting system (paired electrons or whatever) to the entropy of that system.[191] This implies that if the entropy of such a system were reduced and its internal energy increased, then the critical temperature could be significantly increased. One biological area where superconductive phenomena may be of possible relevance is that of the central nervous system and of thought and memory processes in the brain. In this respect it is of interest to note that the mechanism of memory has been described in terms of a decreasing entropy with time in the higher-order centres of the brain.[192]

Despite intensive activity in many laboratories, no organic superconductor has been reported to have been synthesized to date along the lines suggested by Little. Some doubts exist regarding the theoretical basis for the concept,[193] but Little himself is still confident that high temperature superconductivity remains a possibility for certain organic structures.[194] At present, the highest known superconductivity transition temperature of 23.8 K is that exhibited by the Nb_3Ge alloy, and the possibility of synthesizing organic room temperature superconductors offers exciting technological consequences.

Experimental evidence for high temperature superconduction in biological molecules has been reported from at least two laboratories. In a series of magnetic and electrical measurements, Halpern and Wolf have produced extensive evidence to indicate that a class of bile cholates exhibit behaviour resembling superconductivity at high temperatures.[195-199] Below certain transition temperatures, ranging from 7.5 K for sodium dioxycholate to 277 K for sodium cholanate, a range of six cholates were found to behave like perfect diamagnets and the effect was shown not to be due to either ferri-electric, ferro-electric or capacitive effects. The superconductivity was deduced to occur in small domains included in the insulating bulk of the test samples, and so as to distinguish the effects from that normally found for the elemental superconductors, the cholates were designated a *fractional* or *Type III* superconductor. It was considered that the hydrophobic property of the closed four-ring structure common to the cholates, together with the hydrophilic property of their carboxylate groups, were responsible for the domain formations. When small amounts of water are introduced into such materials the hydrophobic groups will tend to cluster together, and on subsequent slow desiccation small micelles will be formed. Such micelles are considered by Halpern and Wolf to form superconducting domains. X-ray diffraction data for these cholate samples indicated that the superconductive-type behaviour was related to electronic rather than structural rearrangements[200] and as such give support to the conclusions of Halpern and Wolf.

Following the suggestion[201,202] that enzymes and other biological materials possess a metastable state with high dipole moment, Ahmed *et al.*[203,204] investigated the dielectric and magnetic susceptibility properties of the dilute solutions of lysozyme. It was found that magnetic fields of the order 0.6 tesla could produce very large changes (~30 per cent) in the relative permittivity of the solutions. To produce such an effect, the magnetic energy available in the solution would have to be sufficiently great to overcome thermal fluctuations, and this in turn required an increment in the magnitude of the magnetic susceptibility of the order 1.5×10^{-6}. This could only have resulted from the cooperative behaviour of at least $10^4 - 10^6$ lysozyme molecules, and was suggestive of superconductive behaviour. This conclusion was supported by measurements of the magnetic susceptibility of a dilute (0.011 per cent) lysozyme solution, which showed that the diamagnetic susceptibility increased with increasing magnetic field strength, until at a critical field of the order 0.6 tesla the susceptibility dropped sharply as would occur from the Meissner effect. It was suggested that in each lysozyme molecule there existed a small superconductive region with linear dimensions smaller than the London penetration depth, and that the collective, superconductor-like, phenomena resulted from the formation of clusters of these small regions. This is similar to the cluster model proposed for the bile cholates. It was also suggested that not only the lysozyme molecules, but also water and ions may have played a rôle in the establishment of the superconducting regions. This could have given rise to a layered structure and also to an a.c. Josephson effect coupled to the electric vibrations thought to occur in biomolecular structures.[205] Later measurements in the same laboratory indicated that magnetic and radio-frequency fields in the frequency range 50 kHz to 300 MHz could alter the enzyme activity of dilute lysozyme solutions.[206] The reported magnetic-field-induced changes in the diamagnetic susceptibility of lysozyme solutions have not been reproduced in other laboratories,[207,208] although it should be stated that the experimental conditions were not exactly those of Ahmed *et al.* and this may be an important factor for the reproducibility of their results. Pohl and Pethig in their investigations[207] found diamagnetic-type effects associated with liquid—air and liquid—glass interfaces, and certain experimental aspects of the work by Ahmed *et al.* would suggest that their interesting results may have involved interfacial effects.

Other indirect evidence to suggest a biological rôle for superconductivity has been discussed by Cope,[209,210] following the suggestion by Ginzburg[193,211] that high temperature superconduction may be expected in a sandwich consisting of a thin conductive film or filament adjacent to a dielectric layer. Cope considers that such superconducting sandwiches may be ubiquitous in biological systems in the form of thin layers of protein and unsaturated lipids and hydrocarbon ring structures (conducting layer) adjacent to layers of water (polarizable dielectric layer). In a search of the biological literature, Cope found six sets of biological data which exhibit behaviour expected from the hypothesis that the rate of the biological process is controlled by single electron tunnelling between micro-regions of such superconducting sandwiches.[209] The biological processes concerned were:

(1) Impulse conduction velocity in frog sciatic nerve.
(2) Junctional electrical resistance of crayfish nerve.
(3) Rate of growth of *E. coli* in trypticase broth.
(4) Rate of growth of *E. coli* in beef peptone broth.
(5) Rate of CO_2 production by growing yeast.
(6) Rate of division of sea urchin eggs.

All of these processes were characterized by the square of the associated activation energies being a linear function of temperature over a moderate range of physiological temperatures. Such an effect can be well described in terms of a model where the rate-limited biological process involves a superconducting tunnelling current of single electrons and/or electron pairs (the Josephson current). It was suggested that as there was an apparent association of superconduction with growth, then the superconductive micro-regions may have been individual purine and pyrimidine rings of DNA and RNA with electron tunnelling between rings along the length of the polymer chain. It was further suggested that superconductive Josephson junctions in living systems may provide a physical mechanism with more than enough sensitivity to explain how many biological organisms are able to respond to weak magnetic fields.[206(b)] Antonowicz[212,213] has observed room temperature current fluctuations, induced magnetically, in thin aluminium—carbon—aluminium sandwiches similar to that expected from the Josephson effect of superconducting electron-pair tunnelling. Microwave irradiation of these carbon films at room temperature caused the appearance of steps in the voltage—current curves, a result again consistent with a Josephson effect. These results support Cope's interesting ideas, and they also led him to suggest that they were also relevant to an understanding of non-thermal biological effects of microwaves.[210]

MELANIN

The melanins are black natural pigments widely present in biological systems, and can be found in skin and hair for example. They have highly stable polymeric chemical structures consisting of condensed, unsaturated, ring structures. They exist in an amorphous rather than crystalline state, and contain a large number of free unpaired electrons of density around 10^{25} m^{-3}. In biological systems melanin does not usually occur in a free form, but is polymerized into the melanosome structure which contains the pigment, lipid, and protein. Melanin in free form can be extracted from malanoma tumour material, and synthetic melanin can be polymerized by the autoxidation of tyrosine or L-dopa (3,4-dihydroxy-phenyl-L-alanine).[214] The melanins have generally been considered to behave as photon absorbers to screen biological systems from harmful ultraviolet radiation and to act as trapping sites for free radical species.[215] This cannot be its only biological rôle however, since apart from occurring in skin, hair, and the retina of the eye, melanin pigmentation can be found in non-illuminated areas of the body such as in the midbrain and the inner ear.[216,217] The loss of melanin from the brain has been related to the onset of Parkinson's disease,[217-219] and has been implicated in such disorders as schizophrenia[218,219] and deafness.[217]

The electronic properties of melanin have been studied in some detail, both theoretically and experimentally. From theoretical considerations of the electronic structure of the monomer unit indole-5,6-quinone, the melanins appear to be able to act as powerful electron acceptors.[220] Blois *et al.*[221-223] investigated the semiconduction and e.s.r. properties of pure melanins and found that the conductivity increased markedly with increasing applied electric field stress, even at low fields, and that no threshold occurred at low temperatures for the e.s.r. signals. Rosenberg *et al.*[11,20] found that the conductivity of the melanins had a protonic as well as electronic component. The properties found by Blois *et al.* led McGinness to consider that the semiconductivity of the melanins can best be understood in terms of the theories developed for amorphous solids, and in particular in terms of the charge carrier 'mobility gap' conduction in the extended and localized electronic states as shown in Figure 8.9.[224] In the theories describing electronic conduction in amorphous solids, particular importance is placed on the effects of the coupling of the electronic states with the lattice vibrational modes (phonons). The black appearance of melanin would suggest that such an electron–phonon coupling mechanism is particularly efficient for this material.[225] The relatively featureless absorption spectrum from the far ultraviolet to the infrared means that the melanins are 'black' over a wider range of energies than just that of the visible spectrum, and the light absorbed over this range is not reradiated but is capable of being converted into heat in the form of vibrations of the atomic structure. The underlying electron–phonon interactions could also be used to absorb the energy of the excited states of neighbouring molecules, and as such this could form another biological rôle for melanin, namely as a device for deactivating potentially disruptive electronically excited molecules.[225]

To extend the analogy and similarity to amorphous electronic solids, the electrical switching behaviour of the melanins was also studied.[214,226] Melanins, of synthetic or natural form, exhibited memory-switching effects with an electric field threshold for the transition from the resistive OFF to conductive ON state occurring at fields of the order 3×10^4 V m^{-1} which is about two to three orders of magnitude lower than that usually found for elemental amorphous materials. No protonic component of the conductivity could be found for the conducting ON state, where the conductivity of the melanin samples was of the order $1-10^{-1}$ mho m^{-1}. The nature of the switching effect and the threshold field value was found to be dependent on the hydration and temperature of the test samples. Switching effects were also explored, without success, in bovine serum albumin, myoglobin, lecithin, polytryptophan, bilirubin, and oxidized cholesterol. Equine cytochrome-c was found to switch from the OFF to the ON state at a field of 4×10^{7} V m^{-1}. Culp et al.[227] confirmed the existence of switching effects in synthetic melanin, although they observed threshold switching and only a pseudomemory effect. The observed switching behaviour was considered to be possibly relevant to melanin's active involvement in biological systems and also in such disorders as Parkinsonism, schizophrenia, and deafness.[214] It should be mentioned here, however, that following the report[228] of memory switching in molecular solids and charge-transfer complexes, Garrett et al.[229] investigated the switching and other high-field effects in various organic films. Whereas switching effects were found, the mechanism involved was considered to involve classical electrical failure and breakdown effects, resulting in irreversible physical damage to the test samples. This would suggest that some caution should perhaps be exercised before considering the possible rôle of electrical switching effects in biological systems. To counteract such a negative attitude it should also be remembered that large electrical fields of value greater than the 3×10^{4} V m^{-1} found to switch the melanin to a highly conductive state do occur across membranes and in other biological systems. For this reason further studies of high field conduction effects in biopolymers may provide rewarding results and lead to a better understanding of the possible relevance of electronic conduction effects in biological systems.

Low temperature specific heat measurements have also been made on melanin samples and tumour melanosomes,[230] and an interesting anomalous feature was observed at 1.9 K which was considered to be possibly associated with a magnetic transition involving unpaired electrons. Such low temperature specific heat measurements should be extended to the study of biopolymers in general, as they should lead to important new knowledge regarding the extent, nature, and electron occupancy of the energy states in these materials.

CONCLUDING REMARKS

In this chapter many examples have been given to show that biological materials exhibit solid-state electronic properties. Much more detailed work remains to be accomplished before the electronic conduction properties can be clearly characterized and understood, but it is now not unrealistic to consider that biopolymers do exhibit electronic effects associated with delocalized mobile electrons. The challenge now is to show that some of the various subtleties and sensitivities that characterize the miracle of the living state have as a basis subtle phenomena involving mobile electrons. Cope has concluded on theoretical grounds that the activity of membrane-bound enzymes such as cytochrome-c oxidase can be well understood by assuming that the enzyme is an electronic conducting particle,[231,232] Indirect evidence, apart from the conduction studies described in this chapter, has been found to show the existence of electronic conduction in biopolymers. Electron spin resonance studies[233] have indicated that unpaired electrons produced by X-ray ionization of solid proteins diffuse through the protein structure to eventually become localized on disulphide bridges, and there is also evidence that such electrons can diffuse across molecular boundaries.[234] Pulse-radiolysis[235] and flash-photolysis[236] studies on lysozyme indicate that electrons migrate from a tryptophan residue to a disulphide bridge, and that some of these electrons must travel through the protein structure. Similar conclusions regarding intra-protein free electron diffusion have also been obtained for cytochrome-c by studying its rate of reduction by hydrated electrons[237] and OH radicals,[238] and nuclear magnetic resonance measurements have given indirect evidence that long-range electron transfer occurs in the polypeptide portion of cytochrome-c.[239,240] The only measurement of electronic conduction effects in living tissue appears to be that obtained by Digby for the cuticle of various crustacea[241,242] and by Pohl and Sauer[243] for the salivary gland of the Lone Star tick (*Amblyomna americanum* L.). There is evidence to suggest that the electronic conduction properties of biopolymers may be involved in preventing the initiation of adverse effects on blood components, such as loss of compatibility and coagulation.[244,245]

Although more direct measurements of the electronic properties of intact biological systems are required, measurements on isolated biopolymers, and especially biopolymeric charge-transfer complexes, will continue to provide important basic data. Evidence is now being collected in several laboratories that cells have an associated 'cytoskeleton' composed of a network of protein fibres, like tubulin and fibrin, that permeate the cytoplasm and possibly even penetrate right through the cellular membrane. Measurements on such protein systems would be of obvious interest. The glycophorins, being proteins known to completely span across cellular membranes, should also be studied. Of particular interest will be studies of such proteins that have been incorporated into an artificial bilipid membrane, and where carefully controlled charge-transfer and redox reactions can be performed on the hydrophilic ends of the protein that protrude into the aqueous medium on either side of the membrane. New measurement techniques must also be developed, especially those capable of dealing with fully hydrated biological samples. Conventional direct current and low frequency alternating current measurements on dried polycrystalline powders can only ever be of limited value, important though they are in establishing a firm base on which to develop future work. More accurate methods must also be developed which can accurately distinguish between electronic, protonic, and ionic conduction effects. Bardelmeyer, for example, has proposed that collagen exhibits electronic conduction below a hydration level of 20 wt per cent, protonic conduction in the range 20–45 wt per cent, and beyond about 65 wt per cent water content small inorganic ions dominate any electrical conductivity.[246] It will be of great interest to determine how the electronic conduction component varies throughout such a hydration range. This in turn will require accurate determinations of the ionic conductivity, which for low ionic concentrations can present serious problems.[247] Complications can also arise as a result of the presence of dissolved gases in the samples, especially if the samples have been compacted at great pressure without high-vacuum conditions. Such dissolved gas molecules may capture free electrons and diffuse as ions through the sample.[248] Measurements on well-characterized ionic conductors may help the experimenter gain experience of such effects, and one such model material could be the solid solutions of perchlorate salts in polyacrylonitrile matrices.[249] Also, the extensive work that now exists on the one-dimensional electronic conductors such as polymeric sulphur nitride[250] will have direct relevance to some of the electronic properties that may be expected for linear polypeptide chains. A knowledge of both the electronic and ionic conduction properties of biological materials will be essential, for example, for the elucidation of the new unifying model of bioenergetics proposed by Blondin and Green,[251] which describes the essential feature of energy coupling in living cells in terms of pairs of moving electrons and ions.

Electron transfer and charge-separating phenomena are known to be essential features in many biological processes. It will certainly not be the case that rapid electron transfer within broad energy bands will always be involved in such processes. Indeed, there is a case for supposing that low electronic mobility may be essential for maintaining localization of the electron's 'wave packet' during transport, as would be appropriate for the microminiaturization that characterizes biological systems.[252] Quantum mechanical electron tunnelling mechanisms have been found to give an adequate explanation of the catalytic activity of various biological electron transport systems including complex I of the mitochondrial electron transport chain,[253,254] although certain features of the optical absorption band of the $Fe-S_{II}$ centre in complex I could be consistent with a significant electron mobility associated with a semiconduction mechanism.[254] The temperature dependence of the electron transport reactions has been explained in terms of the motions of the protein molecules to which the redox functional groups are bound.[253] It should be mentioned here that the theoretical energy band calculations described in Chapter 8 for the proteins are essentially based on infinite structures. For the realistic situation of finite-sized protein molecules, suitable boundary conditions must be introduced, which will result in the appearance of surface energy states. The activation of electrons into and out of these surface states could certainly involve tunnelling mechanisms influenced by conformational motions of the protein molecules. When the photosynthetic *Bacterium chromatium* is optically excited, electron transfer is known to occur from a cytochrome molecule to the chlorophyll reaction centre. This process has been studied in detail[255] and the electron transfer process was found to be characterized by a temperature-dependent activation energy. Below around 100 K the activation energy was lower than 3.5×10^{-3} eV and above this temperature the value rose to 0.14 eV. Jortner[256] has analysed this data in terms of a non-adiabatic multiphonon process which accurately describes the results as a temperature-independent tunnelling process at low temperatures and as a phonon-assisted activated tunnelling mechanism at the higher temperatures. Experimental evidence from optical absorption studies has been presented[257] to show that the fundamental mechanism of electron transfer between the two components of cytochrome-c hexacyanide proceeds via electron tunnelling. Furthermore, from a comparison of the tunnelling parameters derived from measured rates and enthalpies of photosynthetic systems, it was concluded that a tunnelling mechanism is the basis of electron transfer in photosynthesis and mitochondrial respiration. Apart from electron tunnelling mechanisms, it is also possible that proton tunnelling may occur in biopolymers. Yano et al.[276] interpreted their dielectric loss measurements on polyethylene at temperatures below 4.2 K in terms of a proton tunnelling mechanism.

An aspect of the semiconduction data for biopolymers and many organic systems that has aroused curiosity and speculation is the so-called 'compensation effect' whereby for many of the materials a linear relationship is observed between $\log \sigma_0$ and $\mathcal{E}$ of equation (9.1).[5,258] In other words, a large value for the conductivity activation energy is compensated for by an increase in either or both the number of conducting charge carriers and charge carrier mobility. However, for the larger activation energies the observed conductivity can be considered to be anomalously large by comparison with that expected by conventional semiconductor theory. A mechanism was proposed[259] to account for these effects in terms of an interaction between the mobile electrons and the lattice vibrational motions. It was envisaged that on changing its electronic state during the process of conduction, the electron's motion would give rise to an activation entropy associated with a change in the lattice vibration frequency, arising from conformational changes for example. This entropy term would appear in the σ_0 of equation (9.1) and could lead to the observed compensation law and conductivity enhancement. Kaplan and Mahanti[260] have investigated this proposal further, and conclude that the observed effects can only arise if the conformational changes required for the large entropy enhancement are associated with localized holes, and the large mobility required for the conductivity enhancement results from delocalized electrons. This model would also be valid for localized electrons and mobile holes.

Finally, interest is now growing in the concept that high frequency ($10^{11} \sim 10^{12}$ Hz) cooperative vibrational models are of fundamental importance in understanding the organized collective behaviour of biological systems. This concept was introduced by Fröhlich[261-263] and basically it envisages that longitudinal electric modes can exist in many biological systems, which are established by oscillations involving the cellular membranes, delocalized electrons and hydrogen bonds present in biomacromolecules such as DNA and proteins. Fröhlich was able to show[264] that if energy is supplied to these oscillating dipolar modes above a critical rate, then the oscillating subunits will condense into the lowest energy state and give rise to a very strong and coherent excitation of a single mode. Such a condensation of cellular vibrational modes could serve as a method for the efficient storage of energy, and also as a channel for selective biological processes. This concept could also help solve such puzzles as how biological systems, unlike classical physical systems, are able to remain relatively stable whilst at the same time be far from thermal equilibrium. It could also explain the apparent correlated behaviour of cells in governing the growth and final size and shape of the various biostructures. These ideas have received close attention by other theoreticians[265-267] and have gained experimental support from electromagnetic radiation experiments on cells.[268,269] The appearance and breakdown of the coherent modes have also been related to cellular division and the occurrence of cancer.[269,270]

The appearance of delocalized charge carriers in biomacromolecules, produced by charge-transfer or high field charge injection effects for example, can be envisaged as leading to a wide range of submolecular electronic subtleties whose relevance has largely been overlooked by the biological sciences. The electronic effects involved could include not only those relatively simple conduction properties outlined in this Chapter, but also such phenomena as solitons, charge density waves, spin density waves, and the other types of collective electronic modes currently under extensive study in organic conductors, polymeric charge-transfer systems, and in the extremely interesting inorganic sulphur nitride polymer (e.g. see reference 278). Such materials have molecular structures that are much simpler than those of the polypeptides and polynucleotides, and we can expect that some of the electronic concepts applicable to the biopolymers will have first to be fully established in the simpler organic and inorganic systems. One area where the concept of collective electronic modes may be applicable is that concerning the functioning of neuronal networks such as those present in the human brain. A complete investigation of the processes involved in thought and memory may well prove to be the last frontier for the bio-medical sciences, and a good indication of the possible direction such researches may follow has been given by Little.[279] He has been able to show, using certain plausible assumptions, that the problem of the existence of persistent states in a neuronal network closely resembles that of an interacting system of spinning charges. Such a problem is well known in the field of statistical mechanics and solid state physics. It appears that the existence of persistent states in the neuronal network corresponds to the occurrence of long range order in the spin problem, which in turn can be related to the existence of a degeneracy of the maximum eigenvalue of a matrix. This matrix is determined by the topology of the neuronal network, the size and nature of the synaptic junctions and the various electrochemical potentials in the brain. Of significance regarding the plausibility of determining this matrix is the conclusion that whilst the number of possible states in the brain is of the order of 2^N, where N is the number of neurons ($\sim 10^{10}$ in the human brain), the number of states which determine the long term behaviour is a very much smaller number. Of particular interest is the fact that the persistent states are distinguished by the property that a coherence or correlation exists between the neurons throughout the entire brain or large portions of it. These states are thus a property of the brain as a whole rather than a localizable entity. This is exactly the same concept as that regarding the existence of delocalized electronic states in protein structures, where the valence and conduction bands can be a property of practically the whole protein molecule. It should be possible to carry some of the protein conduction concepts directly into the study of the properties of the brain. Extensive networks of interacting protein fibrils are known to exist in the brain and it is of significance to note that we think with our 'grey matter'. This dark colouration of the protein structures involved in memory and thought processes will almost certainly be found to be related to the electronic processes controlling such functions.

The attainment of a deeper understanding of the basic functioning of the processes of the living state at the submolecular level is an important challenge requiring the collaboration of all the sciences. Progress in such interdisciplinary research will lead to new insights in biology, and along with it advances in all the fields of medicine. Szent-Györgyi has written:[129]

> 'We will really approach the understanding of life when all structures and functions, all levels from the electronic to the supramolecular, will merge into one single unit. Until then our distinguishing between structure and function, classic chemical reactions and quantum mechanics, or the sub- and supra-molecular, only shows the limited nature of our approach and understanding'.

References

1. S. Baxter and A. B. D. Cassie, *Nature*, **148**, 408 (1941).
2. S. Baxter, *Trans. Faraday Soc.*, **34**, 207 (1943).
3. A. Szent-Györgyi, *Nature*, **157**, 875 (1946).
4. F. Gutmann and L. E. Lyons, *Organic Semiconductors*, pp. 492–504. J. Wiley and Sons, New York (1967).
5. D. D. Eley, in *Organic Semiconducting Polymers*, pp. 259–94. Ed. J. E. Katon, Edward Arnold, London (1968).
6. L. I. Boguslavskii and A. V. Vannikov, *Organic Semiconductors and Biopolymers*, Plenum Press, New York (1970).
7. E. O. Forster and A. P. Minton, in *Physical Methods in Macromolecular Chemistry*, Vol. 2, pp. 185–237. Ed. B. Carroll, Marcel Dekker, New York (1972).
8. G. King and J. A. Medley, *J. Colloid Sci.*, **4**, 9 (1949).
9. P. Taylor, *Faraday Soc. Disc.*, **27**, 237 (1959).
10. L. Vieth, *Kolloid Z.*, **152**, 36 (1957).
11. M. R. Powell and B. Rosenberg, *J. Bioenergetics*, **1**, 493 (1970).
12. B. Rosenberg, *Nature*, **193**, 364 (1962).
13. S. Maricic, G. Pifat, and V. Pravdic, *Biochem. Biophys. Acta*, **79**, 293 (1964).
14. E. J. Murphy, *Can. J. Phys.*, **41**, 1022 (1963).
15. E. J. Murphy, *Ann. N.Y. Acad. Sci.*, **118**, 725 (1965).
16. D. A. Seanor, *J. Polym. Sci. A-2*, **6**, 463 (1968).
17. M. R. Powell and B. Rosenberg, *Biopolymers*, **9**, 1403 (1970).
18. B. Rosenberg, *J. Chem. Phys.*, **36**, 816 (1962).
19. N. F. Mott and M. J. Littleton, *Trans. Faraday Soc.*, **34**, 485 (1938).
20. B. Rosenberg and E. Postow, *Ann. N.Y. Acad. Sci.*, **158**, 161 (1969).
21. D. D. Eley and R. B. Leslie, in *Electronic Aspects of Biochemistry*, p. 105. Ed. B. Pullman, Academic Press, New York (1964).
22. Y. Miyoshi and N. Saito, *J. Phys. Soc. Japan*, **24**, 1007 (1968).
23. D. D. Eley and R. Pethig, *Disc. Faraday Soc.*, **51**, 164 (1971).
24. K. M. C. Davis, D. D. Eley, and R. S. Snart, *Nature*, **188**, 724 (1960).
25. D. D. Eley and R. B. Leslie, *Advan. Chem. Phys.*, **7**, 238 (1964).

350

26. D. D. Eley and R. B. Leslie, *Trans. Faraday Soc.*, **62**, 1002 (1966).
27. D. D. Eley and R. S. Snart, *Biochim. Biophys. Acta*, **102**, 379 (1965).
28. D. D. Eley and D. I. Spivey, *Nature*, **188**, 725 (1960).
29. N. V. Riehl, *Z. Fiz. Khim. (U.R.S.S.)*, **29**, 1537 (1955).
30. S. Bone, P. R. C. Gascoyne, and R. Pethig, *J.C.S. Faraday I*, **73**, 160 (1977).
31. D. D. Eley and R. Pethig, *J. Bioenergetics*, **2**, 39 (1971).
32. E. M. Trukhan, *Biofizika*, **11**, 412 (1966).
33. D. D. Eley and R. Pethig, in *Conduction in Low-Mobility Materials*, pp. 397–402. Eds. N. Klein, D. S. Tannhauser, and M. Pollak, Taylor and Francis, London (1971).
34. M. H. Cardew and D. D. Eley, *Disc. Faraday Soc.*, **27**, 115 (1959).
35. D. D. Eley and R. Pethig, *J. Bioenergetics*, **1**, 109 (1970).
36. D. D. Eley, G. D. Parfitt, M. J. Perry, and D. H. Taysum, *Trans. Faraday Soc.*, **49**, 79 (1953).
37. A. V. Vannikov and L. I. Boguslavskii, *Biofizika*, **14**, 421 (1969).
38. D. D. Eley and D. I. Spivey, *Trans. Faraday Soc.*, **56**, 1432 (1960).
39. D. D. Eley and P. W. Thomas, *Trans. Faraday Soc.*, **64**, 2459 (1968).
40. M. Pollak, *Phil. Mag.*, **23**, 519 (1971).
41. A. K. Jonscher, *J. Non-Cryst. Solids*, **8–10**, 293 (1972).
42. M. Pollak and T. H. Geballe, *Phys. Rev.*, **122**, 1745 (1961).
43. R. Pethig, *3rd Int. Symp. Chem. Org. Solid State, Glasgow*, pp. 95–100 (1972).
44. D. D. Eley, S. J. Rooker, M. Landon, and R. J. Mayer, *Biochem. Soc. Trans. 534th Meeting, Nottingham*, **1**, 62 (1973).
45. R. Pethig and D. Hayward, *Phys. Status Solidi*, **24**, K23 (1974); **32**, K177 (1975).
46. T. J. Lewis and R. Pethig, in *Excited States of Biological Molecules*, pp. 342–52. Ed. J. B. Birks, Wiley-Interscience, New York (1976).
47. P. Carnochan and R. Pethig, *J.C.S. Faraday I*, **72**, 2355 (1976).
48. S. Bone and R. Pethig, *J.C.S. Faraday I*, **74**, 720 (1978).
49. R. Pethig, *Z. Naturforsch.*, **33a**, 389 (1978).
50. T. J. Lewis, in *Dielectric and Related Molecular Processes*, Vol. 3, pp. 187–218. The Chemical Society, London (1977).
51. V. P. Tomaselli and M. H. Shamos, *Biopolymers*, **13**, 2423 (1974).
52. R. B. Adler, A. C. Smith, and R. L. Longini, *Introduction to Semiconductor Physics*, p. 46, J. Wiley & Sons, New York (1964).
53. R. A. Smith, *Semiconductors*, p. 91, Cambridge Press (1964).
54. Y. Nakahara, K. Kimura, and H. Inokuchi, *Chem. Phys. Lett.*, **47**, 251 (1977).
55. R. Pethig and A. Szent-Györgyi, *Proc. Nat. Acad. Sci. USA*, **74**, 226 (1977).
56. S. Bone, T. J. Lewis, R. Pethig, and A. Szent-Györgyi, *Proc. Nat. Acad. Sci. USA*, **75**, 315 (1978).
57. R. Pethig, *Int. J. Quant. Chem. Quant. Biol. Symp.*, **5**, (1978) (In Print).
58. P. Otto, S. Suhai, and J. Ladik, *Int. J. Quant. Chem. Quant. Biol. Symp.*, **4**, 451 (1977).
59. W. E. Spear and P. G. LeComber, *Proc. 7th Int. Conf. Amorphous and Liquid Semiconductors, Edinburgh (1977)*, p. 309.
60. R. J. P. Williams, in *Current Topics in Bioenergetics*, Vol. 3, pp. 79–156. Ed. D. Rao Sanadi, Academic Press, New York (1969).
61. J. Duchesne, J. Depireux, A. Bertinchamps, N. Cornet, and J. Van Der Kaa, *Nature*, **188**, 405 (1960).
62. D. D. Eley and D. I. Spivey, *Trans. Faraday Soc.*, **470**, 411 (1962).
63. R. S. Snart, *Trans. Faraday Soc.*, **59**, 754 (1963).
64. C. T. O'Konski, P. Moser, and M. Shirai, in *Quantum Aspects of Polypeptides*

and Polynucleotides, p. 479. Ed. M. Weisspluth, John Wiley Interscience, New York (1963).
65. G. Mesnard and D. Vasilescu, *C. R. Acad. Sci., Paris*, **257**, 4177 (1963).
66. M. E. Burnel, D. D. Eley, and V. Subramanyan, *Ann. N.Y. Acad. Sci.*, **158**, 191 (1969).
67. C. Y. Liang and E. G. Scalco, *J. Chem. Phys.*, **40**, 919 (1964).
68. E. Subertova, V. Prosser, and J. Drobnik, *Biopolymers*, **8**, 421 (1969).
69. D. D. Eley and R. B. Leslie, *Nature*, **197**, 898 (1963).
70. J. Ladik, *Int. J. Quant. Chem. Quant. Biol. Symp.*, **1**, 65 (1974); **2**, 133 (1975).
71. G. Mesnard, D. Vasilescu, and Y. Mathon, *Phys. Stat. Solidi*, **11**, K137 (1965).
72. S. Basu and W. J. Moore, *J. Phys. Chem.*, **67**, 1563 (1963).
73. M. E. Burnel, D. D. Eley, and V. Subramanyan, in *The Purines, Theory and Experiment*, p. 244. Jerusalem Symp. Quant. Chem. and Biochem IV., Eds. E. D. Berman and B. Pullman, Israel Academy of Sciences and Humanities (1972).
74. S. Maricic, G. Pifat, and V. Pravdic, *Biochim. Biophys. Acta*, **79**, 293 (1964).
75. S. Maricic and G. Pifat, *Abh. dt. Akad. Wiss, Berl.*, **4**, 63 (1966).
76. C. T. O'Konski, *Rev. Mod. Phys.*, **35**, 723 (1963).
77. N. Riehl, *Kolloid Z.*, **151**, 66 (1957).
78. P. Douzou and J. M. Thullier, *J. Chim. Phys.*, **57**, 96 (1960).
79. G. M. Spruch and C. Perskin, *Science*, **163**, 1350 (1969).
80. R. C. Nelson, *J. Chem. Phys.*, **39**, 112 (1963).
81. Yu. A. Vladimirov and K. N. Timofeyev, *Biofizika*, **11**, 33 (1966).
82. C. Y. Liang, *J. Chem. Phys.*, **43**, 1835 (1965).
83. D. D. Eley and E. Metcalfe, *Nature*, **239**, 344 (1972).
84. D. D. Eley, E. Metcalfe, and R. J. Mayer, *Biochem. Soc. Trans. 534th Meeting, Nottingham*, **1**, 59 (1973).
85. E. Subertova, V. Prosser, and J. Drobnik, *Biopolymers*, **8**, 421 (1969).
86. R. S. Snart, *Biopolymers*, **6**, 293 (1968).
87. B. Gähwiler, I. Zschokke-Gränacher, E. Baldinger, and H. Lüthy, *Phys. Med. Biol.*, **15**, 701 (1970).
88. E. H. Hall, *Am. J. Math.*, **2**, 287 (1879).
89. W. Shockley, *Electrons and Holes in Semiconductors*, p. 339. van Nostrand, New York (1950).
90. L. Friedman, *Phys. Rev.*, **133**, A1668 (1964).
91. D. Emin, *Proc. 7th Int. Conf. Amorphous and Liquid Semiconductors, Edinburgh (1977)*, pp. 249–60.
92. R. R. Rau and M. E. Caspari, *Phys. Rev.*, **100**, 632 (1955).
93. B. Donovan and J. Webster, *Proc. Phys. Soc.*, **79**, 46 (1962); **81**, 90 (1963).
94. A. M. Portis and D. Teaney, *J. Appl. Phys.*, **29**, 1692 (1958).
95. E. M. Trukhan, *Biofizika*, **11**, 412 (1966).
96. R. A. Bogomolni and M. P. Klein, *Nature*, **258**, 88 (1975).
97. M. M. Sayed, C. R. Westgate, and J. H. Perlstein, *Electron. Lett.*, **9**, 529 (1973); **10**, 178 (1974).
98. M. M. Sayed and C. R. Westgate, *Rev. Sci. Instrum.*, **46**, 1074, 1080 (1975).
99. N. P. Ong and A. M. Portis, *Phys. Rev. B*, **15**, 1782 (1977).
100. R. Pethig, *J. Biol. Phys.*, **1**, 193 (1973).
101. D. D. Eley, R. J. Mayer, and R. Pethig, *J. Bioenergetics*, **3**, 271 (1972); **4**, 389 (1972).
102. S. Y. Chai and P. O. Vogelhut, *J. Appl. Phys.*, **38**, 613 (1967).
103. J. R. Fletcher, *J. Phys. E*, **9**, 481 (1976).
104. R. Pethig and R. B. South, *IEEE Trans. Instr. Meas.*, **1M23**, 460 (1975).

352

105. B. Pullman, private communication.
106. S. Winstein and H. J. Lucas, *J. Am. Chem. Soc.*, **60**, 836 (1938).
107. H. A. Benesi and J. H. Hildebrand, *J. Am. Chem. Soc.*, **70**, 2832 (1948).
108. G. M. Bennett and G. H. Willis, *J. Chem. Soc.*, **1929**, 256.
109. G. Briegleb, *Z. Physik. Chem.*, **B16**, 249 (1932).
110. G. Briegleb, *Zwischenmolekulare Kräfte*, G. Braun, Karlsruhe (1949).
111. R. E. Gibson and O. E. Loeffler, *J. Am. Chem. Soc.*, **62**, 1324 (1940).
112. D. L. Hammick and R. B. M. Yule, *J. Chem. Soc.*, **1940**, 1539.
113. H. M. Powell and G. Huse, *Nature*, **144**, 77 (1939).
114. H. M. Powell and G. Huse, *J. Chem. Soc.*, **1943**, 435.
115. J. Weiss, *J. Chem. Soc.*, **1942**, 245.
116. W. Brackmann, *Rec. Trav. Chim.*, **68**, 147 (1949).
117. R. S. Mulliken, *J. Am. Chem. Soc.*, **74**, 811 (1952).
118. R. S. Mulliken, *J. Phys. Chem.*, **56**, 801 (1952).
119. G. N. Lewis, *Valency and the Structure of Atoms and Molecules*, Reinhold, New York (1923).
120. G. Briegleb, *Elektronen-Donator-Acceptor Komplexe*, Springer-Verlag, Berlin (1961).
121. L. J. Andrews and R. M. Keefer, *Molecular Complexes in Organic Chemistry*, Holden-Day Inc., San Francisco (1964).
122. R. Foster, *Organic Charge-Transfer Complexes*, Academic Press, London (1969).
123. R. S. Mulliken and W. B. Person, *Molecular Complexes*, Wiley Interscience, New York (1969).
124. M. A. Slifkin, *Charge Transfer Interactions of Biomolecules*, Academic Press, London (1971).
125. R. Foster (Editor), *Molecular Complexes*, Vols. 1 & 2, Paul Elek (Scientific Books) Ltd., London (1973, 74).
126. J. Yarwood (Editor), *Spectroscopy and Structure of Molecular Complexes*, Plenum Press, London (1973).
127. R. Foster (Editor), *Molecular Association*, Vol. 1, Academic Press, London (1975).
128. A. Szent-Györgyi, *Bioenergetics*, Academic Press, New York (1957).
129. A. Szent-Györgyi, *Introduction to a Submolecular Biology*, Academic Press, New York (1961).
130. A. Szent-Györgyi, *Bioelectronics*, Academic Press, New York (1968).
131. A. Szent-Györgyi, *Electronic Biology and Cancer*, Marcel Dekker, New York (1976).
132. A. Szent-Györgyi, *Science*, **124**, 873 (1956).
133. A. Szent-Györgyi, *Radiation Research, Supplement 2*, 4 (1960).
134. A. Szent-Györgyi, *J. Theoret. Biol.*, **1**, 75 (1961).
135. A. Szent-Györgyi, *Int. J. Quantum Chem.*, **3**, 157 (1969).
136. Reference 122, Table 8.2, p. 233.
137. K. O. Strømme, *Acta Chem. Scand.*, **13**, 268 (1959).
138. O. Hassel and H. Hope, *Acta Chem. Scand.*, **14**, 341 (1960).
139. C. A. Fyfe, reference 20, Vol. 1, pp. 209–99.
140. C. A. Fyfe, *J.C.S. Faraday II*, **70**, 1642 (1974).
141. R. S. Mulliken and W. B. Person, *Ann. Rev. Phys. Chem.*, **13**, 107 (1962).
142. E. M. Voigt and C. Reid, *J. Am. Chem. Soc.*, **86**, 3930 (1964).
143. J. Collin, *J. Electrochem.*, **64**, 936 (1960).
144. M. W. Hanna and J. L. Lippert, reference 125, Vol. 1, pp. 1–48.
145. A. Ottenberg, C. J. Hoffman, and J. Osiecki, *J. Chem. Phys.*, **38**, 1898 (1963).
146. A. Rembaum, *J. Polymer Sci. C.*, **29**, 157 (1970).

147. A. M. Hermann and A. Rembaum, *J. Polymer Sci. C.*, **26**, 107 (1967).
148. A. M. Hermann and A. Rembaum, *J. Polymer Sci. B.*, **26**, 445 (1967).
149. F. Gutmann and L. E. Lyons, *Organic Semiconductors*, pp. 460–67, 720–31. J. Wiley, New York (1967).
150. A. H. Price, reference 126, Chapter 7.
151. N. Kulevsky, reference 127, Chapter 2.
152. V. M. Vincent and J. D. Wright, *J.C.S. Faraday I*, **70** , 58 (1974).
153. R. Pethig and V. Soni, *J.C.S. Faraday I*, **71**, 1534 (1975).
154. J. B. Birks and M. A. Slifkin, *Nature*, **197**, 42 (1963).
155. R. S. Snart, *Biopolymers*, **6**, 73 (1968).
156. J. Nagchaudhuri and R. Suri, *Can. J. Chem.*, **54**, 59 (1976).
157. A. Pullman and B. Pullman in *Quantum Theory of Atoms, Molecules, and the Solid State*, Ed. P. Löwdin, Academic Press, New York (1966).
158. *Molecular Associations in Biology*, Ed. B. Pullman, Academic Press, New York (1968).
159. A. J. P. Martin, *Proc. Phys. Soc.*, **53**, 186 (1941).
160. E. Fukada, R. L. Zimmerman, and S. Mascarenhas, *Biochem. Biophys. Res. Commun.*, **62**, 415 (1975).
161. W. P. Mason, *Piezoelectric Crystals and their Application to Ultrasonics*, D. Van Nostrand, Princeton, N.J. (1950).
162. E. Fukada and I. Yasuda, *Jap. J. Appl. Phys.*, **3**, 117 (1964).
163. Y. Wada and R. Hayakawa, *Jap. J. Appl. Phys.*, **15**, 2041 (1976).
164. H. Kroemer, *R.C.A. Rev.*, **18**, 332 (1957).
165. E. Fukada, *Advan. Biophys.*, **6**, 121 –55 (1974).
166. R. L. Zimmerman, C. Suchicital, and E. Fukada, *J. Appl. Polym. Sci.*, **19**, 1373 (1975).
167. R. L. Zimmerman, *Biophys. J.*, **16**, 1341 (1976).
168. E. Fukada, H. Ueda, and R. Rinaldi, *Biophys. J.*, **16**, 911 (1976).
169. M. Lüscher, R. Giovanoli, and P. Hirter, *Chimia*, **27**, 112 (1973).
170. E. P. Katz and S. T. Li, *J. Molec. Biol.*, **73**, 351 (1973).
171. B. Onaral, *Biophys. J.*, **19**, 91 (1977).
172. E. Baer, R. Kohn, and Y. S. Papir, *J. Macromol. Sci. Phys.*, **B6**, 761 (1972).
173. H. Stefanou, A. E. Woodward, and D. Morrow, *Biophys. J.*, **13**, 772 (1973).
174. K. Hikichi, *J. Phys. Soc. Japan*, **19**, 2169 (1964).
175. E. Menefee, in *Electrets, Charge Storage and Transport in Dielectrics*, p. 661. Ed. M. Perlman, The Electrochemical Society, Princeton (1973).
176. H. Maeda, K. Tsuda, and E. Fukada, *Jap. J. Appl. Phys.*, **15**, 2333 (1976).
177. C. A. L. Bassett, *Calcif. Tissue Res.*, **1**, 252 (1968).
178. M. S. Shamos and L. S. Lavine, *Nature*, **213**, 267 (1967).
179. E. Fukada, *Biorheology*, **5**, 199 (1968).
180. H. Athenstaedt, *Z. Anat. Entwickl. Gesch.*, **136**, 249 (1972).
181. G. Caserta and T. Cervigni, *J. Theoret. Biol.*, **41**, 127 (1973).
182. G. Caserta and T. Cervigni, *Proc. Nat. Acad. Sci. USA*, **71**, 4421 (1974).
183. D. E. Green and S. Ji, *Proc. Nat. Acad. Sci. USA*, **69**, 726 (1972); **70**, 904 (1973).
184. D. E. Green, *Ann. N.Y. Acad. Sci.*, **227**, 6 (1974).
185. K. G. Brown, S. C. Erfurth, E. W. Small, and W. L. Peticolas, *Proc. Nat. Acad. Sci. USA*, **69**, 1467 (1972).
186. D. L. Spears, *Phys. Rev.*, **B2**, 1931 (1970).
187. J. L. Shohet and S. A. Reible, *Ann. N.Y. Acad. Sci.*, **227**, 641 (1974).
188. W. A. Little, *Phys. Rev.*, **134A**, 1416 (1964).
189. F. J. London, *J. Phys. Radium*, **8**, 397 (1937).
190. J. Ladik, G. Biczo, and J. Redley, *Phys. Rev.*, **188**, 710 (1969).
191. A. A. Wolf, *J. Franklin Institute*, **289**, 193 (1970).

192. E. R. John, *Mechanisms of Memory*, Academic Press, New York (1967).
193. V. L. Ginzburg, *Soviet Phys. Uspekhi*, **13**, 335 (1970).
194. W. A. Little, in *Electronic Structure of Polymers and Molecular Crystals*, p. 159. Eds. J. M. Andre and J. Ladik, Plenum Press, New York (1975).
195. E. H. Halpern and A. A. Wolf, *Advan. Cryogenic Engineering*, Ed. K. D. Timmerhaus, **17**, 109 (1972).
196. A. A. Wolf and E. H. Halpern, *Proc. IEEE*, **64**, 357 (1976).
197. A. A. Wolf and E. H. Halpern, *Physiol. Chem. Phys.*, **8**, 31 (1976).
198. A. A. Wolf, E. H. Halpern, and J. Sherman, *Physiol. Chem. Phys.*, **8**, 135 (1976).
199. A. A. Wolf, *Physiol. Chem. Phys.*, **8**, 495 (1976).
200. S. Goldfein, *Physiol. Chem. Phys.*, **6**, 261 (1974).
201. H. Fröhlich, *Nature*, **228**, 1093 (1970).
202. H. Fröhlich, *J. Collective Phen.*, **1**, 101 (1973).
203. N. A. G. Ahmed. J. H. Calderwood, H. Fröhlich, and C. W. Smith, *Phys. Lett.*, **53A**, 129 (1975).
204. N. A. G. Ahmed. C. W. Smith, J. H. Calderwood, and H. Fröhlich, *J. Collective Phen.*, **2**, 155 (1976).
205. H. Fröhlich, *Phys. Lett.*, **51A**, 21 (1975).
206. S. Y. Shaya and C. W. Smith, *J. Collective Phen.*, **2**, 215 (1976).
207. H. A. Pohl and R. Pethig (1975), unpublished data.
208. C. M. Sorensen, F. R. Fickett, R. C. Mockler, W. J. O'Sullivan, and J. F. Scott, *J. Phys. C.*, **9**, L251 (1976).
209. F. W. Cope, *Physiol. Chem. Phys.*, (a) **3**, 403 (1971); (b) **5**, 173 (1973).
210. F. W. Cope, *J. Microwave Power*, **11**, 267 (1976).
211. V. L. Ginzburg, *Contemp. Phys.*, **9**, 355 (1970).
212. K. Antonowicz, *Nature*, **247**, 358 (1974).
213. K. Antonowicz, *Phys. Stat. Sol.*, **28a**, 497 (1975).
214. J. E. McGinness, P. M. Corry, and P. H. Proctor, *Science*, **183**, 853 (1974).
215. B. Commoner and J. L. Fernberg, *Proc. Nat. Acad. Sci. USA*, **47**, 1374 (1961).
216. L. Erway, L. Hurley, and A. Fraser, *Science*, **152**, 1766 (1966).
217. N. G. Lindquist, *ACIA Radiol. Supplement 325*, pp. 1–92 (1973).
218. P. H. Proctor, *Physiol. Chem. Phys.*, **4**, 349 (1972).
219. P. H. Proctor, J. E. McGinness, and P. M. Corry, *J. Theoret. Biol.*, **48**, 19 (1974).
220. B. Pullman and A. Pullman, *Quantum Biochemistry*, Academic Press, New York (1963).
221. M. S. Blois, in *Biology of Normal and Abnormal Melanocytes*, pp. 125–40. Eds. T. Kawamura, T. B. Fitzpatric, and M. Seiji, University Park, Baltimore (1971).
222. M. S. Blois, in *The Biological Effects of Ultraviolet Radiation*, Ed. F. Urbach, Pergamon Press, New York (1969).
223. M. S. Blois, A. B. Zahlen, and J. E. Maling, *Biophysics J.*, **4**, 478 (1964).
224. J. E. McGinness, *Science*, **177**, 896 (1972).
225. J. E. McGinness and P. H. Proctor, *J. Theoret. Biol.*, **39**, 677 (1973).
226. J. Filatovs, J. E. McGinness, and P. M. Corry, *Biopolymers*, **15**, 2309 (1976).
227. C. H. Culp, D. E. Eckels, and P. H. Sidles, *J. Appl. Phys.*, **46**, 3658 (1975).
228. J. Kevorkian, M. M. Labes, D. C. Larson, and D. C. Wu, *Disc. Faraday Soc.*, **51**, 139 (1971).
229. S. G. E. Garrett, R. Pethig, and V. Soni, *J. Chem. Soc. Faraday II*, **70**, 1732 (1974).
230. U. Mizutani, T. B. Massalski, J. E. McGinness, and P. M. Corry, *Nature*, **259**, 505 (1976).

231. F. W. Cope, *Ann. N.Y. Acad. Sci.*, **227**, 636 (1974).
232. F. W. Cope, *J. Biol. Phys.*, **3**, 1 (1975).
233. W. Gordy, W. B. Ard, and H. Shields, *Proc. Nat. Acad. Sci. USA*, **41**, 983 (1955).
234. T. Henricksen, T. Sanner, and A. Pihl, *Radiation Res.*, **18**, 163 (1963).
235. G. E. Adams, R. L. Wilson, J. E. Aldrich, and R. B. Cundall, *Int. J. Radiation Biol.*, **16**, 333 (1969).
236. L. I. Grossweiner and Y. Usui, *Photochem. Photobiol.*, **13**, 195 (1971).
237. I. Pecht and M. Faraggi, *FEBS Letters*, **13**, 221 (1971).
238. N. K. Boardman, J. M. Anderson, and R. G. Hiller, *Biochim. Biophys. Acta*, **234**, 126 (1971).
239. R. K. Gupta, *Biochim. Biophys. Acta*, **292**, 291 (1973).
240. R. K. Gupta and T. Yonetani, *Biochim. Biophys. Acta*, **292**, 502 (1973).
241. P. S. B. Digby, *Proc. Roy. Soc. (London)*, **B161**, 504 (1965).
242. P. S. B. Digby, *Proc. Linn. Soc.*, **178**, 129 (1967).
243. H. A. Pohl and J. R. Sauer, *J. Biol, Phys.*, (1978) (In Print).
244. S. D. Bruck, *Nature*, **243**, 416 (1973).
245. S. D. Bruck, *Polymer*, **16**, 25 (1975).
246. G. H. Bardelmeyer, *Biopolymers*, **12**, 2289 (1973).
247. L. Glasser, *Chem. Rev.*, **75**, 21 (1975).
248. R. E. Barker, *Pure Appl. Chem.*, **46**, 157 (1976).
249. S. Reich and I. Michaeli, *J. Polym. Sci: Polym. Phys.*, **13**, 9 (1975).
250. H. P. Geserich and L. Pintschovius, *Festkörperprobleme*, **XVI**, 65–94 (1976).
251. G. A. Blondin and D. E. Green, *Chem. Engng News*, **53**, 26–42 (Nov. 10, 1975).
252. W. Moorhead, J. E. McGinness, and P. H. Proctor, *Physiol. Chem. Phys.*, **9**, (1977).
253. B. J. Hales, *Biophys. J.*, **16**, 471 (1976).
254. A. van Heuvelen, *Biophys. J.*, **16**, 939 (1976).
255. D. DeVault, J. H. Parkes, and B. Chance, *Nature*, **215**, 642 (1967).
256. J. Jortner, *J. Chem. Phys.*, **64**, 4860 (1976).
257. M. J. Potasek and J. J. Hopfield, *Proc. Nat. Acad. Sci. USA*, **74**, 229 (1977).
258. B. Rosenberg, B. B. Bhowmik, H. C. Harder, and E. Postow, *J. Chem. Phys.*, **49**, 4108 (1968).
259. G. Kemeny and I. Goklany, *J. Theoret. Biol.*, **40**, 107 (1973).
260. T. A. Kaplan and S. D. Mahanti, *J. Chem. Phys.*, **62**, 100 (1975).
261. H. Fröhlich, in *Theoretical Physics and Biology*, pp. 13–22. Ed. M. Marois, North Holland Publ. Co., Amsterdam (1969).
262. H. Fröhlich, *Phys. Lett.*, **26A**, 402 (1968); **39A**, 153 (1972); **51A**, 21 (1975).
263. H. Fröhlich, *Nature*, **228**, 1093 (1970).
264. H. Fröhlich, *Int. J. Quant. Chem.*, **2**, 641 (1968).
265. M. A. Lifshitz, *Biofiz*, **17**, 694 (1972).
266. D. Bhaumik, K. Bhaumik, and B. Dutta-Roy, *Phys. Lett.*, **56A**, 145 (1976); **59A**, 77 (1976); **62A**, 197 (1977).
267. T. M. Wu and S. Austin, *Phys. Lett.*, **64A**, 151 (1977); **65A**, 74 (1978).
268. N. D. Deryatkov, *Sov. Phys. Usp.*, **16**, 568 (1974).
269. S. J. Webb, *Int. J. Quant. Chem. Quant. Biol. Symp.*, **1**, 245 (1974).
270. M. S. Cooper, *Phys. Lett.*, **65A**, 71 (1978).
271. K. Koga, T. Kajiyama, and M. Takayanagi, *J. Polym. Sci: Polym. Phys.*, **14**, 401 (1976).
272. T. Furukawa and E. Fukada, *J. Polym. Sci: Polym. Phys.*, **14**, 1979 (1976).
273. M. Date, *Reports Prog. Polymer Phys. Japan*, **16**, 473 (1973).
274. R. Hayakawa, J. Kusuhara, and Y. Wada, *J. Macromol. Sci. Phys.*, **B8**, 483 (1973).

275. R. Hayakawa, K. Namiki, T. Sakurai, and Y. Wada, *Reports Prog. Polymer Phys. Japan,* **19,** 317 (1976).
276. O. Yano, K. Saiki, S. Tarucha, and Y. Wada, *J. Polym. Sci: Polym. Phys.,* **15,** 93 (1977).
277. P. Otto, J. Ladik, K. Laki and A. Szent-Györgyi, *Proc. Natl. Acad. Sci. USA* , **75,** 3548 (1978).
278. *Organic Conductors and Semiconductors* (L. Pal. G. Grüner, A. Janossy, and J. Solyom, Eds.) Springer-Verlag, Berlin (1977).
279. W. A. Little, *Mathematical Biosciences,* **19,** 101 (1974).
280. E. M. Trukhan, N. F. Perewoschikof and M. A. Ostrowski, *Biofizika,* **15,** 1052 (1970).

Some Physical Constants and Parameters

Avogadro's number	$N_A = 6.0225 \times 10^{26}$ kmole^{-1}
Boltzmann's constant	$k = 1.3805 \times 10^{-23}$ J/K
	$= 8.6171 \times 10^{-5}$ eV/K
Electrical dipole moment	1 debye $= 3.335 \times 10^{-30}$ C m
Electronic charge	$q = 1.60210 \times 10^{-19}$ C
Electron rest mass	$m_e = 9.1091 \times 10^{-31}$ kg
Electron volt	1 eV $= 23.04$ kcal/mole
	$= 96.4$ kJ/mole
Speed of light	$c = 2.997925 \times 10^8$ m/s
Permittivity of free space	$\epsilon_0 = 8.8543 \times 10^{-12}$ F/m
Planck's constant	$h = 6.6256 \times 10^{-34}$ J s
	$= 4.1356 \times 10^{-15}$ eV s
	$= 1.5836 \times 10^{-34}$ cal s

Author Index

(The chapter and reference number is given. For example, W. O. Baker is referred to in Chapter 2, ref. 4, and Chapter 4, ref. 132.)

368

Taysum, D. H., **9**, 36
Teaney, D., **9**, 94
Teixeira-Pinto, A. A., **6**, 26
Teller, E., **4**, 112, 118
Teramoto, A., **2**, 29
Terlecki, J., **7**, 48
Teslenko, V. I., **8**, 64
Thomas, L., **7**, 1
Thomas, P. W., **9**, 39
Thompson, W. D., **7**, 153
Thompson, W. K., **4**, 45
Thomson, J. J., **4**, 105
Thorne, J. M., **4**, 29
Thullier, J. M., **9**, 78
Tiddy, G. J., **4**, 55
Tien, H. T., **7**, 23, 81
Timmerman, J., **4**, 44
Timofeyev, K. N., **9**, 81
Titulaer, U. M., **1**, 51
Tobolsky, A. V., **2**, 42
Tokuoka, S., **4**, 98
Tomaselli, V. P., **9**, 51
Totin, T., **7**, 67.
Townsley, P. M., **6**, 39
Trotter, I. F., **2**, 39
Trukhan, E. M., **5**, 22, 23; **9**, 32, 95, 280
Tsernoglou, D., **4**, 79
Tsien, R. W., **7**, 75
Tsubomura, H., **2**, 22
Tsuda, K., **9**, 176
Tsutsumi, A., **2**, 43
Tyler, A., **7**, 44

Ueda, H., **9**, 168
Ukrainskii, I. I., **8**, 62
Ur, A., **7**, 121−123
Usui, Y., **9**, 236

Valentine, R. C., **6**, 27
Van Beek, L. K. H., **5**, 10, 11
Van Beek, W. M., **4**, 167
Van der Kaa, J., **9**, 61
Van Heuvelen, A., **9**, 254
Van Olphen, H., **5A**, 4
Van Santen, J. H., **5**, 15; **6**, 14
Van Vleck, J. H., **1**, 2
Vanderkooi, G., **2**, 7
Vanderkooi, J., **7**, 18
Vannikov, A. V., **9**, 6, 37
Vasilescu, D., **3**, 52; **9**, 65, 71
Vaughen, W. E., **1**, 10
Velentino, A. R., **7**, 137
Verbruggen, R., **3**, 36
Verveen, A. A., **7**, 53−55, 57
Verway, E. J. W., **5A**, 6

Vieth, L., **9**, 10
Vincent, V. M., **9**, 152
Vladimirov, Yu, A., **9**, 81
Vogel, A. I., **1**, 32
Vogel, H., **5**, 49
Vogelhut, P. O., **9**, 102
Voigt, E. M., **9**, 142
Volger, J., **5**, 9
Volkenstein, M. V., **1**, 21, 22
Von Hippel, A. R., **1**, 6; **4**, 150, 154, 161; **5**, 8; **6**, 13
Von Hippel, P. H., **4**, 39

Waara, I., **4**, 78
Wada, A., **2**, 20, 30, 31; **3**, 45, 46
Wagner, K. W., **5**, 5, 6
Walter, J. A., **4**, 57, 58
Wanke, E., **7**, 58
Wapnick, S., **7**, 132
Warner, D. T., **4**, 91
Warren, B. E., **4**, 12
Watenpaugh, K. D., **4**, 85
Watkins, J. C., **7**, 94, 95
Watson, H. C., **3**, 35
Watson, J. D., **2**, 33; **4**, 141
Webb, E. C., **4**, 50, 64
Webb, S. J., **4**, 178; **9**, 269
Webster, J., **9**, 93
Wei, L. Y., **7**, 90, 92
Weiser, N. B., **4**, 107
Weisman, I. D., **4**, 94
Weiss, J., **9**, 115
Weissbluth, M., **8**, 79
Weissman, G., **7**, 98
Weissman, M. B., **7**, 59, 64
Wen, W. Y., **4**, 37
Westgate, C. R., **9**, 97, 98
Westphal, W. B., **4**, 150, 161
Wever, R., **7**, 156
Wheeler, T. G., **7**, 125
White, S. H., **7**, 84, 88
Whitehead, J. B., **5**, 7
Whitehead, M. A., **2**, 10
Wiech, G., **8**, 7
Wilkins, M. H. F., **4**, 143; **7**, 11
Will, G., **3**, 31
Williams, E. J., **4** 26
Williams, G., **1**, 8, 36, 40, 42
Williams, R. J. P., **9**, 60
Willis, G. H., **9**, 108
Wilson, E. B., **1**, 27; **2**, 9
Wilson, H. R., **4**, 143
Wilson, R. L., **9**, 235
Winstein, S., **9**, 106
Woessner, D. E., **4**, 95

Subject Index